Tourism Planning

Tourism Planning

Basics, Concepts, Cases

Fourth Edition

Clare A. Gunn
with Turgut Var

Routledge
Taylor & Francis Group

NEW YORK AND LONDON

Published in 2002 by
Routledge
29 West 35th Street
New York, New York 10001
www.routledge-ny.com

Published in Great Britain by
Routledge
11 New Fetter Lane
London EC4P 4EE
www.routledge.co.uk

Routledge is an imprint of the Taylor & Francis Group.
Printed in the United States of America on acid-free paper.

10 9 8 7 6 5 4 3 2 1

Library of Congress Cataloging-in-Publication Data

Gunn, Clare A.
    Tourism planning : basics concepts cases / Clare A. Gunn, and Turgut Var.—4th ed.
        p.   cm.
    Includes bibliographical references.
    ISBN 0-415-93268-8 — ISBN 0-415-93269-6 (pbk.)
    1. Tourism. I. Var, Turgut. II. Title.

G155.A1 G86 2002
338.4'791—dc21

2002021965

Contents

List of Figures

List of Tables

Foreword

The decision to publish a fourth edition of *Tourism Planning* could not have come at a more propitious moment. The dramatic impact on tourism at all levels resulting from the September 11, 2001 terror attacks reinforces the case for sound planning of tourism in all settings. This new edition continues to chronicle the change and growth of the subject matter, scope, and methodologies of tourism planning. The text then describes how the planning can be done with concepts, principles, and examples.

The chapter on tourism policy is a welcome addition to the content of the book. This discussion of policy demonstrates the need for serious thinking and action not just by government but by private enterprise as well. The book now offers a deeper consideration of the concept of sustainability on tourism and a closer look at the significance of ecotourism. These two concepts are now and will continue to be major factors in the future planning and development of tourism. There will also be a need to examine the concept of sustainability of the tourist. The rapid decline of the volume of tourism after September 11 revealed, as perhaps no other event could have, the importance of tourism to the economy of all areas. There was no ready response available to the sudden disappearance of so many tourists. In the search for a quick fix for the problem, many authorities tried the old solution of pumping more advertising dollars into the mix in the hope that this would be the key factor that would lead to a revival of tourism. But, as the text emphasizes, tourism is very complicated and requires more thorough planning for change.

Gunn quite correctly points out that there are four goals for tourism planning: enhanced visitor satisfaction, better business, sustainable resource use, and community integration. The attainment of these goals is complicated by the fact that they differ from place to place and from development to development. The integration of the legitimate desires of site developers, of destination developers, and of regional developers constitutes a major challenge in the successful planning of tourism. I would suggest that reaching the goal of visitor satisfaction is achieved only through the other three.

As sustainability and ecotourism become required elements of planning, we must be careful that we do not restrict the choices that planners may present to decision-makers. Full sustainability must remain a key objective. We should recognize that only partial sustainability might be achievable. The planner's task then is to state the degree of sustainability that he is recommending, why it is being recommended, and what the costs of this recommendation are to the area. The decision can then be made to proceed or not to proceed. Tourism should not become so rigid in its planning requirements that it sets up a zero tolerance policy. There must always be flexibility in the system.

All too frequently tourism is used as an add-on justification for many projects whose main purpose is not tourism. The phrase "it will be good for tourism" gets tagged onto the end of the benefit list for many economically questionable projects. Tourism is too important to be looked on as an additional possible benefactor. The depth of the recent losses in tourism clearly illustrates the importance of tourism in its own right.

The end product of tourism is a satisfied customer. Without a succession of satisfied customers the product will wither and die. All of the pieces of a tourism destination, whether it is at the local, area, or regional level are but the means to that end. Tourism has traditionally been measured in economic terms but such a measurement looks at the success or failure of the means and not at the success or failure of the end. It is true that tourism has significant economic benefits, but the continued success of tourism depends upon the level and sustainability of visitor satisfaction.

As the book points out, planning is key to success and as a follow-up there must be studies that measure how well any particular destination is doing. What is it that turns tourists off and on? Can studies determine when the consumers are not getting the satisfaction from a site that brought them there in the first place? Another key issue at this level would be whether the kinds of tourists visiting a destination are changing and does this change have an effect on the satisfaction level and on the future plans for changes to the site.

Discussion outlines at the end of chapters are another welcome addition to this edition of *Tourism Planning*. These discussion outlines will be extremely useful in a classroom situation and can also provide the basis for discussion by planners, developers, and local citizens at public meetings on tourism planning.

As with previous editions the provision of the great number and variety as well as many new case studies is a major strength of this book. These examples demonstrate not only how essential and effective sound planning could be but also the great diversity, complexity of processes, and time required for implementation of tourism plans.

The changes of updating and adding new concepts make this edition an even more valuable volume for scholars, practitioners, governments, planners, and developers than in the past.

Gordon D. Taylor
Tourism Consultant

Preface

Although I don't know of many, one advantage of being an octogenarian is experience. Over these many decades, it has been my good fortune to have fallen into the field of tourism that turned into a lifetime career. My observations of development errors and pitfalls challenged me to seek ways of avoiding them with better planning.

Readers, especially those unfamiliar with my work, deserve introduction to the genesis of this book. Fortunately, my span of observation, study, teaching, and consulting of tourism planning has covered almost sixty years. In this lifetime tourism has grown from an obscure and soft topic to one that is powerful and accepted worldwide. I have witnessed the transition from horse-and-buggy travel to that of automobile, train, plane, and ship; from our home entertainment around the piano to radio, TV, Internet, performing arts, and cinema; the dramatic shift in many countries from an agricultural economy to that of industry and service; and from a narrow and provincial travel geography to that of the world and even space.

From a personal professional perspective, a helpful beginning point was early family autocamping travels as early as 1921 in Model T Fords. The most extensive was an 8,575-mile journey throughout America's West in a Model A Ford in 1929. The next major influence was obtaining a degree in landscape architecture, in land and water conservation, and the first Ph.D. in landscape architecture in the country. This exposure to land use and design principles became an important foundation for further study of environmental–tourism relationships.

Then, a 21-year career in the first adult educational project in tourism provided hands-on work with tourist and resort developers. This was followed by teaching and research in two hotel schools and departments of landscape architecture, recreation and parks, and tourism—University of Massachusetts, Michigan State University, University of Hawaii, Texas A&M University, University of Guelph, Canada.

These experiences, together with the dissertation search for planning principles, stimulated strong interest in the field of tourism planning, resulting in *Vacationscape: Designing Tourist Regions* (1972). Further activity was serving on the boards of The American Society of Landscape

Architects (ASLA), U.S. Travel Data Center, and The Travel & Tourism Research Association (TTRA). A valuable and enriching experience has been consulting work in over 16 countries, fostering cross-cultural breadth of tourism planning principles. Personal awards have included: named "Fellow," ASLA, Emeritus Member of the International Academy for the Study of Tourism, *Who's Who in America,* and Lifetime Achievement Award, TTRA.

When Dave McBride of Routledge/Taylor/Francis publishers called and suggested a fourth edition, I had two reactions. First, it was apparent that this volume had proven its value as a guide toward better tourism planning and development, for real-world practitioners as well as students and educators. Second, the world of tourism has continued to explode and the literature of books and journals has proliferated. Therefore, it seemed that the third edition did indeed need updating. To assist on this task, I sought and obtained the cooperation of a colleague at the Department of Recreation, Park and Tourism Sciences, Texas A&M University, Dr. Turgut Var.

Clare A. Gunn, 2001

I was pleased to be invited to assist in this fourth edition. My experience in the economics of tourism planning complements the physical design work of Dr. Gunn. I began my tourism career as a tourist guide in 1954 and obtained university degrees in business, accounting, finance and economics, with emphasis on tourism. Following the award of the prestigious Ingersoll Milling Machine Scholarship, I was awarded a Fulbright postdoctoral fellowship and returned to my homeland of Turkey.

Consulting and writing as well as study and teaching followed, joining with Charles Gearing and William W. Swart to produce the pioneering work, *Planning for Tourism Development: Quantitative Approaches* (1972). Among my projects were Kusadasi Dilek Peninsula Tourism Development Planning, a major vacation village, Corum Development, and Hittite National Park.

Then, back in the United States, I became involved in tourism studies and teaching in several universities: University of Kansas, Simon Fraser University (Canada), University of Hawaii, Northern Territory University (Australia), and Texas A&M University. This experience led to serving on the editorial boards of *The Journal of Travel Research, Annals of Tourism Research,* and several other scholarly journals. I was pleased to be elected as a Senior Fellow of the Academy of Hospitality and to the International Academy for the Study of Tourism. Most recently I served as joint editor for *VNR Encyclopedia of Hospitality and Tourism* (1992) and *Encyclopedia of Tourism* (2000).

Turgut Var, 2001

It is our hope that this fourth edition will provide even greater acceptance and applicability than before. The intent is not to provide the only answers to tourism planning but rather to stimulate planners, designers, developers, tourism leaders, educators, and students to do further research and develop new concepts for the betterment of this powerful phenomenon called tourism.

Acknowledgments

We genuinely appreciate the constructive criticism and valuable information we have received from a great many people and from many disciplines. This work would not have been done without their input. Because so many have assisted, and we are confident they know who they are, we cannot possibly name them all even though we would like to. Instead we want to recognize some key individuals who have helped greatly on this edition.

Genuinely appreciated were the reviews from those who responded so favorably to the proposal for updating this book. Their constructive comments and encouragement were of great help. Many thanks to these experienced and competent reviewers: J. Kent Stewart (Western Management Consultants), Dr. Stephen L. J. Smith (University of Waterloo, Canada), Dr. Allan Mills (Virginia Commonwealth University), Dr. Mussafer S. Uysal (Virginia Polytechnic Institute), Dr. Edward McWilliams (D. K. Shifflet & Associates, Virginia), and Prof. Michael Fagence (University of Queensland).

Dr. Var wishes especially to thank Dr. Peter Witt, former head, and Dr. Joseph O'Leary, head, Department of Recreation, Park and Tourism Sciences, Texas A&M University, for allowing him release time from regular duties to provide assistance on this book project.

Special thanks are offered here to some of the many suppliers of new research information and cases, including: Dr. Douglas C. Frechtling (George Washington University), Dr. James F. Burke (The Collins School of Hospitality Management), Claudia Kopkowski (Massachusetts Audubon Society), Dr. Graham Brown (Southern Cross University, Australia), Dr. Peter Williams (Simon Fraser University, Canada), Dr. David Wood (Curtin University of Technology, Australia), Dr. Brian Hay (Scottish Tourist Board), Dr. Jafar Jafari (University of Wisconsin-Stout), Rachelle Richard-Collete (Bouctouche Bay, New Brunswick), Laurel J. Reid (University of New Brunswick), Per Nilsen (National Parks Canada), Bruce Rogers (City of Sanibel, Florida), Dr. Richard O. B. Makopondo (University of Illinois), Dr. Gene Brothers (North Carolina State University), Eileen Calveley (AILLST, Scotland), Dr. Esteban Bardolet (University of the Balearics), Dr. Ross Dowling (University of Notre Dame, Australia), Thomas Combrink (Northern Arizona University), Michael Kelly (The

Hopi Tribe, Arizona), Sue Walker (Cape Byron Trust, Australia), Juli Grot (EDAW, London), Carmela Avagliano Argenziano (Genova, Italy), and John and Kristin Denure Hunt, tourism consultants.

As principal author, I am deeply indebted to several people whose special assistance made this volume possible. Over many decades, the depth and breadth of experience in tourism of my colleague and friend, Gordon Taylor, have been very valuable to me. As market researcher and policy director for the national tourism agency of Canada for many years, he shared his special insight into many dimensions of tourism. Our work together in the Travel & Tourism Research Association, our many meetings and programs together, and our exchanges of tourism philosophies have aided me greatly in all my studies and writings. Also threaded through this work are the profound environmental land design policies of genius landscape architect, William Johnson. He was my mentor during my doctoral work and inspired me to develop my concepts of design and spatial relationships for tourism development. His firm was one of the first to practice the integrity of environmental sustainability for land design projects. Much credit also must be attributed to my association with the only state tourism official in the 1970s who advocated that there was more to tourism development than promotion. During his tenure as director of the Texas Tourist Development Agency, Frank Hildebrand gave me great support for creating better tourism development vision and planning for the future. I am extremely grateful for his many years of friendship, collegiality, and professional insight.

This book would not have been possible without the many years of collaboration with Dr. Uel Blank. His expertise in economics, political action, and sensitivity to human involvement stimulated me to broaden my concepts of tourism planning and implementation. I owe a great debt to two very supportive family members. Son Richard has frequently responded to computer SOS signals as well as provided quality photographic techniques to improve the many illustrations in this book. And, finally, without the many hours, days, and months of dedicated word processing by my wife of over sixty years, Mary Alice, this manuscript never would have been completed. Her support, advice, and encouragement are very precious to me.

Introduction

As tourism's great growth worldwide has provided a huge social and economic windfall it has become a high priority agenda item for nations and communities everywhere. Its positive economic impacts continue to be cited for justification of tourism development. But a new awareness of several negative impacts, especially on the environment, is now stimulating a greater demand for fresh ideas and processes that can guide growth without such stress—the need for more and better planning.

Throughout the world today most tourism leaders continue a dominant policy of travel promotion, based on the simple rule of the more tourists the better the economy. Billions of dollars are spent by public and private interests on advertising and publicity to entice greater travel. Mass tourism has become an unprecedented phenomenon. Personal benefits and new economic support have exceeded all expectations.

But there is a growing backlash to this halcyon period of tourism's growth. Environmental awareness is no longer restricted to emotional zealots but is now supported by more tourism policymakers and developers. It has given rise to new tourism jargon—sustainability, ecotourism, nature tourism, green tourism, cultural tourism. More writings and discussions are now focused on capacities and oversaturation, and even demarketing. Travelers themselves have become more aware and vocal about congestion, foul air, polluted waters, wildlife extinction, deforestation, and conflict with residents. Tourism's wanton growth is taking its toll. Tourism is too important to mankind to let it continue to drift.

In other fields, such as education, health, and even defense, when such drifting creates chaos, there is a cry for remedy. New curricular plans, new health plans, and new defense plans are put forward. If tourism is to ameliorate its negative impacts and yet provide its proven values, new plans must be created and implemented by those responsible for its development.

The purpose of this book, as has been true of earlier editions, is to present plans and examples that illustrate a balance between development and environmental protection. It is based not only on current theories and concepts but also on practical applications. It also describes the difficulties as well as successes of planning processes. Its writing and organization are directed to public and private planners and developers, as well as students seeking an understanding of the topic of planning tourism development.

For ease of understanding, the book is presented in two parts. Part I includes four chapters that summarize the basics of tourism that are essential to planning—the purpose of planning; tourism as a system; growth, sustainability, and ecotourism; and policy related to planning.

Experience has demonstrated that planning concepts and methods must vary depending upon the geographic scale. Therefore Part II is divided into three scales—regional (national, provincial, state), destination (community and surrounding area), and site (specific land area for facilities and services). For each scale, a chapter of cases illustrates both difficulties and accomplishments of tourism planning.

As an aid to students, teachers, and community leaders each chapter is followed by several questions and discussion topics. This addition to this new edition is intended to stimulate thinking and discussion beyond the information provided in the text.

An important corollary purpose of the book is to stimulate readers to go beyond the information presented here and seek new approaches to the complicated field of tourism planning. Because of its complexity and dynamic nature, tourism deserves the very best study, judgment, and conceptual thinking for its future development. Researchers, scholars, developers, residents, and the body politic must give priority to tourism's development issues if its future is to continue to make a positive and quality impact on humanity.

Part I

The Basics

Before one can understand and engage in tourism planning, at any scale, certain basic fundamentals need clarification. The purpose of this part of the book is to summarize some of the more important foundations. Because public and private designers, planners, and developers vary in their experience and understanding, these chapters may refresh their present knowledge or offer new insight into better design and planning.

Chapter 1, "The Purpose of Tourism Planning," is intended to establish the necessity of planning and also dispel past fears and misunderstandings of planning. If tourism is to reach toward better economic impact, it must be planned as well toward goals of enhanced visitor satisfaction, community integration, and above all, greater resource protection. Planning today has a new look, worthy of attention by all sectors involved in development.

Chapter 2, "Tourism as a System," endorses the need for all developers, planners and managers to understand the dynamic interrelation among all functioning parts of tourism as a system. Essential is new understanding of the need for balancing development with demand even though market interests are constantly changing. Only by all sectors planning toward better integration of all parts will tourism avoid difficulties and meet desired objectives.

Chapter 3, "Growth, Sustainability, Ecotourism" calls attention to today's need for greater environmental awareness and planning sensitivity in all tourism development. Instead of limiting opportunities, when properly planned, tourism can even foster both positive objectives and strengthened conservation of resources. New market demand for travel to destinations and attractions of natural and cultural resource significance is giving rise to greater environmental sensitivity.

Chapter 4, "Policy" addresses the important role of policy making and implementation. Included are private development policies as well as

1

Tourism Planning

those by public agencies. And, all scales, from regional to local, are involved in preparing and taking action on tourism policies.

These four chapters lay the base for the second part—concepts and examples of planning tourism at three scales (regional, destination, site) and concepts and principles valuable to future tourism development.

Chapter 1

The Purpose of Tourism Planning

INTRODUCTION

As millions more people travel and seek personal rewards from their experiences, massive development of resources is the consequence. This pervasive tourism growth is a significant part of the global expression of the new services economy. With the weakening of many other aspects of the economy, nations and communities see tourism as a quick and easy solution.

Closer examination of this trend shows mixed results. At the same time that improved technologies of automobile and air transportation have given more people the opportunity to travel, destinations often have been glutted with congestion and overburdened facilities. As scenery and other natural resources are touted in promotion, the visitor often experiences more ugly commercialization than scenery. As coastal areas are promoted for clear water and pristine beaches, the visitor often discovers a concrete jungle of hotels and sometimes waterfront sewage pollution. As communities seek the economic benefits of foreign visitors, these often come at the cost of cultural conflict and unfulfilled promises. Many scholars and observers have well documented the negative as well as the positive impacts of tourism development.

The truth is that tourism development is being done by those who focus primarily on individual parts rather than tourism as a whole. Tourism can enrich people's lives, can expand an economy, can be sensitive and protective of environments, and can be integrated into a community with minimum impact. But a new mind set is called for, that demands more and better planning and design of all tourism development, especially how the many parts fit together.

The purpose of any planning is to create plans of action for a foreseeable future and implement these actions. For example, an owner wishing a new home cannot expect the structure to be completed without a building plan and a contractor to build it. Such a plan is the result of a designer visualizing the completed building and supervising its construction. It must be acceptable to the goals and objectives of the owner—meet functional needs, suit aesthetic desires, and meet financial limitations. Application of this metaphor to planning of tourism development demonstrates the fundamental fact that tourism in its full breadth is much more complicated and for many reasons.

First of all, tourism itself is an abstraction. It doesn't exist, at least not in the same sense as a residence. Tourism is not even a discipline, such as chemistry or geography. Tourism is a field made up of many physical, program, and action parts. It is only the pieces of tourism and their aggregation that can be planned.

Tourism is not under the control of one owner, it has no CEO. It is controlled by a multitude of owners, and mostly within three categories—government, nonprofit organizations, and private commercial enterprise. In many nations even these divisions become blurred, such as quasi-governmental corporations. Furthermore, tourism is influenced by a great many other factors that can make or break the planning process and its implementation, such as local residents, financial institutions, and market demand.

Tourism has no solitary goal or objective but many. A popular governmental goal of tourism is economic improvement. Sometimes this is defined with specific annual objectives, such as an estimated number of visitors and a dollar amount of their expenditures. Some governmental agencies that impinge on tourism but have their own policies direct their management to other goals. For example, park and recreation departments generally base their existence on social goals, not economic.

Even though the overall goal of the private commercial sector of tourism is profitmaking, this varies with businesses. Not all hotels or restaurants expect the same profits because their products and services as well as costs of operation vary. The magnitude of debt service and other overhead can influence objectives of pricing. And, the costs of physical plant updating vary greatly across tourist service businesses.

The concept of tourism planning is almost an oxymoron. The bulk of tourism is rooted in *voluntary travel* (except for business, often linked with personal). Planning tourism, therefore, seems contrary to such an unplanned phenomenon. This presents a major complication for tourism planning. Travel preference remains elusive in spite of much market research on demographics, psychographics, life styles, opportunity, experience, and a great many other factors that influence what travelers like for their travels. Tourism developers can foster but not force these important reactions.

Even the physical foundation for tourism is dynamic. Generally, the mountains, rivers, lakes, plains, deserts, seas, and forests are where they were centuries ago. But from both natural and man-made causes, these are constantly changing—damming rivers has degraded the geology and biology of free-flowing waters, deforesting has caused desertification, agricultural practices have polluted rivers and ground waters, wildlife growth plagues suburbia and yet some species are facing extinction, and settlement and irrigation have increased humidity on previously arid lands. Certainly not all negative impacts are caused by tourism, but its share is substantial and growing. Planning must cope with these mounting changes.

A further complication of trying to plan tourism is paradoxically its very strength—creativity. The ingenuity of creative people regularly makes dramatic changes in tourism but is difficult to predict. Many of the names of tourist activities have remained the same over time—resorting, fishing, hunting, camping, boating, cruising, attending events. But how these are expressed is very different in this century from how they were in the past. The magnificent hotels suited to the rich and famous carriage and train travelers of the 1800s hardly suit the similar market category of modern air and automobile travelers. The ingenuity and creativity of scientists, designers, and technicians regularly introduce innovations that upset past formulas for planning tourism. Modern technology, such as computers and the Internet, has added to the opportunities and also the complexity of visualizing tourism's future. New plans for theme parks, public parks, aquariums, zoos, interpretive visitor centers, hotels, resorts, and other features regularly introduce changes.

An increasing complication, especially with worldwide proliferation of ecotourism, is acculturation, the dramatic evolutionary changes in many aboriginal societies. Native populations, based on a non-industrial foundation, face the dilemma of introducing tourists to their special cultural uniqueness and yet moving themselves into newer areas of industrial societies. It is difficult for them to plan for tourism development that might give them new economic advantages without encountering great internal stress for protection of their cultural heritage. Acculturation, sparked by tourism, may require many years of planning and adaptation.

In spite of these complications, those who are involved in tourism in its diverse dimensions are active in planning, if only from their own narrow perspective or on broad economic and social scales. Planning futures is not a new concept around the world.

THE CONCEPT OF PLANNING

Physical planning, as a concept and practice, has taken place for centuries. As Branch (1985, 12) points out, cities in India as early as 3000
B.C. were divided into square blocks, oriented to the cardinal points, and
laid out to allow circulation between them. Medieval cities frequently
were planned with encircling walls for fortification. Even building codes
and zoning date back to ancient times. But such order in city planning
resulted from strong centralized authoritative control. Elsewhere, unregulated development had little order. Even in more recent times, socialistic
and planned-economy nations exacted strong land use controls from a
central authority.

Town planning has been practiced in the United Kingdom for two centuries (Cherry 1984, 187) and physical layout planning reaches back to
early Greek and Roman times. For England, interest in planning was
stimulated by the physical and social ills resulting from industrialization.
Visionaries and philanthropists dreamed of utopian cities. The bias for
many years was toward physical planning—the visual appearance of
architecture and patterns of land use. This concept was followed by
trends toward comprehensive planning set into law. In recent decades two
dimensions have been added to planning—social and economic.

> Planning is a multidimensional activity and seeks to be integrative. It em-
> braces social, economic, political, psychological, anthropological, and tech-
> nological factors. It is concerned with the past, present and future. (Rose
> 1984, 45)

Although such lofty goals are at the heart of the planning concept, carrying them out has not been simple or easy. For many reasons, including
the complexity of thousands of decisions made by individuals, corporations, and governments all over the world, planning has not been as effective as planners might have wished. Professional planners have agreed
upon some general directions—a better place to live and the like—but
there is no neat body of theory for planning such as can be found in other
disciplines. In fact, planners generally agree that planning is not a distinct
discipline but an amalgam of many.

In capitalistic market-economy countries, planning has become nebu-
lously blurred in meaning, often even a pejorative term. Even though
much planning is known to take place, several negative connotations
have become strong enough in public opinion to block many planning
efforts. A few of these are worth noting so that individuals, communities,
and areas can take steps to cope with them.

Many people feel that the idea of planning places too much *power in a
governmental bureaucracy.* Because urban planning departments have

become a legally sanctioned institution of many city governments, many people resent bureaucratic control over what they believe to be their freedoms, especially for land use and development.

Much of this resentment is based on ascribing power and titles to *professional planners*. In the past, university programs of urban planning focused on professional elitism patterned after such other professions as medicine and law. This was based on the educational criterion that there was a technical body of knowledge that was exclusive and essential to a profession.

Another negative public reaction has resulted from so many *plans being aborted.* It has happened so often that the phrase "plans collecting dust on the shelf" has become a popular planning cliché.

In spite of these known images of planning, it continues to grow as a process for visualizing and guiding action to avert pitfalls and meet challenges of the future.

In the context of history, the planning of tourism development is relatively recent. Although a few geographers had written about tourism planning in the 1940s, the first major works appeared in the 1970s. In 1972, the book *Vacationscape: Designing Tourist Regions* (Gunn 1972) was published and included a model of the tourism system and a participatory process derived from a case for tourism-recreation planning in Michigan's Upper Peninsula, perhaps the first regional tourism plan in the United States. For several years, tourism researchers Charles E. Gearing, William W. Swart, and Turgut Var collaborated on studies of economic models of tourism planning. In their book, *Planning for Tourism Development* (Gearing et al. 1976), a comprehensive overview of many economic models and their authors is presented. Mathematical formulas of demand, attractiveness, supply variables, and the selection of alternative plans are included. The principle of economic efficiency is emphasized. In 1977, a recreation-tourism process called Product's Analysis Sequence for Outdoor Leisure Planning (PASOLP) by Fred Lawson and Manuel Baud-Bovy was published in Europe. Later, Baud-Bovy elaborated on the approach based on experience of application in several European countries. Elements of this process—resource analysis, objectives, conclusions, plans, recommendations, and impacts—have continued through many iterations by others over the last several decades. Planners and scholars who have put forward planning theories and applications have been summarized by Sandro Formica (2000), and include Gunn (1979), Getz (1986), and Inskeep (1987).

Generally, most tourism planning approaches have also been influenced by the field of urban and rural planning. For example, in the United States, the National League of Cities (Dodge and Montgomery 1995) published a guide to be used by city officials. Its recommended process, equally applicable to tourism, begins with agreement on a process and analyzing the region's resources and anticipated changes. This is followed by the important step of creating a vision of a best future that avoids pitfalls and reaches toward desirable goals. A final and very important step is selection of strategies and implementation. (More discussion of tourism planning and applications is elaborated in Part II.)

The need for tourism planning may not be as blatantly conspicuous as other development concerns. But, nevertheless it is real. Those involved in the tourist business today realize that tourism is *more competitive* than every before. In the past, growth of demand was so great that competition was not a concern. Today, throughout the world, thousands of investors, public and private, are developing new tourism areas. At no time in history has the *proliferation of promotion* of travel places been so massive. Market specialists are discovering that travel markets are not as simple as they once were considered. Now they talk of market segmentation and much greater sophistication levels of travelers. New markets are burgeoning—adventure travel, cultural travel, ecotourism, intellectual travel, and travel to spectacular events. And they are dynamic, demanding continuing market research. Tourism, once considered the sole responsibility of hotels and travel promoters, is now being recognized as encompassing much more; it is a *very complicated phenomenon.* The supply side involves nearly every citizen, every public agency, and every organization because the visitor is exposed to everything. How the voters support the development and maintenance of streets, safe and attractive neighborhoods, educational museums, beautiful and well-kept parks, and even health services—all are as important to tourism as to the welfare of local citizens. Added to these concerns is a new awareness of *environmental impacts of tourism.* Many of the foundations of tourism have been and continue to be eroded. Natural and cultural resources with poor tourism planning have produced beach pollution, soil erosion, threats to wildlife, and scenic spoliation. These and many other issues are sufficient cause for mounting new planning initiatives and actions.

TOURISM DEFINED

If one is to plan tourism, one must have some understanding of how tourism is being defined. As Jafari points out, the popular and professional meanings have changed over time (Jafari 2000, xvii). Whereas earlier meanings focused mostly on economics—tourism as an economic generator—newer meanings have a much broader scope. This new perspective is probably the result of greater knowledge in recent decades of the immense dimensions of tourism. It is truly a very complicated phenomenon.

In the first major writing about tourism, the *Travel and Tourism Encyclopaedia,* J. G. Bridges cites several historic dimensions of tourism (Bridges 1959). Human movement is an innate characteristic demonstrated by mankind throughout history. Early travel took place for many reasons—visit friends, curiosity, better living, pilgrimage, commerce, trade, and general wanderlust. Evidence of the great power of travel motivation is its emergence and growth centuries ago in spite of many obstacles, such as discomfort, unreliability, threat from bandits, and many health hazards.

As travel modes and quality increased dramatically and as research expanded, new interpretations of tourism emerged. Mathieson and Wall (1982, 1) recognized the breadth of tourism, much beyond economics, and offered this definition:

> Tourism is the temporary movement of people to destinations outside their normal places of work and residence, the activities undertaken during their stay in those destinations, and the facilities created to cater to their needs.

A similar meaning was provided by Chadwick (1994), who identified three main concepts: the movement of people; a sector of the economy or industry; and a broad system of interacting relationships of people, their needs, and services that respond to these needs. As travel economists sought to identify economic impacts of tourism, more precise definitions of trips emerged—distance, length of stay, purpose, expenditures. There was a tendency among some researchers to exclude business travel because it was not believed to be promotable.

In recent decades, the concept of tourism has broadened into holistic interpretations that have given rise to the modeling of tourism as a system (Leiper 1993). He cites the works of several who have proposed models of the tourism system: Cuervo (1967), Gunn (1972), Leiper (1979), Mill and Morrison (1985), and Jafari (1989). Such a systems approach is the focus of this book. Key elements of this holistic and interrelated model include:

- Tourism is not a discipline; instead it is a multidisciplinary field.
- Tourism is generated by two major powers—demand, supply.
- Within demand is a diversity of traveler interests and abilities.
- Within supply are all the physical and program developments required to serve tourists.
- Tourism includes many geographic, economic, environmental, social, and political dimensions.
- Tourism is not an industry; it is made up of a great many entities as well as business.

These factors represent the definition of tourism most closely related to planning challenges and opportunities in the world today and to the concepts and cases provided in this book.

THE PLANNERS

For tourism, the maze of those who guide and direct development for the future encompasses a huge array of planners. For the purpose of understanding the planners of tourism, they could be categorized in four main groups.

The Business Sector

Every hotelier and restaurateur makes plans for future development and operation. The new investor searches for the best location. Site factors, relevance to other businesses, size, and land cost are only a few factors that influence the location of a future business. After establishment, the planning of products, services, and quality level to meet travel market segments is a continuing obligation. Key planners are owners and managers. The group of main tourist service businesses are often called "the hospitality industry." However, there are several reasons why the tourism business sector does not fit the definition of industry.

Tourism does not produce a singular product, such as the automobile. Tourism involves a tremendous diversity of "products." Tourism products—satisfactory visitor experiences—are quite in contrast to goods and services produced by industries. The tourism distribution system of transportation services moves the markets to the products at destinations. This is the complete opposite of manufactured products that are distributed to market areas.

This truth, the significance of destinations, demands entirely different planning strategies for tourism as compared to manufacturing industries. Plant locations have very little importance to product markets, whereas destination locations have everything to do with meeting the needs of tourism markets.

All destination places for travel may have some similar characteristics but it is because of each one's uniqueness that people travel to it. Places are different. They have different geographic positions, geographic settings, development patterns, histories, traditions, and societies. Planners have the obligation to discover the special qualities that make them different and to plan for the development of these special features that will appeal to markets.

Tourism "products" and their foundations are far more "perishable" than are manufactured products. If the manufacturer of children's clothing discovers that a community has become primarily a place for retirees, the products may be redistributed to communities with a high birth rate. However, when a large capital investment is made in resort hotels, marinas, and theme parks, and the travel market decides to go elsewhere, it is costly if not impossible to move the physical plant.

When the industrial committee of a community seeks new development, efforts are directed toward the chief executive officers of specific plants. Not so with tourism. Tourism development requires a large number and a diverse mix of decision makers. The task for tourism is far more complicated.

Tourist businesses at the planning stage are obligated to consider not only potential profits but also the many implications of their decisions— on the environment, on competition, the relevance to other businesses, and on the infrastructure and social values of a community.

The Public Sector as Planners of Tourism

Contrary to being an industry or dominated only by the business sector, tourism is also developed and managed by another very important group—the public sector. Generally, the primary role of government is governance—enactment and implementation of laws and regulations and provision for defense. However, for tourism most countries have assumed many very important functions other than regulation.

In a great many countries, tourism promotion and marketing have been accepted as roles of government. Billions of dollars are spent by governments annually on promotion to attract visitors to their areas. For most of these agencies, advertising makes up the largest share of the budget. Only a few include money for research, planning, training, and information systems.

Throughout the world, governments provide a great number of visitor attractions, such as museums, archeological sites, and reserves. National parks not only protect valuable natural and cultural assets but also attract millions of visitors. Parks at state, provincial, county, and community levels also provide a great many outdoor recreation functions for travelers. In some countries, government has taken the initiative to invest in resorts and destination enclaves.

The governmental sector owns and manages much of the infrastructure upon which tourism depends. From community to the national level, this often includes water supply, sewage disposal, police and fire protection, streets and lighting, as well as electrical power and communications.

Although the governing agencies may set policies and exercise practices primarily for residents, these utilities are of critical importance to travelers. For example, some destinations are not popular with travelers because water supplies and police protection are inadequate. Official city planning, building codes, and zoning have much to do with how tourism is developed.

Certainly, understanding the many policies and practices of governmental agencies is as essential to all tourism planning and development as decisions of the business sector.

The Nonprofit Sector as Planners

Often omitted as a prime planning and development sector of tourism is the very important category of nonprofit organizations. Throughout the world there has been great growth of tourism development by voluntary organizations. Many health, religious, recreation, historic, ethnic, professional, archeological, and youth organizations plan, develop, and manage land and services for visitors.

In recent years, both cultural and natural resource attractions have increased in numbers due largely to nonprofit organizational support. Historical societies have recognized the value of mounting campaigns to preserve historic sites and buildings. In addition to protecting lands and structures, they have rebuilt and modified structures to adapt them to tourism. Retaining the historic patina of architecture, adaptations for visitors have been made—adding heat, air conditioning, special protection of floors and walks, exhibits, toilet facilities, descriptive materials, and guided tours. Conservation organizations and ecotourism societies have sponsored stronger natural and cultural resource protection alongside tourism development. Tourism trade organizations such as those for hotels and restaurants not only have policies for internal operation, such as labor, taxes, insurance, and governmental regulation but also for conservation of energy and recycling.

The voluntary, informal, family sector holds great promise for tourism expansion, especially in developing countries. Rather than inviting the large multinational firm to invest outside capital and labor, local talent can be harnessed for many indigenous and small-scale tourism developments. Because the goal is less profit than ideology, many cultural benefits can accrue from nonprofit tourism development.

Although laws vary, the nonprofit sector generally is one in which charges can be made for products and services but all revenues must be expended for operational and capital costs, not as investment. Nonprofit tourist attractions typically receive revenues from contributions, admission receipts, and foundation grants.

As planners and developers, those in the nonprofit sector are as important to overall tourism as the business and governmental sectors. Their visions and plans generally contain strong social and ethical elements and their work must be considered at both the macro and micro levels.

Professional Consultants

Today, several kinds of professionals are engaged in planning many aspects of tourism. Their roles are not investment or management and therefore can provide unbiased information and plans. Their functions range from specific buildable site development to consulting services on many tourism planning and development topics. Even though most of their functions are popularly known, they are reviewed here to encourage more developers to utilize their services. No other group can provide planning assistance as effectively.

The most popularly recognized are *building architects*. Plentiful worldwide are their completed projects—hotels, resorts, restaurants, attractions, monuments, airport terminals, theaters, arenas, and aquariums. The typical planning process begins with consultation with owners, development of a design program, creation of sketch plans, completion of working drawings and construction details, and writing a construction contract. Architects can be found in both the public and private sectors. Their training, experience, and artistic talent contribute to their concepts of how a project will appear when completed.

Closely related are *landscape architects*. They are specialists in site and land planning. Their projects utilize their talent combined with understanding of land resources and their utilization. Tourism has been influenced greatly by their work, including site planning for resorts, marinas, roads, historic complexes, and community plans. Most parks around the world have been designed by landscape architects. As with architects, they utilize a process that begins with client consultation and follows through analysis, design concepts, and final details of construction plans. They also are involved in public and private practice.

The field of *urban planning* has provided guidance on services, utilities, transportation, and land use for communities. Most often, this work has been the foundation for regulations and zoning. Their role has focused primarily on residents because local taxes provide revenues for their activities. Planning for visitor use is not yet a major function but holds great opportunities for providing a valuable integration of resident and tourist use.

Civil engineers provide the design and details for many projects related to tourism—bridges, harbors, airports, waterfronts, and water and waste systems.

For tourism, the major change among these specialists is their collaboration as *teams*. Many professionals combine the services of architects, landscape architects, engineers, and others in order to provide the needed mix for projects. Often other specialists are added—historians, archeologists, wildlife specialists, foresters, golf designers. Not only do these teams work on specific land development projects but also provide consulting services. They are often called upon to resolve planning conflicts because they have no vested interest and can offer objective solutions.

As *universities* have expanded their curricula to include tourism, many researchers and teachers are frequently engaged in tourism planning consulting. They often have a mixed background of disciplines such as one or two of the following—forestry, parks, recreation, history, architecture, landscape architecture, planning, anthropology, economics, and marketing. Their main contribution is the latest technical or scientific information, important for many tourism planning projects. Many universities are engaged in adult and extension educational programs directed toward tourism that include planning. They work in the field holding seminars, meetings, creating publications, and providing demonstrations. Because they have no vested interest and can be objective, they often serve as catalysts to bring many factions together to resolve community planning conflicts.

There are great opportunities for developers of tourism to make greater use of these planning specialists. They can provide detailed analyses of resources, determine threats to the environment, and bring innovative and creative solutions to difficult tourism planning tasks. However, two aspects of their work must be emphasized. Because their contractual agreement is with an owner, the tourist's interests and actions are often left out or false assumptions are made. Every agenda of planning should include detailed present and potential travel market descriptions. Their consulting work usually ends before implementation. The productive recommendations are of little value unless the action agents—tourism policymakers, developers—put the guidelines in place. It is critical that implementation steps are built into all planning projects from the beginning.

Finally, the important role of lay people must be added to that of the planning and design professionals. With the universal growth of travel has come a much more sophisticated travel public. Personal travel experience by these nonprofessionals is valuable input for guiding the direction of physical development, environmental impact, and quality of service.

GOALS FOR DEVELOPMENT

For tourism, as an overall comprehensive activity, planning can provide betterment if directed toward several major goals. Here, goals are defined as different from objectives. Objectives are specific, real, and actual activities that can be accomplished within a given time. Goals are ideals or aims that one strives for but may never completely accomplish. Goals provide the framework for the identification of policies and accomplishment of specific objectives. For example, a nation may have the goal of economic improvement within which there may be objectives for specific roads, hotels, and attractions to be built by a certain date.

Put forward here are four suggested goals for today's vision of better tourism development: enhanced visitor satisfactions, improved economy and business success, sustainable resource use, and community and area integration. These go beyond the present level and are aims that if accepted by all involved in tourism planning and development will eliminate many of today's ills and at the same time provide sustainable tourism.

Enhanced Visitor Satisfactions

Tourism begins with the desires of travelers to travel and ends with their satisfactions derived from such travel. But, as has been pointed out, the complicated characteristics of modern tourism tend to reduce these satisfactions from the possible level desired. Mere volumes of mass participation, a popular method of evaluating "success," do not necessarily translate into satisfaction. Visitors often waste much time, money, and energy attempting to weave the myriad of development into a meaningful whole. Exaggerations in promotion can result in disappointment upon actual participation. Visitors are too often frustrated finding their way about, seeking the appropriate services, and understanding the attractions when they do find them. Furthermore, developers' lack of understanding of the tourist market results in less than satisfactory services, such as poor accommodations, food services, and interpretive tours. Fragmentation of transportation services creates irritation and sometimes disappointment, especially when packaged tours are aborted midstream. Recommendations of decades ago are equally applicable today:

> The amateur tourist needs help. He needs confidence that the strange world to which affluence admits him deals candidly with him and that there are standards of price, service and facility he can rely upon. He needs assurance that if somehow things do not turn out as expected, he has a recourse. (*Destination USA* 1973, vol.1\95)

Planning should not only eliminate the problems outlined above but also provide the positive mechanism whereby land acquisition, design, development, and management have the greatest chance of providing user satisfactions. In this sense, planning is not only user problem-solving, it is user problem-avoiding. Planning should provide a check on interrelationships of development to make sure the participant's desires, habits, wishes and needs are satisfied insofar as physical development and management can do so. The worth of the planned development is not to be judged solely by the owner nor the planner but by the visitor. This standard demands a user-oriented planning policy that must be flexible to meet market changes.

The importance of tourism to the individual and society was identified in the "Manila Declaration on World Tourism," (*Records* 1981, 118) resulting from the World Tourism Conference of 1980. The "spiritual" elements that must be given high priority were cited:

> The total fulfillment of the human being.
>
> A constantly increasing contribution to education.
>
> Equality of destiny of nations.
>
> The liberation of man in a spirit of respect for identity and dignity.
>
> The affirmation of the originality of cultures and respect for the moral heritage of peoples.

This expression by 107 national delegations and 91 observer delegations from international, governmental and non-governmental organizations was evidence of the need for considering the human dimension in all tourism planning.

All planners, developers, and managers of tourism must never lose sight of a major fundamental—the intrinsic value of travel for the individual and society. Although stakeholders cannot guarantee experiential satisfactions, they are bound to provide the best setting and the programs most likely to make them happen.

The child's first thrill of paddling through foam and sand along a beach or observing wonders of nature, such as a colorful butterfly on a beautiful flower, are important outcomes of tourism. For families of all ages, seeing and photographing elk and moose in the north, the exotic animals and birdlife of the tropics, the snow-capped and glaciered mountain ranges, the brilliance of autumn foliage, or the full palette of sunsets over water, are the real substance of much of travel. At the same time that business proprietors are concerned about airline schedules, hotel room maintenance, and labor payrolls, they must always maintain high sensitivity to visitor interests. The thrill rides of theme parks or whitewater rafting that push human endurance to the limit may be the most important travel experiences for some. The fascination and revelation of scientific endeavor capture the interest of masses of tourists as they visit museums and science centers. New understandings of other cultures—their back-

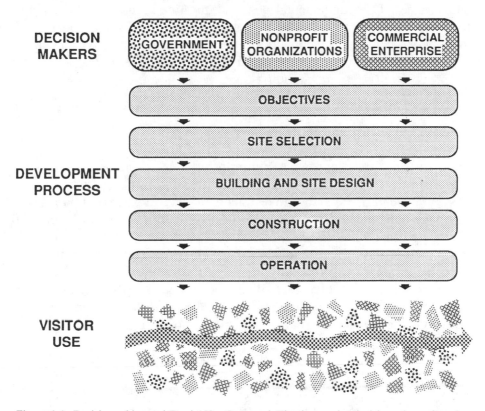

Figure 1-1. Decisionmaking and Tourist Use Compared. The three major decisionmakers of tourism create a mosaic of independent developments. Travelers, in contrast, are the only ones to view and use this mosaic as a whole. Needed is planning that integrates the many parts for visitor use.

ground, customs, crafts, philosophies, human values—are the rewards for many who travel. For others, the change of venue for common local activities—socializing, games, parties—is sufficient reason for travel to a distant location. Fantasizing the life style, play, and battles of ancient peoples has no stimulus equal to actually being in the theaters and coliseums of Greeks and Romans. These and hundreds of other experiences fill the rich and abundant cornucopia of travel values that must be considered in all tourism planning.

Planners and managers of tourism are challenged by the fundamental principle that the visitor is the only one to observe and use the full range of development, as illustrated in Figure 1-1. Every tourist trip cuts across hundreds, perhaps thousands, of developments resulting from independent plans and decisions by innumerable actors of all sectors. The traveler's interest spans a sequence of many experiences from leaving to returning home. The primary concern of the traveler is how satisfying these experiences are, not who owns or manages the individual places and services. This fundamental begs the question of how well the planners and developers of this complicated system understand and implement better planning and cooperative development of tourism as a whole.

Therefore, one major goal of collaborative tourism planning should be the *provision of user satisfactions.*

Improved Economy and Business Success

Most areas and nations place a high priority on an improved economy as a tourism development goal. Too frequently this is interpreted to mean merely a search for new investment in facilities and advertising. Although both are important, questions of what should be developed and where are equally significant. For planning purposes, one must reach into many factors that influence tourism's success. Such factors as geographical relationship to markets, attractions and attractiveness, resources for development, and involvement of all sectors need to be examined for their potential for developing a tourist economy. It is equally important to determine the social, environmental, and economic impact of every new investment.

The positive impact of tourism, no matter how measured, shows that tourism does strengthen the economy of many areas. This economic benefit is best understood as a "gross increase in the wealth or income, measured in monetary terms, of people located in an area over and above the levels that would prevail in the absence of the activity under study" (Frechtling 1987, 328). Economic benefits can be expressed by both primary and secondary effects as shown in Table 1-1.

Just now being recognized in the United States (and well accepted in other countries) is the inducement of foreign tourism through the expansion of domestic tourism. Evidence of the economic impact of tourism is so overwhelming that it is no wonder that undeveloped countries seek it and industrialized nations wish to protect it.

Frequently, in their desire to improve the tourism economy, governments constrain rather than assist the process. Lack of communication and even antagonism between governmental agencies and private enterprise frequently do not allow private investors adequate freedom to get an important commercial job done. When private development becomes "so involved in red tape and so expensive that it has no appeal to a potential investor," something is wrong (Brown 1976, 24). Successful private enterprise must make profits. Likewise, successful bureaucratic development, as in local, state, and federal parks and reserves, must meet those objectives of the common weal important to public goals. This kind of "success" is important for government agencies. Only adequate rewards

TABLE 1-1

ECONOMIC BENEFITS OF TRAVEL AND TOURISM

A. Primary or Direct Benefits
  1. Business receipts
  2. Income
    a. Labor and proprietor's income
    b. Corporate profits, dividends, interest, and rent
  3. Employment
    a. Private employment
    b. Public employment
  4. Government receipts
    a. Federal
    b. State
    c. Local
B. Secondary Benefits
  1. Indirect benefits generated by primary business outlays, including investment
    a. Business receipts
    b. Income
    c. Employment
    d. Government receipts
  2. Induced benefits generated by spending of primary income
    a. Business receipts
    b. Income
    c. Employment
    d. Goverment receipts

Source: Frechtling 1987, 330

of this nature can provide the incentive to develop. The desired economic impact then follows. Fragmentation and isolation of policies, regulations, and managerial practices tend to reduce greatly the potential of rewards.

Planning should not only address itself to the elimination of the problems above but also to the provision—the insistent provision—of positive rewards to those who identify, design, develop, and manage areas for tourism. In other words, cooperation, collaboration, and coordination must foster, not destroy, individual creativity and innovation in development to meet new needs. Public land and resource agencies should be guided into locations and programs that meet their special governmental mandates and yet are compatible with commercial enterprise and non-profit organizations.

Therefore, another goal of collaborative tourism planning is the provision of *increased rewards to ownership and development but only in an environmentally sustainable manner.*

Sustainable Resource Use

The controversy between environmentalism and development continues. This stems mainly from development other than tourism but carries over to it as an ideological conflict. This paradox is difficult to explain because so much of tourism depends upon resource protection.

Traditionally, travelers have sought destinations with attractive scenery, recreational waters, aesthetic landscapes, undeveloped mountain slopes and peaks, and protected wildlife. More recently, there has been a surge of interest in historic and archeological sites. All of these require environmental protection if they are to serve as foundations for development desired by visitors. Businesses must support measures that protect wildlife from endangerment, reduce soil erosion, reduce runoff from pesticides and other agricultural additives, and prevent toxic wastes and other pollutants from destroying valuable water resources.

Both a tourism economy and visitor satisfactions depend upon the absolute necessity of stopping resource degradation so flagrant around the world. Tourism businesses, governmental agencies, and nonprofit organizations developing tourism—all will have little to promote in the future unless attitudes and policies change. For some areas, it may already be too late. In order to develop tourism, the resources may already have been so severely diminished that no one will wish to invest and no traveler will wish to come.

The new trends toward sustainable tourism development (see chapter 3) are double-edged. They demonstrate new recognition of a large market segment interested in natural and cultural attractions. The trend has also encouraged greater energy conservation and recycling of waste. At the same time, excessive marketing that uses labels of "ecotourism," "green tourism," and "nature tourism" only as hucksterism is stimulating overdevelopment in fragile and rare resource areas. If it is used only as a tool for enticing more visitors, it reduces chances of true sustainable tourism development.

Essential then to all tourism planning is new commitment toward the goal of *sustainable resource use.*

Community and Area Integration

Many communities and regions view tourism as a separate layer that is simply added to a community. Engaging in tourism from this viewpoint is always disappointing because it fails to integrate tourism into the social and economic life of the community.

A study in Cairns, Australia, by Ross (1992) found that residents perceived both positive and negative impacts from tourism development. Concerns centered on cost of buying land and houses, cost of renting a house, cost of living, and crime levels. Positive results included more job opportunities, more business opportunities and greater entertainment, parks, shopping, and hotels and restaurants. As tourism developed, there appeared to be a lessening of friendliness and degradation of social life among residents.

Probably no other economic development of communities has so many far-reaching tentacles as does tourism. Tourism involves all the city's businesses, agencies, organizations, and segments of the public. It involves the local society and its ability to host masses of outsiders, often creating congestion, litter, and even competition for goods and services. It involves competition for land. It competes for amenities, such as parks, museums, and cultural events. And tourism often demands extra utility infrastructure, such as water supply, waste disposal, police and fire protection, streets, lighting, and maintenance.

Integrating tourism planning into official community planning has been slow to take place. The majority of planning goals for legal planning agencies have been directed toward the citizenry, not visitors. Although this is logical, it ignores the role of all city departments to cooperate in satisfying the needs of citizens as they host travelers. As Branch (1985, 76) points out, the city planning department should be the catalyst (but often is not) to coordinate the actions of the many city operating units. These units often include airport, animal regulation, art, attorney, building, controller, data services, engineering, health, fire protection, harbor, library, museum, parks, pensions, personnel, police, power plant, public works, purchasing, recreation, sanitation, social services, transportation, treasurer, and water supply. Certainly, the majority of these, with their separate policies and practices, have much to do with how a community is able to provide all the supply side development so necessary for long range tourism success.

However, the political and private organizational structure in most countries, such as the United States, tends to mitigate against long range planning. Branch (1985) cites as one cause the reelection process whereby politicians hesitate to make commitments beyond their tenure in office. He states as another reason that, to politicians, long range planning seems too difficult and demanding. These may be valid reasons why official urban planning tends to be short range. Because of increasing social, environmental, and economic ills of cities, the populace may be more supportive of long range and coordinated planning in the future—for all development including tourism.

An example of incorporating tourism development into official planning is the case of Viborg County, Denmark (Munk 1991). Both the physical comprehensive planning and business development program of the county include tourism as well as agriculture, recreation, and extraction of raw material. The aims of the overall plan include increased work opportunities, high quality of life, and sufficient public and private services.

The official plan is renewed every four years. All local authorities are involved in discussions of needed changes before revisions are made.

The aims of this county tourism plan follow those of the regional plan. The policy for tourism is to improve local conditions so inhabitants of the county, the trade, and institutions can receive Danish and foreign guests and give those guests experiences of great value involving the county's qualities and characteristics. At the same time, this policy of tourism will promote those forms that, on one hand can contribute to the economic and qualitative development of the region, and on the other, may thrive and develop together with local people as well as protect the county's natural, cultural, and environmental resources.

This is to be accomplished through five strategies: product development, marketing and information, education, public planning and administration, organization, and economy.

Generally, tourism's positive economic impact is believed to be of great enough value to offset the costs of integrating tourism into a community. But in order to make sure of this, tourism must be planned with the specific goal of fusing tourism with the social and economic life of a region and its communities.

Long ago, the rule of economic diversity was proven to be of value to communities. Diversity in the kinds of economic bases provides the best hedge against major drops in support from any one kind. The same policy is true for tourism. Today, as many communities and nations look to tourism for economic strengthening, there is danger if tourism becomes the sole economic provider. This danger is especially true with tourism because travel markets are less secure than local markets. Travelers are located some distance away, making each area vulnerable to competition from other destinations.

In recent years, awareness of tourism's potential negative impacts— social, environmental, economic—has increased. The tradeoffs in terms of employment, incomes, and tax revenues are often considered of greater value. But, in order to make sure of this, developers must *beforehand* analyze the potential threats and initiate plans and action programs to ameliorate them. Described in Part II are concepts and cases of participatory planning whereby people of a community are intimately involved in all tourism planning and development.

An important goal of tourism planning is *to integrate all tourism development into the social and economic life of a community.*

These four goals—*enhanced visitor satisfactions, better business, sustainable resource use, and community integration*—should be the motivating forces for all stakeholders in tourism to plan and develop the needed objectives and the strategies to carry them out.

TOURISM PLANNING SCALES

Experience with tourism planning demonstrates that even though the goals may be similar, there are differences in objectives and processes from the macro to the micro scale. The most popular today is at the *site* scale—individual property development for hotels, restaurants, resorts, roads, and attractions. However, when tourism functions are better understood, it becomes clear that there are many opportunities for better tourism success by planning at the scale of the *destination zone,* here defined as a community (or several) and the surrounding area. In order to determine greatest potential for a larger area, *a regional scale* (nation, province, state) of planning is needed. Even though planning is slightly different at these separate scales, integrating tourism development at all scales holds greatest promise for guiding development toward the desired goals.

The Site Scale

All three sectors—businesses, nonprofit organizations, government agencies—have increasingly employed professionals to plan and design their properties. Gradually, professional firms of architecture, landscape architecture, engineering, and planning have found opportunities in the tourism field. Owners of land have increasingly engaged their services to plan a wide range of facilities, services, and attractions. Perhaps the most popular approach is for the larger firms to bring together several specialists depending upon the needs of the project. For example, some projects require design and planning teams that include historians, golf course specialists, and exhibit experts as well as the traditional design specialists.

For better tourism site planning in the future, a major change in the design process is called for. Because the contractual agreement is between designer and client, important information about the eventual user, the tourist, is often omitted. Every site project for tourism must include detailed and current travel market information. Market researchers can provide pertinent market segment guidance for specific kinds of tourism development and location. They should be part of the design team.

At the lower end of the economic development scale, many entrepreneurs do not avail themselves of professional designers. Sometimes this results in unusually creative tourism developments, such as for bed-and-breakfast, craft stores, gift shops, farm vacations, and outdoor sports guiding, such as hunting, fishing, and trekking. These are especially successful when the owner-planner understands all factors of market demand as well as location.

However, this individuality often produces aesthetically ugly and non-economic tourism development. Just because an owner holds a low-cost piece of property is an insufficient reason for locating a tourist business. It may not meet the needs of travelers or gain from related businesses and attractions nearby. If built in an unattractive and shoddy fashion, it may be within the owner's budget but fail to meet the needs and desires of the traveler.

Too frequently, a series of such individual developments strung along a highway can be damaging on many counts. They destroy attractive roadside scenery. They create marginal business operations and often leave derelict monuments to business failure because they are improperly located for most favorable site and market factors. In a sense, nonprofit and public sector developments can create similar problems.

Governments, however, pose another issue. Park agencies, for example, generally follow one of two policy directions for site planning and design. Some maintain full-time staff with these responsibilities. The advantage cited is that they can respond more quickly to planning needs and that the planners are familiar with the agency's policies. Delays in letting out design and planning proposals to consultants are avoided. Some claim that this also saves public moneys by using staff rather than paying consultants with their overhead costs. Critics, however, believe that this "in-house" policy tends to prevent innovation and creativity. Furthermore, they remind the agency that there is wasted overhead at times when staff are not busy on projects. Some believe that public agency policies, such as the U.S. National Park Service agreements with concessions, place excessive constraints on what can be built as well as how it is to be designed, priced, and managed. Their policy of a concession monopoly for each park prevents competition, thus inhibiting the diversity of services and products desired by the travel markets. Although it may be desirable to have aesthetic and management controls over businesses within the park, nearby communities should be free to develop services for the complete diversity of travel markets.

There is little question that there is need for new cooperation and new guidelines between the decision makers and the designers-planners at the site scale for tourism development. The individual's freedom to develop tourism properties can be protected at the same time such owners can become enlightened on greater opportunities for success when exercising worthwhile principles of planning. Perhaps the most important principle is that of balancing internal with external influences on success. For example, a hotelier may be implementing state-of-the-art in-house management practices and still fail because external factors were not considered. These often include surrounding developments for traveler attractors and attractions, transportation, information systems, relevance to resident needs and desires, and the economic health of the community. Planning for every tourism facility and service must encompass many relationships if they are to be most successful. (Site scale tourism concepts and examples are described in chapters 9 and 10.)

The Destination Scale

As used in this book, "destination" refers to the "community-attraction complex" and "destination zone" as described elsewhere. If tourism development is to reach toward improved social, economic, and environmental goals, the *destination scale* (community and surrounding area) is as important to plan as the regional and site scales. The basic elements of planning at the destination scale are:

- Transportation and access from travel markets to one or more communities.
- One or more communities with adequate public utilities and management.
- Attraction complexes (clusters) that meet market needs.
- Efficient and attractive transportation links between cities and attractions.

Because all travel modes terminate at communities, they become very important for all tourism development. Yet few planning specialists focus on the community level. Most urban planners are concerned only about the resident needs because with their taxes they support a great many physical developments and services. Essential is the need for communities to plan and design transportation services for *both* visitor and local needs and preferences.

Communities are focal points in destinations and perform key roles for tourism. As a result of mass travel market growth and rapid expansion of transportation systems, today's tourist impact on communities has often exceeded their ability to accept mass tourists. Meeting travelers' needs of getting about, using services, and enjoying the attractions they came to see, compounds the challenge for planners and managers. For some cities, these new demands have exceeded capacities of basic infrastructure—water, waste, police, fire control, utilities, and streets. For others, prime natural and cultural resources that were basic to the area's appeal and were not adequately protected have been overrun by tourists.

A desirable planning principle at destinations is attraction clustering rather than dissemination. Clusters of attractions within and in the surrounding area are more efficient for development, visitor use, and operational management than when they are scattered about. Frequently neglected are the design and appearance of the travelway links between the community and attraction clusters. Travel through the most rundown part of town does not give a tourist the most aesthetic first impression of the attractions.

Proper planning and management can resolve most of these issues but require concerted action. Most have the power of governance that can control and direct growth for better land use. Because tourism impacts the entire community, participatory planning is essential. Residents deserve to know how new tourism development will affect them. A first step is to make a complete study of existing attractions and resources with future potential. Their size, location, quality, and capacities are key to future planning and development. Further steps include decisions on all other elements of tourism development, their planning and management.

These steps are similar to community planning approaches outlined by the National League of Cities (Dodge and Montgomery 1995). After organizing a process and analyzing the region's resources and anticipating changes, a critical step is the creation of a vision of the desired future. Such a vision must include action needed to overcome the existing pitfalls and a view of a better future. Next is a selection of strategies and implementation.

Destination zones must be planned with sensitivity to social, environmental, and economic impacts. Because a destination zone is a collection of a great many sites and encompasses numerous jurisdictions, considerable public-private cooperation is needed. Because of tourism's complexity, planners at the destination scale should encompass several interest groups—residents, businesses, arts and humanities, cultural and natural resource protection advocates, civic leaders, and professional designer/planners. Because of this diversity, controversies often arise. A catalytic role to resolve issues is often carried out by designer/planners. When a zone has been planned for attractions, access, services, and facilities, it can stimulate and guide developers into projects that provide the best relationships between all parts at the same time it fosters individual project success. (Destination scale concepts and examples are described in chapters 7 and 8.)

The Regional Scale

As important as planning tourism at the destination and site scales, is integrated planning at the regional scale. Most governments are preoccupied with the role of promotion, inducing more visitors rather than guiding development of the supply side to handle more visitors. Opportunities are abundant for greater involvement at the federal level, especially to avoid negative social and environmental impacts of tourism growth.

Today the public sector has the opportunity and even obligation of protecting and enhancing basic resources. Worldwide the environmental alert has been sounded—for example, the European Community established over 160 directives on environmental policy between 1973 and 1991 (Collins 1991). Key issues addressed include prevention of further degradation of water, air, wildlife, and other resource quality.

Planning at the regional scale is even more comprehensive than at the site and destination zone scales. Many more resource areas are involved; a greater number of political jurisdictions are included; and the time periods of accomplishment are much longer. Nevertheless, the main reason for planning at this macro scale is better integration of the whole. The better that each site and destination zone relates to others the better it will be for individual tourism success and for the region as a whole.

For example, today it is increasingly difficult for the isolated resort to succeed. It cannot benefit from related developments and programs. It has great difficulty in meeting market needs for the diversity of activities required at all seasons. Infrastructure costs may reduce feasibility because they are not shared with other developments.

An important planning effort would be greater collaboration among public sector units at the federal level, often restricted by tradition and turf protection. Tourism's future could be assured if public agencies of customs, transportation, agriculture, forestry, water, health, human services, and even defense would initiate new programs of integration on tourism matters. For example, overall regional transportation plans by both private and public sectors can benefit by knowing the needs of tourism. All purposes of personal movement to and within the region can be planned together. Too often tourism objectives are omitted from such plans.

Tourism planning efforts can benefit from the creation of a regional tourism plan. When a nation seeks investors, public and private, for tourism development, it is difficult to attract them with only generalizations about the attractiveness of the region. When regional plans identify the zones where opportunities are well founded on study of tourism development factors, investment feasibility is more appealing to investors and developers. As an example, the federal government of Canada took the initiative to provide all provinces with tourism planning methodology for guiding development (Gunn 1982). Recommended was a three-pronged annual process of study and distribution throughout every province: (1) a report on market-economic foundations for tourism development, (2) resource foundations for tourism development, and

(3) destination zone analysis and potential (based on parts 1 and 2). (Regional scale tourism concepts and examples are described in chapters 5 and 6.)

CONCLUSIONS

This chapter focuses on six fundamentals essential to planning tourism. First, tourism is defined as all travel not just pleasure travel—so that all aspects related to planning can be included. Second, the pluralistic nature of decision makers emphasizes that tourism is run by more than business. Third, all sectors need to understand the difference between their perspective of development and that of travelers. Fourth, experience has demonstrated that planning can and should be directed toward goals of visitor satisfactions, protected resource assets, and community and area integration, as well as the more commonly accepted goals of improved economy and business success. Fifth, the important changes in philosophies and processes of planning itself are described, emphasizing today's greater public involvement in planning. Finally, when planning for tourism development, there are differences in scope and content for regional, destination zone, or site scales. It is from these six principles that the following conclusions were derived.

Planning should encompass all travel.

In the past, the term tourism has often been applied only to pleasure travel. When tourism is viewed from the perspective of the individual firm, this definition is too limited. Hotels, transportation, food services, and shops are interested in sales to all travelers, no matter their purpose. Furthermore, if planning is to be comprehensive, it should include all the elements that need to be integrated for best success of all involved.

Planning must predict a better future.

Planning tourism is not merely a perfunctory or bureaucratic exercise. Its main purpose must be the long-term betterment of all involved. This means not only greater individual success but overall betterment through greater team action. Tourism involves so many individuals and organizations that it must be planned with greater unity of purpose. Unless planning can predict a better future it will be ineffective.

Both planning and plans are needed for tourism today.

It is valuable for a nation, state, or area to make specific plans for tourism development from time to time. These plans give focus and direct action to specific project and program development. However, to be most effective these plans should be coupled with a system of ongoing planning. Too often plans have been seen only as documents and not integrated with action on a regular basis.

Economic development must not be an exclusive goal of planning.

Tourism planning efforts are most popularly directed toward improving the economy—more jobs, income, and taxes generated. Although this continues to be an important goal it will not be achieved unless planning for the economy is accompanied by three other goals—enhanced visitor satisfactions, protected resource assets, and integration with community social and economic life.

Planning must incorporate all three sectors of tourism.

Well known is the business sector of tourism. Generally, it produces the greatest economic impact. However, tourism cannot function without the equally important sectors of nonprofit organizations and governments (as developers), especially for attractions, infrastructure, and transportation. Planning that includes only business will not succeed in reaching desired tourism development objectives.

Planning processes today are becoming much more interactive.

The new planning processes require involvement of decision makers and other influences at each step in the process. The experience, training, and conceptual ideas of professional planners are as essential as ever. But, the older perception of "top-down" planning is being replaced by "bottom-up" planning under coordinated professional leadership. The role of the planner/designer is becoming more strongly one of a catalyst or facilitator.

Three scales of planning need integration.

Site planning for tourism development, no matter how well done, falls short if it is not related to more macro-scale planning. How sites fit into community, destination, and regional development is critical to individual as well as collective nationwide success of tourism. Today, all three levels of tourism planning are in great need.

DISCUSSION

1. Discuss whether tourism can be planned, and if so, by whom.
2. What is the difference between plans and planning as applied to tourism development?
3. As compared to short-range planning, why is long-range planning so difficult?
4. How best can professional planners/designers accomplish better planning for tourism?
5. What are the barriers to reaching planning goals other than the economic, and how can they be overcome?
6. Why should communities and local residents be involved in tourism planning?
7. What role can national governments play in tourism planning?

REFERENCES

Branch, Melville C. (1985). *Comprehensive City Planning; Introduction & Explanation.* Washington, DC: American Planning Association, Planners Press.

Bridges, J. G. (1959). "A Short History of Tourism." *Travel and Tourism Encyclopaedia.* H. P. Sales, ed. London: *Travel World.*

Brown, D.R.C. (1976). "The Developer's View of Ski Area Development." *Man, Leisure, and Wildlands,* proceedings, Eisenhower Consortium, September 14–19. Vail, CO. Springfield, VA: National Technical Information Service.

Chadwick, Robin A. (1994). "Concepts, Definitions, and Measures Used in Travel and Tourism Research." *Travel, Tourism and Hospitality Research,* 2nd ed. J. R. B. Ritchie and C. R. Goeldner, eds. New York: John Wiley & Sons.

Cherry, Gordon E. (1984). "Town Planning: An Overview." In *The Spirit and Purpose of Planning,* 2nd ed. M. J. Burton, ed. London: Hutchinson, pp. 170–188.

Collins, Ken (1991). "Keynote Address," pp. 1–5. *The Planning Balance in the 1990s.* Papers of a Conference. London: Sweet & Maxwell.

Cuervo, Raimondo (1967). *Tourism as a Medium for Human Communication.* Itaxapalapa, Mexico: Mexican Government Department of Tourism.

Destination U.S.A. (1973). Report of the National Tourism Resources Review Commission, Vols. 1–6. Washington, DC: U.S. GPO.

Dodge, William R. and Kim Montgomery (1995). *Shaping a Region's Future: A Guide to Strategic Decision Making for Regions.* Washington, DC: National League of Cities.

Formica, Sandro (2000). "Tourism Planning," pp. 235–242, *Annual Conference Proceedings,* Travel and Tourism Research Association, June 11–14.

Frechtling, Douglas C. (1987). "Assessing the Impact of Travel and Tourism—Introduction to Travel Impact Estimation," Chap. 17, *Travel, Tourism and Hospitality Research.* J. R. B. Ritchie and C. R. Goeldner, eds. New York: John Wiley & Sons, pp. 325–332.

Gearing, Charles E. et al. (1976). *Planning for Tourism Development.* New York: Praeger.

Getz, Donald (1986). "Models in Tourism Planning: Towards Integration of Research and Practice," *Tourism Management* 7(1), pp. 21–32.

Gunn, Clare A. (1979). *Tourism Planning.* 1st. ed. New York: Crane Russak.

Gunn, Clare A. (1972). *Vacationscape: Designing Tourist Regions.* 1st ed. Austin: Bureau of Business Research, University of Texas.

Gunn, Clare A. (1982). *A Proposed Methodology for Identifying Areas of Tourism Development Potential in Canada.* Ottawa: Canadian Government Office of Tourism.

Inskeep, Edward (1987). "Environmental Planning for Tourism." *Annals of Tourism Research,* (14) pp. 1, 118–135.

Jafari, Jafar, ed. (2000). *Encyclopedia of Tourism.* London: Routledge.

Lawson, Fred and Manuel Baud-Bovy (1977). *Tourism and Recreation Development.* London: Architectural Press.

Leiper, Neil (1979). "The Framework of Tourism: Towards a Definition of Tourism, Tourist, and the Tourist Industry." *Annals of Tourism Research* (6) pp. 390–407.

Leiper, Neil (1993). "Defining Tourism and Related Concepts: Tourist, Market, Industry, and Tourism System." *VNR's Encyclopedia of Hospitality and Tourism.* M. A. Kahn, M. D. Olson, and T. Var, eds. New York: Van Nostrand Reinhold.

Mathieson, Alister and Geoffrey Wall (1982). *Tourism: Economic, Physical, and Social Impacts.* London: Longman.

Mill, Robert Christie and Alastair Morrison (1985). *The Tourism Systsem.* Englewood Cliffs, NJ: Prentice-Hall.

Munk, Inger (1991). "Quality of Life—Planning Processes and Models of Cooperation." (Viborg County) presentation, IVLA meeting, Paros, Denmark.

Records of the World Tourism Conference (1981). Madrid: World Tourism Organization.

Rose, Edgar A. (1984). "Philosophy and Purpose in Planning." In *The Spirit and Purpose of Planning,* 2nd ed. M. J. Burton, ed. London: Hutchinson, pp. 31–65.

Ross, Glenn F. (1992). "Resident Perceptions of the Impact of Tourism on an Australian City." *Journal of Travel Research,* 30 (3), pp. 13–17.

Chapter 2

Tourism as a System

INTRODUCTION

From the standpoint of the planners and developers of tourism, the many parts and actors are so numerous and complicated that they seem to defy any order. Each element rightly approaches tourism from its own perspective. Owner-developers of traveler lodging, for example, create a vision of the ultimate finished project and its management. This vision includes type of facility, such as health spa, beach resort, mountain lodge, urban conference hotel, motel, ecolodge, RV park, campground, hunting lodge, bed-and-breakfast, or some other. For every type, many factors are considered. Management characteristics, design and construction costs, buildability, and availability of finance influence the final decision. Today, it is likely that a multinational corporation rather than an independent entrepreneur will be influenced greatly by past successes of their projects. When estimated costs and returns are added, these factors will produce a feasibility study for the project.

Although this approach has been used in hotel education and practice, it is part, but not the whole, of tourism. After being established, such a business may fail or succeed depending upon several other factors. Internal operational management may include a good product, high level of service, and excellent managerial practices and yet not fulfill its role in tourism.

Part of this problem is confusion over the "product." Hoteliers believe their product is selling rooms. The restaurant focus is on selling food. Airlines direct their attention to selling passenger seats. These are necessary but miss the influence of many external factors involved in the overall tourism functioning system.

For tourism, a hidden hand guides an important interrelationship among the many parts that helps spell their success. When all these relationships are complementary, the system functions smoothly; when they are not, it breaks down. If, by means of integrated planning, these relationships are understood and fostered, tourism has a better chance of gaining its desirable goals of better visitor satisfactions, improved economy and business success, sustainable resource use, and community and area integration. This functional truth complicates planning but helps to explain why it is so necessary to view and plan tourism as an overall system.

The purpose of this chapter is to demonstrate that every part of tourism is related to every other part. No owner or manager has complete control of his own destiny. But the more each one learns about the others, the more successful he can be in his own enterprise no matter whether it is run by commercial business, nonprofit organization, or government. Tourism cannot be planned without understanding the interrelationships among the several parts of the supply side, especially as they relate to market demand.

THE TOURISM SYSTEM

One way of modeling the functioning tourism system is illustrated in Figure 2-1. The two main drivers of tourism consist of a Demand and a

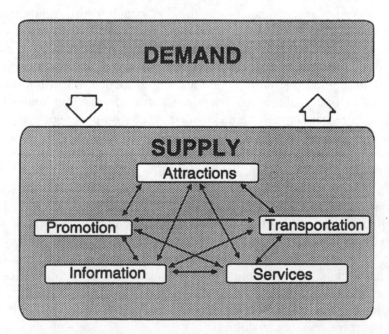

Figure 2-1. The Tourism Functioning System. Virtually all of the elements of tourism can be modeled as an interrelated demand and supply side. The five supply side components are interdependent and require planning that relates to market trends as well as to physical characteristics of land and resources.

Supply Side. Within these major forces are many details that all planner/developers must deal with for success. Although others may use different terms, this relationship is now much the same as it was in 1972 (Gunn, 21). Leiper (1979) describes the system in a similar manner with "tourist generating regions" connected to "tourist destination regions" by means of "transit routes." Boniface and Cooper (1987) called this a system of generating areas connected to destinations by routes traveled between these two sets of locations. No matter how it is labeled or described, tourism is not made up only of hotels, airlines or the so-called "tourist industry" but rather a system of major components linked together in an intimate and interdependent relationship. This model is one way of describing the *functioning tourism system.*

DEMAND-SUPPLY MATCH

In order to satisfy the market demand, a nation, region, or community must be able to provide a variety of development and services—the "supply side." How well this supply side matches the market is the key to reaching the ultimate in correct tourism development. Taylor (1980, 56) has called this the market-plant match and his model is illustrated in Figure 2-2.

Taylor based this on his observations in Canada that "the characteristics of tourism demand are changing rapidly and these changes outstrip the present ability of the plant to adjust and that a measurement system can be devised that will permit the plant to adapt to changing demands in a rational manner." Although the search for such a measurement system continues, there is fundamental logic in always striving for a balance between demand and supply. An Australian tourism research guide recommends steps for a "gap analysis," determining the difference between what travel markets seek and what is provided for them in the region (*Tourism Research* 1985, 14).

All government agencies related to tourism have the obligation of making sure their individual policies and practices provide the opportunity for linking travel market preferences with supply development. For example, national parks that have a dual policy of resource protection and visitor use need a full understanding of travel market interests and needs. Such a policy requires recognition of the traveler's complete needs for accommodation, food service, travel services, and perhaps entertainment but not necessarily within the park boundaries. Called for is cooperation with the surrounding communities to provide these services so that park management can guide visitor use that does not impair the environment.

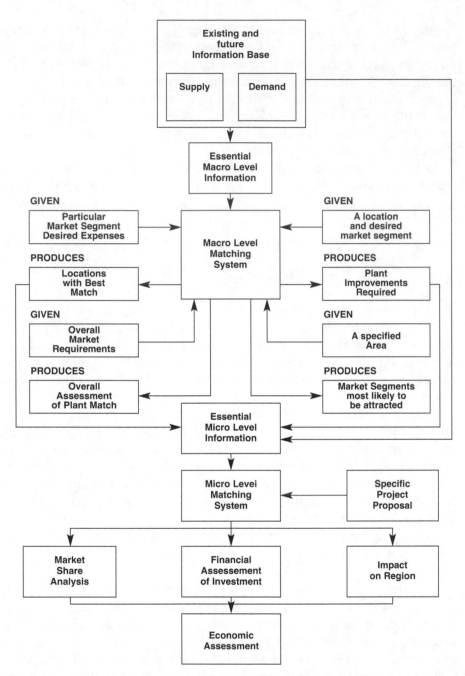

Figure 2-2. Plant-Market Match Model. This macro-micro systems model of tourism planning is directed toward matching appropriate supply development with travel market segment demand. It is a useful guide for designers of specific tourism projects (Taylor 1980, 58).

It is at the destination and site levels that demand-supply linkage is especially important. It is here that land use planning and controls can guide development into those zones best adapted to attractions and services. In the case of a small historic community, civic officials in cooperation with residents can identify tourist services in harmony with historic protection. At the site level designers/developers must be cognizant of market segment requirements and yet adapt development in balance with the local resources.

From this discussion it is clear that preferred demand-supply match *must be evaluated case-by-case.* Examples shown in Part II illustrate this point. Finally, because development policies and market trends continue to change, *the creation of demand-supply match is dynamic, not static.* All public and private actions must remain flexible so that annual adaptation may take place.

DEMAND

As any manufacturer knows, the best product to manufacture is one that is appealing or preferred by the market. This is equally true with tourism. An examination of the Demand Side of tourism reveals four major factors.

A tourist (for business or pleasure) must be *motivated* to travel. Those who do not have the interest, desire, and purpose for travel and stay home cannot be counted in the demand side of tourism. Tourism scholars and marketing specialists continue to probe the elusive human characteristics that influence one's desire to travel.

Within total populations, unless one has the *financial ability* to pay for services and facilities, travel is less likely to occur. Although a significant number still autocamp, stay in free hostels, or with friends and relatives, the larger market of tourist demand occurs in the middle to higher socioeconomic brackets. Air, automobile, and cruise ship travelers have the money required for the bulk of tourism demand.

As the population lives longer and as medical science and practice improve, more people are able to travel. However, unless they have the *time and physical ability* to travel they will not be counted in tourism demand. This factor is not limited to the elderly. Younger brackets may be restricted in their travel time by work schedules or physical disability.

These and many other factors make *predictions* of travel very difficult. *Forecasting* of travel demand is desired by the planner but is one of the most difficult things to accomplish. Forecasting is defined as the art of predicting the occurrence of events before they actually take place (Archer 1980, 5). As the complexities of travel increase—personal taste, environmental policies, international currency exchange, diversity of destinations—projections become more difficult and less reliable. Because planners, developers, and promoters are in constant need of forecasting, the concept continues to occupy an important place in market evaluation as well as development of supply. Although scientific research methods and statistical projections are used increasingly, forecasting, as defined, remains an art based on experience and judgment.

Uysal and Crompton (1985, 7) have provided helpful descriptions of qualitative and quantitative approaches to tourism forecasting of demand. Under qualitative approaches, three methods used by experts are described. *Traditional approaches* include review of survey reports to observe consistent trends and changes. Sometimes surveys within originating market sources are made to obtain the past history of travel as well as opinions of future trends. The *Delphi Method* is an iterative type of research inquiry using opinion of knowledgeable experts. It consists of several iterations by a panel responding to specific questions about trends. Each panel member operates in a manner unknown to the others. Of course this method relies heavily on the extent of expertise of the panel members and the influence of the director. But it is a useful tool, especially when used alongside other measures of prediction. A *judgment-aided model* (JAM) uses a panel in face-to-face contact and debate to gain consensus on several scenarios of the future. Each scenario is based on a different set of assumptions, such as political factors, economic tourism development, promotion, and transportation.

Among quantitative approaches, Uysal and Crompton describe three kinds. *Time series* studies are often statistical measures repeated year after year. Here it is assumed that all variables are working equally over the course of time. In order to reflect changes in influential variables, transfer function models have been developed but involve complex mathematical and statistical techniques. *Gravity and trip generation models* assume that the number of visits from each origin is influenced by factors impinging upon those origins. The primary factors are distance and population. Some researchers criticize gravity models on the basis of not reflecting price, not accounting for shrinking of distance perception by new modes of transportation, and other difficult variables. *Multivariate regression models* allow the use of many variables in predicting travel. Income, population, travel cost, international context, and other variables can be introduced.

This brief discussion is offered only to suggest that much experimentation in methods for forecasting demand is taking place. Some quantitative and statistical approaches can provide clues to future tourist flows. Although professional market analysis may be required for major planning projects, less complicated study by local people can be productive. As a guide, the Western Australian Tourism Commission has issued an excellent self-help publication, *Tourism Research for Nonresearchers*. In any case, understanding travel markets is essential to all planning for tourism development.

Market Segmentation

In the past any tourist was considered to be like all other tourists and all planning and management strategies treated tourists as a homogeneous whole. As has been found in marketing other products, there is much merit to dividing the totality of tourists into groups with similarities.

Market segmentation has been defined by Kotler (1988) as "the subdividing of a market into homogeneous subsets of customers, where any subset may conceivably be selected as a market target." He further offers three basic conditions which should be met for segmentation. First, there must be great enough numbers in each segment to warrant special attention. Second, there must be sufficient similarity of characteristics within each group to give them distinction. Third, the subsets must be viable—worthy of attention. When planning for physical development, as well as assessing social, economic, and environmental impact, it should be very helpful to have segmented refinements of potential tourist groups who might travel to the area.

For tourist market segmentation, many researchers have in the past put forward models of classification. The intent has been to help both marketers and developers make decisions on marketing techniques and creation of physical development. Experience has shown that market segments may vary greatly at regional, destination, and site scales.

A recent model, put forward by James Burke and Barry Resnick (2000), divides tourism markets into four segments—demographic, geographic, psychographic, and behavioristic. With local modification, these are helpful in planning tourism development. For all planning projects and processes, this conclusion suggests that the planners/designers need market characteristic input from professional travel market specialists.

Demographic segmenting refers to measurable personal characteristics, such as age, income, occupation, family size/life cycle, and educational level. Marketers seek to determine how these variables influence travel and the development of facilities and services. Each situation requires a mix of these factors to determine their significance.

Geographic segmenting is used to determine differences in similarities in travel preference due to traveler location. Location factors are important for decisions on air routes, attraction development within given travel distances, and decisions on destination development due to weather conditions.

Psychographic segmenting is a more recent method of grouping travel markets according to their values, attitudes, lifestyles, interests, activities, and personalities. Such grouping can help both marketers and developers direct their programs and projects to meet the interests of these groups.

Behavioristic segmenting divides the travel market into groups that have similar buying habits. Included are travel habits and preferences, purpose of travel, and benefits sought. By making this grouping, those who promote as well as those who plan and develop tourism can focus their efforts more precisely.

In addition to this classification of market segments, other researchers have considered other groupings that may be helpful to the planners. For example, anthropologist Smith (1992) has put forth a possible distinction between the *pilgrim* and *tourist.* Pilgrimages, travel with primary religious motivation, have become especially significant worldwide in recent years. Nolan and Nolan (1989) have described pilgrimages in three categories: centers of interest for religious tourism; shrines; and events related to religion, folklore, or ethnicity. Other scholars have documented the many forms of pilgrimages today and throughout history. However, Smith (1992, 4) points out that secular tourist travel has become increasingly intertwined with pilgrimage travel.

A generalized market segmentation, especially important to physical tourism planning, is by activities dependent upon development using *natural* or *cultural* resources. This has been the foundation for geographic assessment of destinations with tourism potential, as described in chapter 5. Forbes and Forbes (1992, 141) emphasize "special interest travel" as a growing segment, including adventure travel and ecotravel. They characterize these travelers as interactive, highly involved and interested in quality experiences, focusing on in-depth activities within destinations.

Planners and developers—public and private—must have current information on travel market characteristics in order to understand why, where, and what development is most appropriate.

COMPONENTS OF SUPPLY

Equally important in the functioning tourism system is the driving force of the Supply Side—all the objects and services that are provided to meet demand. The supply side includes all those programs and land uses that are designed and managed to provide for receiving visitors. Again, these are under the control of the policies and practices of all three sectors— private enterprise, nonprofit organizations, and governments. For purposes of planning, the supply side could be described as including five major components, as shown in Figure 2-1. Although others have described these with different labels, it is generally agreed that these represent the supply side of tourism. Jafari (1982, 2) refers to these as the "market basket of goods and services, including accommodations, food service, transportation, travel agencies, recreation and entertainment, and other travel trade services." Murphy (1985, 10) also includes similar components of the supply side. Mill and Morrison (1989, 2) combine attractions and services into a "destination" component. Focusing on community tourism, Blank (1989, 6) combines transportation, communications, attractors, services, and other community components for the supply side. But no matter how they are labeled, these are the components that *together* make up tourism supply.

Because planning tourism involves all the components in concert with each other, it is incumbent upon stakeholders to have current knowledge of these components. Hoteliers should be aware of trends in attractions, transportation, information, and promotion. Attraction developers must be aware of market trends as well as activity taking place in the other components of supply. So the following discussion is not meant to be exhaustive but rather to stimulate awareness of the major components of the supply side of tourism and their interdependence. For each component, based on critical observation of present trends, the authors have offered a key issue facing tourism planning.

ATTRACTIONS

The attractions of a destination constitute the most powerful component of the supply side of tourism. They make up the energizing power unit of the tourism system. If the market provides the "push" of traveler movement, attractions provide the major "pull." Service businesses are facilitators, not major causes of travel. Without attractions, these services may not be needed except for local trade. Attractions provide two major functions. First, they *entice, lure, and stimulate* interest in travel. As people in their residential locations learn about attractions of destinations, they make decisions on those that appeal the most. Or, for business travel, the trade center, convention center, or industrial complex may provide the pulling power. Second, attractions *provide visitor satisfactions,* the rewards from travel—the true travel "product."

Scope

Attractions are those developed locations that are planned and managed for visitor interest, activity, and enjoyment. Even though a destination may have an abundance of resources that are attractors, they are not functioning as true attractions until they are ready to receive visitors. Attractors and attractions have stimulated travel throughout the world for centuries. Attractions are numerous and extremely diverse.

Although the potpourri of current attractions seemingly lacks any similarity or definition of help to planning, attractions could be classified in several ways. Such classification may be of assistance to individual enterprises and other stakeholders in tourism when they plan for the future. Offered here are three classifications.

By ownership. Attractions are owned and managed by all three sectors—government agencies, nonprofit organizations, commercial enterprises. Table 2-1 lists examples of attractions classified by ownership.

By resource foundation. Attractions can be grouped according to the basic resource foundation, natural or cultural, as listed in Table 2-2.

By touring/long-stay. Although recent market trends have shown a striking reduction of time devoted to each trip, attractions could be classified by whether they are best adapted to touring circuit travel or long-stay in-place travel. Some examples are listed in Table 2-3.

Other classifications might be made, such as *outdoor versus indoor, by primary or secondary,* and by *market segmentation.*

Planning Considerations

Experience is demonstrating several planning considerations related to attractions. These are conceptual as well as based on research.

Attractions are created and managed. A popular error practiced by promoters of travel is listing attractive features prematurely. Until a site has been identified, designed, built, and managed for visitors, it cannot function as an attraction and should not be promoted. Historic homes as well as natural resource sites can be damaged greatly if hordes of visitors come too soon, needing parking, tours, and interpretation. Without proper design and management, valuable assets may be eroded.

Attractions are places in which the entire array of physical features and services are provided for an assumed capacity of visitors. Again, market and supply are the two sides of tourism that require close examination for attraction planning. For whom are plans being made and what are their interests? What are the features most critical for the site and how can visitors gain an experience without undermining the resource? What design and operational techniques are appropriate for solving these questions? An estimate of peak visitor volume is essential to the planning of every feature of the attraction—parking, trails, walks, exhibits, lectures, toilet facilities, tour guidance, spectator seating, and possibly food service and

TABLE 2-1

CLASSIFICATION OF ATTRACTIONS BY OWNERSHIP

Governments	Nonprofit Organizations	Business Sector
National parks	Historic sites	Theme parks
State parks	Festivals	Cruises
Wildlife reserves	Organization camps	Shopping centers
Scenic/historic roads	Elderhostels	Specialty food
Recreation areas	Historic architecture	Resorts
National monuments	Theaters	Golf courses
Wildlife sanctuaries	Gardens	Theaters
Zoos	Museums	Craft shops
Bike/hike trails	Parades	Plant tours
Sports arenas	Nature reserves	Race tracks

TABLE 2-2

CLASSIFICATION OF ATTRACTIONS BY RESOURCE

Natural Resource Foundation	Cultural Resource Foundation
Beach resorts	Historic sites
Campgrounds	Archeological sites
Parks	Museums
Ski resorts	Ethnic areas
Cruises	Festivals
Golf courses	Medical centers
Nature reserves	Trade centers
Organization camps	Theaters
Bike/hike trails	Plant tours
Scenic roads	Convention centers

TABLE 2-3

CLASSIFICATION OF ATTRACTIONS BY TOURING/LONG STAY

Touring	Long Stay
Roadside scenic areas	Resorts
Natural areas	Organization campsites
Historic buildings, sites	Vacation home complexes
Specialty food places	Gaming centers
Shrines	Dude ranches
Zoos	Convention centers

souvenir sales. The attractor may be the ecosystem, rare plant, landmark, or animal but the attraction is a developed and managed entity.

Attractions gain by being clustered. In today's mass tourism, the minor and isolated attraction requires so much time and effort by the visitor to reach that it is seldom worth it. Mass travel systems, such as fast trains, expressways, and air routes, necessitate stopping and walking before attractions can be enjoyed. This transportation factor supports the planning principle of several attraction features close by.

Attraction themes are best carried out when attractions are grouped together, physically or by tour. Evidence occurs in "garden tours," "historic tours," "architectural tours," and cruises. National parks are examples of attraction clusters, offering many complementary nature attractions, such as beautiful scenery, hiking trails, wildlife conservation parks, challenging topographic features, and outdoor recreation sites. Winter sports resorts frequently contain a combination of attractions: snow and ice sports areas, cross-country ski trails, competitions, indoor entertainment, and sometimes summer attractions for greater revenue production.

Clustered attractions have greater promotional impact and are more efficiently serviced with infrastructure of water, waste disposal, police, fire protection, and power.

Linkage between attractions and services is important. Attractions, although fulfilling a major portion of the travel experience, need support by travel services. Park plans, for example, are incomplete if the non-attraction needs of travelers are ignored. Food service, lodging, and supplementary services (purchases of film, drugs, souvenirs) must be within reasonable time and distance reach of travelers. This fact has posed a policy and planning dilemma for park planners for quite some time. It suggests that many attractions need to be planned for "day-use" only, providing the majority of services in nearby communities where they can be serviced more efficiently and gain from local trade as well as travelers. More remote attraction features, however, may require minimum services within the attraction, such as food service, toilets, and visitor centers.

Attraction locations are both rural and urban. Rural areas and small towns have their own assets to support attraction development. Some market segments prefer the homeyness and lower congestion of these areas. Table 2-4 lists some of the more popular tourist activities in rural areas and small towns (Gunn 1986, 2). Vernacular landscapes, such as farmsteads and rural scenic roads, demand special planning and control to assure scenic appeal in the future. Great growth has taken place with cruise ships. These represent floating resorts with complete entertainment, food service, and accommodations. Enjoying port city experiencees is an important part of the trip for many. Oceangoing ships are no longer merely a means of transportation.

Urban locations are equally viable for tourism development of both cultural and natural resource attractions. Urban rivers, parks, and nature centers as well as museums, theaters, arenas, auditoriums, universities, convention centers, and industries are foundations for attractions. Often, urban and rural attractions can be planned with complementary themes and linked together with tours.

TABLE 2-4

TRAVELER ACTIVITIES IN RURAL AREAS

Picnicking	Canoeing
Camping	Cross-country skiing
Hiking	Swimming
Horseback riding	Resorting
Bicycling	Historic touring
Hunting	Rural festivals
Fishing	Scenic touring
Boating	Visiting friends/relatives
Waterskiing	Nature appreciation

Issue: *As new attractions are developed by public and private sponsorship, a planning requirement is evaluation of potential negative as well as positive impacts—social, environmental, economic, community.*

SERVICES

Scope

Greatest economic impact from travel occurs through the travel service businesses. Accommodations, food service, transportation, travel agencies, and other travel businesses provide the greatest amount of employment, income, and taxes generated. This category is most frequently called the "hospitality service industry." Economists point to not only the direct impact but the multiplier effect. For example, hotels, restaurants, and retail shops offer specific products and services. But revenues received, in turn, provide economic support for contract food services, contract laundries, and indirect services such as housing, food, medical service, and transportation of employees.

Service Sponsorship

Fundamentally, commercial tourist services and facilities operate with the same purpose as all other business—to make a profit. However, there seems to be continuing misunderstanding of the term "profits," some believing that in the tourism, recreation, and resource development field, profitmaking is evil. Many, especially those sponsoring government recreation and park areas, seek a more altruistic and expansive social responsibility from business. But first and foremost is the responsibility for private enterprise to remain economically viable. Such economic viability comes from "profits," which in reality are costs of doing business. According to Drucker (1975),

> There is no conflict between "profit" and "social responsibility." To earn enough to cover the genuine costs which only the so-called "profit" can cover, is economic and social responsibility—indeed it is the specific social and economic responsibility of business. It is not the business that earns a profit adequate to its genuine costs of capital, to the risks of tomorrow and the needs of tomorrow's worker and pensioner, that "rips off" society. It is the business that fails to do so.

In a sense, all owner-managers of tourist services and facilities (governments as well as business) have similar ultimate goals: the satisfaction of needs of tourists. Crudely stated, a motel owner would not sell rooms if travelers did not arrive at that location seeking overnight accommodation. The businessperson has to be creative enough to develop the facility and service and offer it at a price acceptable to the public.

> It is the customer who determines what a business is. It is the customer alone whose willingness to pay for a good or a service converts economic resources into wealth, things into goods.... What the customer thinks he is buying, what he considers value is decisive—it determines what a business is, what it produces, and whether it will prosper. (Drucker 1973, 61)

For the planning of tourist services and facilities, it may be helpful to recognize differences among four types of ownership-management.

First, the *independent ownership and management*, typical of the "mom-pop" category of business, operates on its own forms of personal enterprise policies—market segmentation, pricing, range of services and facilities. One researcher (Bevins: 1971, 3) found that the economics of operation among the small outdoor recreation owner-managers varied greatly according to their goals. He grouped them into three categories: (1) those who did not wish to maximize financial returns but are in business because of belief in conservation, because of recreational values for family members, or for retirees to keep busy, (2) those who seek supplementary income for unemployed or underemployed family labor, and (3) those who seek the more typical economic goals of all business—revenues that will return on the investment. The trend of small tourist business continues to dominate in spite of the more conspicuous large and multinational firms.

Second, the *franchise, chain* and other multiple-establishment organizations have grown greatly in recent years. The advantages cited are greater marketing through single image and toll-free reservations, increased buying power, and uniform standards. Arrangements vary from those in which the properties (land and buildings) are owned and managed by the corporation to those that are independently owned but agree to certain operational standards and advertising logos for promotional advantages. Best Western in 1975 became the largest organization of travel lodging in the world (Best Western 1976, 1). A popular mode is that of the Holiday Inns, which use similar design of buildings, central purchasing, uniform signs and logos, and uniform operational standards. Most of the Inns are owned by local people who have a franchise arrangement with Holiday Inn.

Franchising, born in the United States, has promise for tourist service development elsewhere. It provides for local control but has the advantages of a larger organization. Experience has shown that there may be some difficulties in adapting franchising to other countries. Ashman (1986, 41) identifies a few: lack of understanding of its function, arbitrary governmental restrictions against it, labor requirements, property ownership laws, and quality control.

Franchising demands the right balance between centralized control and unit management. Kaplan (1984, 20) cautions against the application of centralized manufacturing techniques to food service organizations. Too much decision making at the top with few rewards to unit managers can divorce a company from the realities of consumers and service. Production speed, product quality, freshness, speed of service, sales promotion, as well as administration and training of personnel, operations, and quality control are best handled at the unit level. The more successful franchise operations recognize the importance of unit level decision making with adequate rewards at the same time efficiencies of large scale are obtained.

Third, *quasi-governmental* commercial operations, usually called concessions, are of increasing importance. In the United States, many federal and state resource and land agencies have concession agreements with private businesses to provide services to the public on government land. These include hotels, motels, trailer and other camping facilities, restaurants, stores, service stations, and marinas. Reasons cited for a concession arrangement—private profit-making operations on government land—are:

1. Private investment and arrangement reduces the need for public financing.
2. Revenues can accrue to public agencies from concession operations.
3. Innovation and economy may result from management responsibility shared with the private sector.
4. Local economies may be strengthened through profit opportunities for the private sector.
5. Greater recreation opportunity for the public may result from the provision of facilities or services which the managing agencies could not provide.

At the same time, several barriers exist to limit greater use of concession arrangements on public lands:

1. Concession interests may conflict with the management purposes for the public lands.

2. Such businesses are highly seasonal and profits are affected substantially by weather conditions.

3. Concessioners do not hold title to the land, making loans difficult to secure and tenure uncertain.

4. Federal and state civil service regulations may create difficulties in contracting for personal service.

5. Inconsistent or shifting public policies create uncertainty for entrepreneurs.

6. A high degree of onsite supervision by the public agency is generally needed in order to ensure acceptable standards of public service (Bureau of Outdoor Recreation 1973, 82).

Because each concession usually holds a monopoly as a business, it is not subject to the same competition as other businesses outside the control of the agency. Sometimes the political and managerial constraints can perpetuate bad service.

Fourth, *nonprofit organizations,* such as youth clubs and churches, often own mess halls, lodging, and campground facilities with extremely varying policies. Some are of poor quality due to weak financing and incompetent management and depend solely on donations for support. Others are virtually palatial resorts that are "profitable" in the sense that the revenues far exceed their immediate operating expenses. Some are oriented to conservation-resource protection whereas others are strongly program-oriented. Each depends upon the policies of its parent institution.

One area of contention between private enterprise and government is the problem of control. Participants of the public and private sector hearings of the U.S. National Tourism Policy Study (Senate Committee 1977, 30) cited several problems including time-consuming bureaucratic procedures, inadequate and unimaginative strategies for implementing programs, and a lack of continuity in implementing programs. Participants felt that ineffective implementation of federal programs had exacerbated inadequacies in tourism development activities, created difficulties for small business survival, conflicts between environmental and developmental goals, energy constraints on development, and inadequacies in promotion of travel opportunities in the United States, both domestically and internationally.

Fundamentals of Free Enterprise

Throughout the world, private businesses dominate the tourist services. The degree to which these are "free enterprise" depends upon the extent to which the free market system is not interrupted. It was the Scotsman Adam Smith, who, in 1776 in his *Wealth of Nations,* identified the basic principle of free enterprise as dependent on a *voluntary* exchange between buyer and seller. While pure free enterprise business may not exist, even in market-economy countries, the more it strives toward certain fundamentals, the more successful it is.

Allen, et al. (1979) has identified the following five fundamentals as essential to a free enterprise economy.

1. Private property

In a free enterprise system all property is owned by private individuals. This is based on several premises. First is the premise that individuals know best how to manage their property. The individual is believed to have a strong interest in not littering his property and conserving its resources because he is responsible for the consequences. The property owner has certain rights.

- The owner's right to determine how his property is used.
- The owner's right to transfer ownership to someone else.
- The owner's right to enjoy income and other benefits that come his way as a result of his ownership of the property.

2. Economic Freedom

By voluntarily cooperating with each other at the same time, individual interests are pursued and the freedom of individual choice is protected. No outside force, such as government, dictates this choice. The following rights are important but do not guarantee business success.

- The right to start or discontinue businesses.
- The right to purchase any resource they can pay for.
- The right to use any technology.
- The right to produce any product and to offer it for sale at any price.
- The right to invest in any way.

The seller and buyer make a voluntary exchange. The market, by its own selection, tells the producer what to produce and at what price. Of course, total economic freedom must be conditioned by the rights of society as a whole.

3. Economic Incentives

When there are incentives to work efficiently and productively, business becomes more efficient and productive. Workers receive incentives through wages and other rewards for doing good work. Businesses receive their incentives through profits. The more productive and the better a business meets market needs, the more profitable it usually becomes. However, punishments, in the form of business loss or failure, can come when the questions of what to produce and how to produce are not answered properly. For this system to function properly there must be a minimum of outside interference. Economic incentives serve to direct scarce resources to the production of goods and services the market values the most.

4. Competitive Markets

In a free enterprise system, the individual can choose and people vary in their preferences. These preferences of markets are expressed to producers by means of what is purchased. This means that there must be competitive businesses rather than monopolies. Each business can then strive for its market share. If it becomes very profitable, it invites competitors who seek their market share through even better products or services. This competition stimulates greater efficiency and lower prices. Competition spreads the decision of what to produce over many producers rather than by governmental decree.

5. Limited Role of Government

The greatest role of government in a free enterprise system is to stimulate business freedom and provide only basic rules and regulations for the good of society. Governmental intervention into day-to-day economic decision making is not part of a free enterprise system. It does not interfere with what or how to produce. Its role is to keep the system free and competitive.

At the same time that tourism has provided the opportunity for many entrepreneurs to create new travel-oriented service businesses, the field has been plagued with a high percentage of business failures. Some would argue that too many amateurs are attracted to these businesses. The business appears simple and glamorous to amateurs, who soon become disillusioned by the long hours, greater responsibilities, and lower profits than anticipated (Lundberg, 1979).

In order to remedy the tendency for excessive failures and poor service in many tourist businesses, many educational programs at all levels have been provided worldwide. Governments and business associations provide such guidelines for successful operations as *The Inn Business* (Minister

1982), produced by the Minister of Supply and Services, Canada. This guide provides constructive information on pertinent topics, such as entering the business, planning and development, operation (staff, repairs, maintenance, marketing, and other sources of help). In the United States, business advice and guidance is offered by agencies such as the Small Business Administration and U.S.D.A. Cooperative Extension Service.

Planning Considerations

Service businesses for tourism have both traditional and special planning needs.

Location and service are influenced by two markets. All the businesses providing basic and supportive services for travelers also serve local resident markets. Restaurants, shops, entertainment, and local transportation businesses receive much of their trade and revenues from residents as well as travelers. This fundamental influences location. Remote locations generally are much less successful compared to city and even small town locations.

A balanced economic base is more stable. When tourism and travel businesses provide the major economic input, the economy can fluctuate greatly with changes in travel markets. Industry and trade, combined with tourist businesses, provide the best balance of diversity.

Tourist businesses depend on urban infrastructure. Isolated locations require greater investment to provide for water supply, waste disposal, police, fire protection, and sometimes electrical power, compared to urban settings. This tends to encourage the location of lodging, food service, and other travel services in communities.

Businesses gain from clustering. At one time, entrepreneurs believed that they should locate away from their competition. Today the prevailing belief is that food services or lodging accommodations are best adapted to traveler demand when grouped together. When the traveler begins to think of needing food service, it seems best to be located near other kinds of food service.

Fragile environments should be avoided. Care in location is essential to avoid damaging the very reason for providing a service. Too often, tourist businesses have been located too close sites that have important value for attractions—n resource sites.

Services depend on attractions. Service business f ly related to attractions. Therefore, the business sect on plans for increased development of attractions.

attractions are based on natural and cultural resources, these businesses should exercise strong environmental protection advocacy.

Entrepreneurship is critical to tourism planning. Because of the dynamics of tourism, opportunities for innovative service businesses continue to appear. But if a culture does not have a tradition of entrepreneurship, it may have difficulty in creating new businesses. There needs to be a volume of business people interested in and able to see opportunity, obtain a site, gather the financial support, plan, build, and operate a new business. Small business continues to offer the greatest opportunity in spite of the many risks and obstacles.

Issue: *Because visitor service has frequently deteriorated in recent years, it must be improved wherever the investment-management hierarchy has placed greater emphasis on profits than on service.*

TRANSPORTATION

Scope

Passenger transportation is a vital component of the tourism system. It provides the critical linkage between market source and destination. Transportation between cities and attractions within urban areas and within attraction complexes requires special planning consideration. Except when touring is used as an attraction, transportation is not usually a goal; it is a necessary evil of tourist travel. Therefore, in the planning for tourism development, it is essential to consider all travel modes for people-movement throughout the circuit in order to reduce its friction as much as possible.

In contrast to a person's work transportation, which usually employs only one mode, it is not unusual for a modern tourist to utilize several modes on one trip. Planning increasingly requires intermodal considerations. It is not unusual for a tourist to utilize air, automobile, taxi, cable car, and horse carriage (in historic districts) on one trip. Because the several modes are designed, built, and managed by many different owners, a great amount of confusion and uncertainty can upset the traveler. Probably the increased use of motorcoach and cruise ship tours is due to their handling of all transportation arrangements, thereby reducing confusion. Poorly understood is the role of pedestrian movement in tourism. Increased traffic management has solved mass movement of vehicles but often destroyed personal amenities in the process. Attractions and tourist service businesses do need access but the final and most important mode is on foot.

Planning Considerations

All owners and managers of tourist attractions and services
stake in all transportation development policies and practi
routes, pricing, schedules, convenience, and interfacing ᵇᵉᵗ
can foster or spell disaster for tourism. This issue is further complicated
by the different needs of local as compared to long distance travelers.
Some highway planners design routes and capacities for business com-
muting only. Following are a few key planning considerations for the
important component of transportation.

The transportation sector must include tourism in its plans. Because
tourism has grown to major significance internationally, the transportation role must be strengthened. Modern engineering and technology have
greatly increased the quality of construction of highways, bridges, airports, railways, and harbors. But closer input from tourist service businesses and attraction leaders is needed in the transportation
decision-making process. Both can gain from greater integration.

Intermodal travel requires new planning cooperation. Tourist demand
is seldom directed toward a single transportation mode as created by
business and government. Increased availability (price, scheduling, airline options) of air travel has introduced many more destination choices
to the prospective traveler. But access to the specific attractions and circulation within a destination frequently put several other modes into play.
Increased popularity of package tours forces greater integration of travel
modes. If any one travel link fails to provide the quality of service
desired, the entire trip may be spoiled. The planning of intermodal transportation centers is needed for domestic local as well as outside visitor
markets. Today's floating resorts (cruise ships) still require planning linkage with port cities and their attractions even though most activities are
provided on board ship.

Transportation is more than engineering. Greater sensitivity to the
human dimensions of travelers is required for transportation planning.
Finding one's way is increasingly complicated with freeways which tend
to disorient and isolate travelers from their objectives. Better signage and
traveler information continues to present a planning challenge. The experience of travel, especially on byways and rural routes, is often a part of
the tourism product. Scenic routes require special controls on adjacent
land use.

Highways require greater sensitivity to the environment. Although
highway design and planning techniques have advanced in recent years,
special care to protect natural and cultural resources is needed. Although
traveler access is very important, the building of a major highway into

virgin territory can drastically upset the local social and physical environment. National park planners have experienced excessive road kills of wildlife along improperly placed highways.

Pedestrianism. Critical to planning for all travel targets is pedestrianism. The great majority of travel attractions are enjoyed on foot, outside the automobile, train, ship, or plane. Exception are the safari tour, helicopter and plane tour, and cruise ship tour. New design and planning are needed for handling greater volumes of travelers after they leave the mass transport vehicle. New routing, new surface preparations, and new visitor interpretation are challenges for newer, safer, and more satisfying pedestrianism.

Issue: *As demonstrated by the massacre of thousands of civilians on September 11, 2001, tourism is extremely vulnerable to major catastrophes. This event further endorses the principle of the interdependency of all components of the tourism system.*

INFORMATION

Scope

An increasingly important component of the tourism system is traveler information. Many public tourism agencies still confuse information with promotion. Advertising is intended to attract whereas information is to describe—maps, guidebooks, videos, magazine articles, tour guide narratives, brochures, Internet, and traveler anecdotes.

Although much of the provision of information is outside the realm of physical planning, one form of traveler information linkage—the visitor center—is growing rapidly. As ecotourism evolves and as visiting historic sites increases, there is danger of environmental damage. A popular solution is the creation of major visitor centers where masses of tourists can be managed and where they can gain a great experience without destruction of the environment.

Increasingly, planners of zoos, aquariums, nature centers, museums, interpretive centers, and visitor centers are developing facilities and programs to provide a surrogate attraction and richer visitor experience. For example, the Canadian Museum of Civilization, Hull, Quebec, stimulates, educates, and entertains the visitor with a variety of exhibits, live presentations, a children's museum, and Cineplus (a dramatic video experience). Over 900 years of history and cultural development are depicted in ways impossible by any other technique (Lancashire 1990). The Leid Jungle, the largest of its kind, at the Henry Doorly Zoo, Omaha, provides

61,000 square feet of exhibit space and contains 3,000 species of plants and 125 species of animals. Visitors gain an enriching experience without damage to fragile environments in their native settings (Cunningham 1992, 41). Several major aquariums give millions of visitors close exposure to marine life and capsuled environmental information with no impact upon real water resource settings. These and many other large attractions are beginning to respond to the plea of the Smithsonian Institution's Secretary (Adams 1992, 13) for greater responsiveness of museum specialists to the diversity and dynamics of social change. Museums need to be publicly recognized as important institutional means by which every travel group in our pluralistic society can define itself and represent its place within the complex, dynamic circumstances of contemporary life.

Simonelli (1992) has been working toward improved interpretive programs in the Canyon de Chelly, a Navajo area designated as a National Monument in northeastern Arizona. She emphasizes the need to balance the interests of residents, preservation managers, and visitors. This is not an easy task in an area occupied by the ancient Anasazi for two thousand years and by the Navajo for the last 250 years. She states: "The chief aim of interpretation is provocation, not instruction; it is revelation based on information" (Simonelli 1992, 20). Potential conflict arises between archeologists who wish to designate important prehistoric sites and the Indians now using these lands for agriculture and homes. Planning must consider the differing cultural values held by Indians and other Americans. Indians do not support competition, scientific foundations, the work ethic, and private ownership (Lew 1999). Plans for limiting visitor use, both spatially and quantitatively are being considered in order to protect the Indian culture—sustainable tourism.

Directly related to land planning and development for tourism are the behavior and attitudes of visitors. Social and environmental conflicts often result from lack of destination understanding on the part of visitors. Needed are ethics that help prepare travelers for their travel. Such behavioral education should include information on all topics that might cause conflict or reduce the likelihood of gaining anticipated experiential satisfactions. These topics could include:

- Weather conditions—needed clothing
- Physical demands—travel rigor
- Customs—host mores on dress, language, gestures
- Social contact—host-guest taboos
- Host privacy—trespass rules, regulations

- Foods—differences among
- Etiquette—behavioral mannerisms
- Religious beliefs—avoidance of conflict
- History—understanding backgrounds
- Politics—avoidance of conflict
- Communication—how to approach natives
- Facilities, services—different features, standards
- Health—avoidance of problems

Seminars, videos, books, Internet, Web sites, and tour guidance on these topics need not diminish the traveler's ability to obtain adventure.

Planning Considerations

Information segmentation is needed. In the past, some informational literature and guidance has been so generalized that no one really benefits. Instead, special places require special descriptive information and guidance. Historic sites, for example, may require several options related to the sophistication and interest of the visitor. Busloads of school children led by history teachers require different talks and exhibits from the casual visitor coming by personal automobile.

Information systems are not promotion. Much of promotion is directed toward the market before travel decisions are made whereas information is needed both before and during travel. Generally, roadside billboards are of greater scenic destructive value than either informative or promotional value for travelers. A mix of maps, guidebooks, well-marked highways, geographic positioning systems, Internet access, and visitor centers can assist the traveler in finding his way and understanding what he is seeing and doing.

Visitor centers are essential. A well-designed visitor center complex adjacent to resource-based attractions promises to solve many issues. First, it can be designed to accommodate personal cars as well as tour buses. It can handle masses of visitors without environmental damage to the primary resources of importance to the attraction. It can provide a vicarious resource experience for the visitor as a surrogate for direct contact and its accompanying noise, litter, and physical wear and tear. Cooperative planning between public agencies, nonprofit organizations, and commercial business can take much of the financial burden away from public agencies. A visitor center complex could include food service, retail sales (crafts, souvenirs), pageantry, museum, exhibits, and demonstrations as well as lectures, videos, and publications.

A national guidance system is ideal. Some nations, such as Australia and Great Britain, have planned and established networks of information centers of great value to the traveler. Tourist maps and roadside signs identify with a uniform symbol, such as an "I," where such centers are located. Here the traveler can obtain additional literature, maps, and personal counseling. In order to reduce costs, these may be incorporated into local businesses, such as restaurants and shops. This system offers better directions, is better liked by visitors, and creates less visual clutter than excessive roadside signs. The increased popularity of geographic positioning systems (GPS) is assisting greatly in finding one's way.

Local hospitality training is needed. Planning for improved tourism information for visitors requires an adequately informed local citizenry. Too often travelers are given no information or even misguidance when asking a local citizen for aid. Local hospitality training programs can be effective for improving knowledge of services and attractions as well as the ability to properly greet visitors.

Issue: *For better visitor understanding and resource protection, there are great opportunities for new interpretive centers, traveler guidance, and computer technology, a challenge to tourism planners and developers.*

PROMOTION

Although promotion is dominantly programs rather than physical development, it is an important component with strong linkage with all other components. Tourism promotion is a major policy and program activity of many nations, provinces, states, governmental developments, and businesses. Promotion for tourism usually encompasses four activities: advertising (paid), publicity (unpaid), public relations, and incentives (gifts, discounts).

Of great aid in promoting tourism is the recent explosion in the use of the Internet and e-business (electronic business). Very helpful to planners, developers, and managers are two publications produced by the World Tourism Organization: *Marketing Tourism Destinations Online* (Richer and Carter 1999) and *E-Business for Tourism* (Carter and Bedard 2001). Available to chambers of commerce and tourism organizations are electronic networks with linkages to supply side development. Web sites today are proliferating. Following entry to a Web site through a home page, topical displays can provide linkage to a variety of choices. These choices might include: attraction features, a trip planner, a brochure, maps, linkages with services and facilities, even booking. For the sophisticated traveler, detailed descriptions of sites may include photographs, historical background, physical features, and relationship to other attractions and destinations. It can link a site to a specific travel segment. Because Web sites are becoming so popular, destinations and sites can obtain a great amount of information about their competitors. The electronic age is here and can be of great aid in tourism planning, development, and management.

Because so much money is spent on promotion, the important planning linkage is the matter of *what* is promoted. All promotional planning must be closely integrated with all other supply-side planning and development.

For example, Baker (1992, 1) found that the marketing program of the U.S. National Park Service to promote off-scale seasonal use did accomplish that objective but exacerbated the overall use problem. Whereas over 95 percent of the park use was in June, July, and August before their campaign, the percentage dropped to 78 percent in 1990, after the program. But the program resulted in a significant *increase* in total visitors. This effort increased the burden on National Park staff, already overextended because of budget reductions.

Even though promotion is an important tool for increasing economic impact, it must be used with great sensitivity to the goal of user satisfaction, closely related to land development. For example, the planning and management of attractions within destinations may not allow the visitor to experience the view or the objective illustrated in promotional literature. A professional photographer, hired by the promoters, may have required special access permission and many days of waiting for ideal weather to obtain the beautiful and enticing image of a scenic or historic attraction—all of which were unavailable to the visitor. Understanding visitor use and site management are especially important for all tourism promotion.

Closely related to land planning for tourism is the use of billboards and signs along highways. Although informative signs at exits of freeways and at highway intersections may be needed, promotional signs have questionable value. In most instances today, they are less effective for luring visitors than other media—tour guidebooks, radio-TV spots, publicity, Internet, magazine articles, and word-of-mouth from friends and relatives. Furthermore, scenic appreciation of roadsides is such a strong desire among travel markets that defacing the landscape by billboards hardly seems desirable. Many regions, such as Hawaii, a very successful travel destination, have banned billboards and severely limited the use of signs.

Issue: *Misleading and unproductive promotion must be replaced by new ethical standards for better quality visitor experiences.*

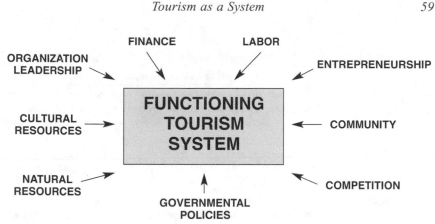

Figure 2-3. External Influences on Tourism System. All development of the supply side of tourism is influenced greatly by several externalities: natural resources, cultural resources, organizations/leadership, finance, labor, entrepreneurship, community, competition, and governmental policies.

EXTERNAL FACTORS

Such a core of functioning components of the supply side of tourism is greatly influenced by many external factors (Figure 2-3). Planning cannot be concerned solely with the core of the tourism system because all sectors may be as subject to outside influences as those under their own control. Several factors can have great influence on how tourism is developed. A brief examination of these may help in understanding this complicated reality of tourism, critical to planning, and the proper functioning of the tourism system.

Natural Resources

The popular emphasis on tourism economics and businesses tends to divert attention from important foundations for tourism development. Again, the *causes* of travel to a destination are grounded in the destination's resources, natural and cultural, and the attractions that relate to them. Even destinations such as Walt Disney World that seemingly are contrived and unrelated to the resource base in fact benefit greatly from it. Nearby Orlando and surrounding area have many complementing attractions—art museum, science center, a 72–building historic district, Lake Eola, wildfowl (ducks, geese, herons, anhingas, cormorants, moor hens), Leu Botanical Gardens, Florida Audubon Society Center for Birds of Prey, Bok Tower Gardens, and Wekiwa Springs State Park (Whitman, 1992). Other attractions nearby are A World of Orchids, Airboat tours,

TABLE 2-5

TOURISM DEVELOPMENT RELATED TO NATURAL RESOURCES

Resource	Typical Development
Water	Resorts, campgrounds, parks, fishing sites, marinas, boat cruises, river float trips, picnic areas, water scenic areas, shell collecting areas, water festival sites, waterfront areas, scuba diving sites, water photographic sites
Topography	Mountain resorts, winter sports areas, mountain climbing, hang gliding areas, parks, scenic sites, glacier sites, plains, ranch resorts, scenic drives, vista photography
Vegetation	Parks, campgrounds, wildflower sites, autumn foliage areas, scenic overlooks, scenic drives, vacation homes, scenic photography sites, habitat for wildlife
Wildlife	Nature centers, nature interpretive centers, hunting, wildlife observations, wildlife photographic sites, hunting resorts
Climate	Sites suited to sunbathing, beach use, summer and winter resorts, sites with temperature and precipitation suited to specific activity development

Green Meadow Petting Farm, riding stables, citrus grove tours, Jungle-Land Zoo, Lakeridge Winery & Vineyards, and hundreds of lakes, conservation areas, state parks, wildlife preserves, and nature trails.

Natural and cultural resources identify the uniqueness of place, very important to travelers and their objectives. Even a cursory review of publicity and advertising of travel today demonstrates the high value that promoters place on attractions related to natural resources. Generally, the term natural resources refers to five basic natural features especially important to tourism: water, topographic changes, vegetation, wildlife, and climate. Table 2-5 summarizes the relationship between these factors and tourism development.

Outdoor recreation has been a major travel purpose for many years. Although promoted primarily for its health and social values, outdoor recreation is very important to tourism economics. For example, a study in Texas (Texas Parks 1984, 6) revealed that Texas travelers spend approximately $9 billion annually on only 20 outdoor recreation activities. Critical, then, for future tourism development is the location and quality of the natural resources that support these activities sought by travel markets.

Probably the most popularly developed natural resource for tourism is *water.* Surface water is magnetic and has appealed to travelers for many years, stimulating many kinds of waterfront development. Ancient fresco paintings of the Egyptian dynasties include generous illustrations of water's attractiveness. Brittain (1958, 124) has aptly stated that in addition to commerce and defense, historically, water

... drew men together in common pleasures, strengthening, no doubt, a sense of individual participation in a larger life that enhances neighbors and strangers, and even foreigners from distant lands wearing their exotic clothes and clacking away in incomprehensible languages.

Reflection pools, ponds, fountains, rivers, lakes, waterfalls, and the seas continue to provide appeals that have no substitute. The appeal of water to both residents and visitors is bound up in cultures throughout the world. "We still like to go beachcombing, returning to primitive act and mood. When all the lands will be filled with people and machines, perhaps the last need and observance of man will be, as it was at the beginning, to come down and experience the sea" (Sauer 1967, 310–311). It is for its great value to tourism that water quality and its protection must be seen by all sectors as essential to tourism's success—economically as well as socially and environmentally.

Historically, and even today, *topographic change*—hills, mountains, and valleys—provides the physical setting for much of tourism. Land relief is an essential ingredient in contemporary culture's assessment of landscape scenery, now heightened by the boundless popularity of photography. Hillsides and mountaintops offer spectacular vistas, near and far. Mountain resorts, winter and summer, retain their appeal for contemporary travel market segments. Related to topography are soils, of significance to tourism development—construction stability, landscape modification, erodibility. Because some mountainsides and slopes are highly erodible, resource protection must be part of the catechism of tourism development. Also related is the geological foundation, often influencing the stability of land and lakes, the absorptive capacity of sewage, and reliability of water supply.

For many kinds of tourism development, from the tundra of the north to the rain forests of the tropics, *vegetative cover* is an important natural resource for tourism development. While deserts appeal to some tourists, much more popular are verdant landscapes. Forests create appealing scenic vistas, support wildlife, offer dramatic panoramas of color in autumn, and aid greatly in preventing soil erosion. Often special plant areas (redwoods, the Big Thicket, silverswords in Hawaii, Michigan jackpine for Kirtland warbler) are singularly important travel destinations for some market segments. Wildflowers are spectacularly attractive in forests in the North and over open fields in the South in springtime. But forested and vegetated regions are extensive and are subject to varying policies by owners and managers. Some timber harvest practices, such as clear-cutting, destroy landscape scenery and stimulate soil erosion. Vegetation is dynamic; trees sprout, grow and die and may be damaged by disease and fire. Management for tourism requires special policies and practices if this resource is to maintain its value to tourism.

Once primarily of interest only to travel segments interested in game hunting, *wildlife* today is even of greater importance for non-consumptive tourist markets. Viewing and photographing wildlife have grown significantly in recent years. It is estimated that about $31 billion was spent by travelers on observing, feeding, and photographing wildlife in the United States in 1996 (USFWS 1997). Photo safaris are far more important today in Africa than hunting ever was. Color and digital photographs and videos are becoming important tourist trophies. Animal habitat management is necessary if the resource is to continue for tourism. Some wildlife is extremely sensitive to human intrusions, requiring special design and management techniques if visitors are to be enriched by this resource. An issue today is conflict between urban sprawl and invasion of wildlife.

Climate and weather are qualities of place that greatly influence the planning and development of tourism. Travelers generally prefer sunny weather, even in winter sports areas and certainly for beach activities. For example, for many of the national parks of the U.S., peak visitation occurs during sunniest weather. Some northern countries, such as Canada, do not try to promote travelers seeking sunny and warm beaches but other attractions more appropriate to their climate. Without doubt, climate plays an important role in the popularity of Hawaii and Caribbean islands. There is little evidence to suggest that storm hazards—lightning, tornadoes, hurricanes—have more than a temporary impact on travel. In fact, some fishing in the Gulf of Mexico is stimulated during periods of hurricanes. Related to climate are conditions of air quality. Although air quality controls are lessening air pollution in some parts of the world, travelers do object to areas where odor, manufacturing gases, and automobile pollution are prevalent. The new wave of "sustainability," "green tourism," and "ecotourism" (as presented in chapter 3) are evidence of new awareness of the importance of natural resources and their protection for tourism's success.

This brief review should be sufficient to endorse the need for vigorous natural resource protection advocacy for all tourism sponsors and developers in order for the tourism system to function at its best.

Cultural Resources

In recent years, several travel market segments have increasingly sought destinations with abundant cultural resources. This category of resource base includes prehistoric sites; historic sites; places of ethnicity, lore,

TABLE 2-6

TOURISM DEVELOPMENT RELATED TO CULTURAL RESOURCES

Resource	Typical Development
Prehistory, Archeology	Visitor interpretive centers, archeological digs, prehistory parks and preserves, nautical archeological sites, festival sites related to prehistory, exhibits and customs related to prehistory
History	Historic sites, historic architecture, historic shrines, museums depicting eras of human history, cultural centers, historic pageants, festivals, landmarks, historic parks
Ethnicity, Lore, Education	Places important to legends and lore, places of ethnic importance (customs, art, foods, dress, beliefs), ethnic and national cultural centers, pageants, festivals, dude ranches, gardens, elderhostels, universities
Industry, Trade, Professionalism	Manufacturing and processing plants, retail and wholesale businesses, conference centers, educational and research institutions, convention centers, performing arts, museums, galleries
Entertainment, Health, Religion, Sports	Spas, health centers, fitness resorts, health specialty restaurants, religious meccas, shrines, sports arenas, night clubs, gaming casinos, theaters, museums (history, art, natural history, applied science, children's, folk), art galleries

education; industries, trade centers, professional centers; places for performing arts, museums, galleries; and sites important for entertainment, health, sports, and religion. It is estimated that cultural tourist development in the U.S. attracted 214.1 million person-trips from domestic travelers in 1999 (TIAA 2001). Examples of development related to cultural resources are shown in Table 2-6.

Peterson's research (1990, 209) has categorized cultural travelers as aficionados (sophisticated, professional), casual visitors (urban backyard visitors), event visitors (activities at sites), and travel-tourists (historic site visitors). She cites three reasons for visiting cultural sites: experiencing a different time or place, learning, and sharing knowledge with others. A major international conference on cultural and heritage tourism (Hall and Zeppel 1990, 55) concluded that, in spite of the surge of interest within the travel market, there are major gaps in planning and operation of such attractions. Stressed was the need for greater public-private cooperation. (Twenty papers presented at this ICOMOS conference are contained in "Cultural Heritage and Tourism," (1990) *Historic Environment*, (7), Department of Environment and Planning, Adelaide, South Australia: 3–4). The field of cultural resources spans virtually all resources except those that can be called "natural."

The travel market interest in *prehistory* and archeology has stimulated development of these resources for visitors. Locations where scientists are discovering structures and artifacts of ancient peoples are of increasing interest to travelers. Nautical archeology (discovery and analysis of ancient ship transport and ways of life) is becoming as important as terrestrial archeological digs, but, because of their rarity, these sites must be under rigid control to prevent their destruction by visitors. It has been estimated that illegal trade in antiquities is approximately $4.5 billion a year in the U.S., ranking fourth in illicit activities after drugs, guns, and money laundering (Black 2001). Archeologists emphasize the fact that the context (relationship to setting and other artifacts) is more important than the artifact. Documentation of what these clues suggest for ancient peoples—dates, foods, customs—is more important than collecting. Special design and management policies, such as restrictions against collectors and treasure hunters and the establishment of interpretive visitor centers and museums, are needed to handle volumes of visitors to prehistoric sites, terrestrial as well as marine.

As travelers have become more sophisticated, they have much greater interest in *historic* areas. The topic of history deals with the documented past. Even though every place has a history, places of local significance are of less interest to visitors than those of state, provincial, national, or world importance. Generally governmental agencies and nonprofit organizations have been the leaders in preserving, restoring, and developing sites important to history. For tourism, sites, structures, and events related to places are the foundations for historic attractions. As with archeological sites, historic sites require very special control, design, and management so that the resource is protected at the same time visitors gain historic appreciation and enriching experiences. It is important for tourist businesses to support the development and maintenance of historic sites because they stimulate the market for services.

For discussion purposes, places important for *ethnicity* lore and *education* have been grouped together as a category of cultural resource foundations for tourism development. Travel interest in the exotic and special customs, foods, costumes, arts, and entertainment of ethnic groups continues to rise. As an example, forty-two percent of the visitors to South Dakota want to see Indians (Mills 1991). Because native resources are rooted in the past, they are prone to disappear because of the social and economic desire of localities to progress and modernize. Many cultural organizations have established programs to protect early cultural elements and special design and management of places is required to develop such places for tourism. For example, Barry Parker, executive director, First Nations Tourism Association of Canada, has identified goals and objectives for that organization (Parker 1991, 11):

Goals:

> To position native tourism business as a major player in the Canadian tourism industry.
>
> To preserve, protect and promote cultural uniqueness in the tourism industry.
>
> To facilitate growth in the Canadian native tourism industry.

Objectives:

> Communications—to enhance image/perception by establishing a data base and networking system.
>
> Human resource development—to coordinate national level training to ensure cultural integrity through standards, quality, certification.
>
> Advocacy—to influence policy development at the federal, provincial and territorial levels.
>
> Marketing—to develop a national marketing strategy.

Evidence of progress on this topic is illustrated in Part II, especially the planning and policies of the Hopi in Arizona. Close cooperation between planners and ethnic groups is essential in order to avoid misinterpretation that may demean a past society. Often legends and lore are as important to visitors as true ethnic culture.

Universities, colleges, technical institutions, and research centers are of interest to many travelers but require special access, exhibits, and tour guidance for tourism.

Travel objectives of *industry, trade,* and *professionalism* continue to be very important for several travel segments. These objectives are often combined with pleasure. Manufacturing and processing plants are not only of interest to business travelers but also to pleasure travelers if the sites provide tours, facilities, and services for visitors. Trade and business centers are important cultural sites for many travelers. Places that establish meeting services and convention centers are attracting many travelers for professional and technical seminars, meetings, and conventions. In spite of the growth of Internet and telecommunications, face-to-face meetings and conferences are still important. And many areas are major tourist objectives because of the diversity of shops. Shopping is a very important activity for a great many travelers.

Places for *performing arts, museums,* and *galleries* are very important for a great many travelers. Tighe (1988) cites many examples of the significance of cultural tourism. Aspen, Colorado, known primarily for its skiing, also hosts over 55,000 people annually for a music festival and other performing arts activities. The Spoleto Festival of Charleston, South Carolina, holds 125 performances a year with over 90,000 in attendance. The Port Authority of New York-New Jersey reports arts institutions contribute $5.6 billion annually to the economy. In all instances, a high percentage of attenders are tourists. The United States Travel and Tourism Association (USTTA)'s in-flight surveys have indicated that about 27 percent of all overseas visitors to the United States went to an art gallery or museum and some 21 percent went to a concert, play, or musical. In 1984 the Los Angeles Olympic Arts Festival drew 1,276,000 people. The Travel Industry Association of America reported that domestic travelers took 214.1 million person-trips to cultural attractions in 1999.

Finally, cultural resources also include places that provide for *entertainment, health, sports,* and *religion.* Health spas, centers for physical fitness and weight reduction, and special medical treatment become travel objectives for many travelers. Sports arenas throughout the world attract millions of visitors to special events such as the Olympic Games. Some communities are known as centers for certain religious groups. Others attract many visitors because of cultural resources such as gaming casinos, music halls, opera houses, and nightclubs.

Entrepreneurship

Because tourism is dynamic, needed are entrepreneurs who visualize opportunities for new developments and creative ways of managing existing developments. The ability to see an opportunity, to obtain needed financing, to obtain the proper location and sites, to engage designers to create physical settings, to gather the human resources needed to manage the physical plant and services is important for travel development. For industrialized nations, entrepreneurship is a part of the culture. The lack of this factor in many undeveloped countries is a major handicap that increases the difficulty of creating and expanding tourism.

Finance

Certainly, for the development of tourism, capital is required. But the ease of obtaining the financial backing for tourism varies greatly. Public and private lenders are often skeptical, have a negative image of the financial stability of tourism generally, and are slow to accept innovations. Because so much of the tourism physical plant is small business and has attracted many inexperienced developers, some of this reputation is justified. However, recent trends have demanded much greater business sophistication and higher capital investment. Tourism does take considerably more capital than is popularly believed. Investors are more likely to support projects that demonstrate sound feasibility. Financial backing is an important factor for both public and private tourism development.

Labor

The availability of adequately trained workers in an area can have considerable influence on tourism development. As markets demand higher levels of service, well-trained and competent people are in greater need. The popular view that the untrained can perform all tasks needed in the diversity of tourism development is false. When the economic base of any area shifts, those taken out of industrial employment may be retrainable but are not truly available for tourism jobs unless such training is provided. Remote locations become more costly for development because employees must be housed on site. The labor capacity of an area has much to do with tourism development.

Competition

The freedom to compete is a postulate of the free enterprise system. If a business can develop and offer a better product, it should be allowed to do so in order to satisfy market demand. However, before an area begins tourism expansion it must research the competition—what other areas can provide the same opportunities with less cost and with greater ease. Is there evidence that tourism plant has already saturated a market segment? Certainly, competition is an important influence upon the tourism system.

Community

A much more important factor influencing tourism development than has been considered in the past is the attitude toward tourism by the several community sectors. While the business sector may favor greater growth of tourism, other groups of the local citizenry may oppose it on the grounds of increased social, environmental, and economic competition for resources and other negative impacts. Political, environmental, religious, cultural, ethnic, and other groups in an area can make or break the proper functioning of the tourism system.

Governmental Policies

From federal to local governing levels, statutory requirements may foster or hinder tourism development. How the laws and regulations are administered—loosely or rigidly—can influence the amount and quality of tourism development. Policies on infrastructure by public agencies may favor one area over another. The policies of the many departments and bureaus can have a great bearing on how human, physical, and cultural resources are utilized. Smooth or erratic functioning of the tourism system is greatly influenced by governmental policies.

Organization, Leadership

Only recently being recognized is the great need for leadership and organization for tourism development. All planning is subject to implementation by many sectors. Many areas have hired consultants to identify tourism opportunities but frequently such plans for development have not materialized due to lack of organization and leadership.

Without doubt, as tourism development research and experience broadens, more influential factors will be found. Any planning for tourism in the future must take into account the core of the tourism functional system and the many factors influencing it.

CONCLUSIONS

Every stakeholder of tourism will *gain,* not lose, by making plans in the context of tourism as a system. Governmental agencies can gain because their plans and decisions on parks, highways, infrastructure, and promotion will be more supportive of development by the other sectors. As capitalistic and market economies grow, privatization can be integrated to a higher degree with public agency activities. Nonprofit organization development of tourism can fulfill goals and objectives more successfully if it is designed and managed in the context of the overall tourism system. Certainly, the business sector of tourism will benefit greatly when it takes advantage of the complementary action by the other two sectors. And, finally, the tourist and the travel experience, the true product and purpose of all tourism development, will gain because the system is working in greater harmony. Travelers benefit when all parts of all supply side components make their travels easier, more comfortable, and more rewarding. Difficult and challenging as system planning for tourism may be, it holds promise of greatest rewards for everyone. All parts depend upon one another for smoothest functioning. By considering tourism functions as a system, several conclusions can be drawn.

Tourism planning at all scales is most productive when done in the context of the tourism functioning system.

Because every part of tourism is related to and influenced by every other part, its design/planning must consider this relationship. No hotel, airline, or theme park can be properly planned without consideration of market demand, other supply side components, and many externalities. Public park and recreation areas function not only as protected resource lands but also as attractions within the overall tourism system.

Markets, as well as supply, drive tourism development.

Critical to all tourism development and its planning are the many characteristics of travelers' tourism demands. All physical development and programs must meet the interests and needs of travelers. If not, economic rewards may not be obtained, the environment may be eroded, and local conflict may ensue. Planning for visitor interests can ameliorate or prevent these negative impacts. All sectors seeking improved tourism must be fully cognizant of market characteristics and trends.

Supply side components are owned and managed by all three sectors.

Supply side development is not exclusively under control of the business sector. All five major components of supply—attractions, transportation, services, information, and promotion—are created and managed by governments and nonprofit organizations as well as business. This means that for tourism to function properly, planning should integrate policies and actions of all three sectors.

External factors impinge on the functioning of the tourism system.

The tourism system does not operate in an isolated manner. Several factors need to be analyzed and worked into plans for best future operation of the system. These external factors include: natural resources, cultural resources, entrepreneurship, finance, labor, competition, community, governmental policies, and organization and leadership.

Business success depends on resources and their protection.

Tourist business enterprises are as dependent upon natural and cultural resources as internal management. Good business practice is not the only cause of travel. Equally important are the attractions nearby that, in turn, depend primarily on basic natural and cultural assets. Without protection, restoration, and visitor development of these assets, business cannot thrive.

Tourist business location depends on two markets.

All tourist businesses gain revenues from sales of products and services to local as well as travel markets. Therefore, their business operations, and especially site locations, must be planned to serve both markets. It is important for all community planning to recognize this fundamental for best economic input.

Matching development of supply with demand is a constant challenge.

Because travel demand and supply development are dynamic, plans must be updated regularly. In addition to project plans, required is a continuing planning process, by the public as well as the private sector.

The tourism system requires integrated planning.

Even though private and independent decision making are cherished by most enterprises in all tourism sectors, each will gain by better understanding the trends and plans by others. The public sector can plan for better highways, water supply, waste disposal, parks, and other amenities when private sector plans for attractions and services are known. Conversely, the private sector can plan and develop more effectively when public sector plans are known.

DISCUSSION

1. How can developers of the supply side of tourism gain better information on market characteristics?
2. Discuss why travel markets are dynamic, not static, and what this has to do with planning.

3. How can individual tourism designers/planners and developers communicate better with members of the several components?

4. Speculate on how a destination can avoid obsolescence if its travel market chooses other destinations.

5. Consider creative solutions to the several planning and development issues raised in this chapter.

6. What should be the role of government in fostering rather than interfering with the success of free enterprise?

7. How can tourist service businesses become more proactive for resource protection and how would they gain?

8. Discuss why urban planners have not incorporated the tourism system and its implications in their village, city, and town plans.

REFERENCES

Adams, R. McC. (1992). "Smithsonian Horizons." *Smithsonian*, 23 (1) April: pp. 13–14.

Allen, J.W. et al. (1979). *The Foundation of Free Enterprise*. Center for Education and Research in Free Enterprise. College Station, TX: Texas A&M University.

Archer, Brian H. (1980). "Forecasting Demand: Quantitative and Intuitive Techniques." *Tourism Management,* 1 (1), pp. 5–12.

Ashman, R. (1986). "Born in the U.S.A." *Nation's Business*, 74 (11) November: 41ff.

Baker, P. (1991–1992). "The National Park's Unique Marketing Phenomenon." *Arizona Hospitality Trends*, 6 (1) Winter: 1ff.

Best Western 1975 Annual Report (1976). Phoenix: Best Western.

Bevins, M. I. (1971). *Private Recreation Enterprise Economics,* proceedings of the Forest Recreation Symposium, Pinchot Institute Consortium for Environmental Research.

Black, Kent (2001). "The Case of the Purloined Pots." *Smithsonian* 32 (6), pp. 34–44.

Blank, Uel (1989). *The Community Tourism Industry Imperative: The Necessity, The Opportunities, Its Potential*. State College, PA: Venture.

Boniface, Brian G. and Christopher P. Cooper (1987). *The Geography of Travel and Tourism*. London: Heinemann.

Brittain, Robert (1958). *Rivers, Man and Myths*. Garden City, NY: Doubleday.

Bureau of Outdoor Recreation (1973). *Outdoor Recreation for America: A Legacy for America*. Washington, D.C.: USGPO.

Burke, James F. and Barry P. Resnick (2000). *Marketing and Selling the Travel Product*. 2nd ed. Albany, NY: Delmar/Thompson Learning.

Carter, Roger and Francois Bedard (2001). *E-Business for Tourism*. Madrid: World Tourism Organization, WTO Business Council.

"Cultural Heritage and Tourism" (1990). Department of Environment and Planning, Adelaide, South Australia. *Historic Environment*, (7): pp. 3–4.

Cunningham, D. (1992). "The Lied Jungle." *Nebraskaland*, 70 (2) March: 40ff.

Drucker, Peter F. (1975). "The Delusion of Profits." *Wall Street Journal*, February.

Drucker, Peter F. (1973). *Management*. New York: Harper & Row.

Forbes, Robert J. and Maree S. Forbes (1992). "Special Interest Travel." *World Travel and Tourism Review*, pp. 141–144. Oxon, UK: C.A.B. International.

Gunn, Clare A. (1986). "Small Town and Rural Tourism Planning." In *Integrated Rural Planning and Development*, Floyd W. Dykeman, ed. Sackville, NB: Mount Allison University. pp. 237–254.

Gunn. Clare A. (1972). *Vacationscape: Designing Tourist Regions*. Austin: Bureau of Business Research, University of Texas.

Hall, C. M., and Heather Zeppel (1990). "History, Architecture, Environment: Cultural Heritage and Tourism." *Journal of Travel Research*, 29 (2) Fall: pp. 54–55.

Jafari, Jafar (1982). "The Tourism Market Basket of Goods and Services." In *Studies in Tourism, Wildlife, Parks, Conservation*, Tej Vir Singh et al., eds. New Delhi: Metropolitan..

Kaplan, A. (1984). "Overworked and Undertrained; Unit Managers Need Attention." *Nation's Restaurant News*, November 5.

Kotler, Philip (1988). *Marketing Management: Analysis, Planning and Control*, 2nd ed. Englewood Cliffs, NJ: Prentice-Hall.

Lancashire, D. (1990). "Canada's Cultural Dynamo Enlightens While it Entertains." *Smithsonian*, 20 (12) March: 114ff.

Leiper, N. (1979). "The Framework of Tourism." *Annals of Tourism Research*, 6 (1): pp. 390–407. *Marketing, Management* (1986). (Marketing management program, Maclean Hunter Ltd.) Toronto: Canadian Hotel and Restaurant.

Lew, Alan A. (1999). "Managing Tourism-Induced Acculturation Through Environmental Design on Pueblo Indian Villages in the U.S.," pp. 120–136. *Tourism Development in Critical Environments*. T. V. Singh and S. Singh, eds. New York: Cognizant Communications.

Lundberg, Donald E. (1979). *The Hotel and Restaurant Business*. 3rd ed. Boston: CBI.

Mill, Robert Christie, and Alastair Morrison (1989). *The Tourism System*. Englewood Cliffs, NJ: Prentice-Hall.

Mills, R. (1991). "The U.S.A. Experience," presentation, National Native Tourism Conference, Winnipeg, Manitoba, May 22–23.

Minister of Supply and Services Canada (1982). *The Inn Business*. Ottawa: Canadian Government Publishing Centre.

Murphy, Peter (1985). *Tourism: A Community Approach*. New York: Methuen.

Nolan, Mary Lee, and Sidney Nolan (1989). *Christian Pilgrimage in Modern Western Europe*. Chapel Hill: University of North Carolina Press.

Norvell, H. (1986). "Outlook for Retired/Older Traveler Market Segments." *1985–1986 Outlook for Travel and Tourism*, proceedings of the Eleventh Annual Travel Outlook Forum. Washington, DC: U.S. Travel Data Center.

Parker, Barry (1991). Proceedings of the National Native Tourism Conference, Winnipeg, Manitoba, May 22–23.

Peterson, K. I. (1990). "The Heritage Resource as Seen by the Tourist: The Heritage Connection," proceedings of the Twenty-First Annual Conference, Travel and Tourism Research Association. Salt Lake City.

Richer, Paul and Roger Carter (1999). *Marketing Tourism Destinations Online.* Madrid: World Tourism Organization, WTO Business Council.

Ryan, Chris (1992). "The Child as a Visitor." *World Travel and Tourism Review,* pp. 135–139. Oxon, UK: C.A.B. International.

Sauer, Carl O. (1967). "Seashore—Primitive Home of Man." In *Land and Life*, John Leighly, ed. Berkley: University of California Press.

Senate Committee on Commerce, Science, and Transportation (1977). *National Tourism Policy Study: Ascertainment Phase.* Washington, DC: USGPO.

Simonelli, J. M. (1992). "Tradition and Tourism at Canyon de Chelly." *Practicing Anthropology*, 14 (2) Spring: pp. 18–22.

Smith, V. L. (1992). "Introduction: The Quest in Guest." *Annals of Tourism Research*, 19: pp. 1–17.

Spotts, D. M., and Edward M. Mahoncy (1991). "Segmenting Visitors to a Destination Region Based on the Volume of Their Expenditures." *Journal of Travel Research*, 29 (4) Spring: pp. 24–31.

Taylor, G. D. (1980). "How to Match Plant with Demand: A Matrix for Marketing." *Tourism Management*, 1 (1) March: pp. 56–60.

Texas Parks and Wildlife Department (1984). *1983 Outdoor Recreation Trips Expenditures in Texas.* Austin: Texas Parks and Wildlife Department.

Travel Industry Association of America (2001). Communication with director, Suzanne Cook, August 29.

Tighe, A. J. (1988). "The Arts and Tourism: A Growing Partnership." *1988 Outlook Forum for Travel and Tourism.* Washington, DC: U.S. Travel Data Center, pp. 247–252.

Tourism Research for Non-Researchers (1985). Perth: Western Australian Tourism Commission.

US Fish and Wildlife Service (1997). *1996 National Survey of Fishing, Hunting & Wildlife Associated Recreation.* Washington, DC: USFWS.

Uysal, M., and J. L. Crompton (1985). "An Overview of Approaches Used to Forecast Tourism Demand." *Journal of Travel Research,* 23 (4) Spring: pp. 7–14.

Whitman, S. (1992). "Discovering the Real Orlando." *Rotarian,* 160 (3) March: 28ff.

Chapter 3

Growth, Sustainability, Ecotourism

INTRODUCTION

Even with an understanding of basic fundamentals of tourism planning, several major issues of implementation must be considered, primarily the development of land. All functioning components of the tourism system are related to land resources. Most attractions, even events, are anchored to place settings. Transportation facilities, even air and sea, have important land and site requirements. Services require land and location factors important to both local and travel markets. Information and promotion are about land development for tourism. Because land resources are finite and now show great wear and tear, from tourism growth as well as other economic development, the urgency of planning is becoming more and more evident.

Environmental alarms, once the prerogative of minor environmental groups, are now being sounded by tourism interests. The sacred cow of growth is now being questioned by observers and researchers of tourism. The search for the positive economic impacts of tourism continues but the pressures of mass tourism are revealing many pitfalls such as oversaturation and travel glut. Reactions to these pressures have spawned a new tourism vocabulary—ecotourism, sustainability, green tourism, soft and appropriate tourism, and alternative tourism. In spite of the faddishness of these terms, they have serious implications for the future planning and development of tourism. For many communities and nations these concerns are stimulating new tourism development policies. The following discussion examines concerns over growth, sustainability, and ecotourism, especially as related to tourism planning.

GROWTH

In the past, most planning, promotion, and management of tourism has been focused toward growth. Nations and communities have accepted the popular belief in expanded tourism as an unlimited economic good. To this end, they have invited and accepted virtually any and all new tourism development as worthwhile. Well documented has been the increased jobs, incomes, and tax revenues from tourism growth. For most undeveloped regions of the world today, tourism growth is seen as necessary to economic salvation. Governments and businesses invest heavily in promotion to increase visitor volumes.

But, in all these generalizations about growth, it is not clear what kind of growth is most desirable. For example, it requires a relatively small number of the hotel-staying visitors to equal the impact of large numbers of campers or visitors staying with friends and relatives. However, if an area pushes only for one travel market segment, it will not enjoy a diversity of trade that is necessary for greater stability. When its resources are studied, there may be opportunities for growth in a variety of attractions that will stimulate a better balance of visitor markets.

Promotion and tourism economics have dominated tourist interests for many years. As early as 1937, the Council of the League of Nations recommended that statistics be developed for international tourists (TSAs 2001). In 1984, Canada created a National Task Force to hold the International Conference on Travel and Tourism Statistics, Ottawa 1991. From this, the World Tourism Organization (WTO) created a set of statistical definitions that were adopted by the United Nations Statistical Commission in 1993. Canada announced its Tourism Satellite Account (TSA) in 1994. This was then followed by the Enzo Paci World Conference on the Measurement of the Economic Impact of Tourism, Nice, France, 1999. Participants included representatives from many organizations throughout the world, including the Organization for Economic Cooperation and Development (OECD), Statistical Office of European Communities (EUROSTAT), World Travel and Tourism Council (WTTC), WTO, and other tourism oriented organizations and agencies. A conference on refinements of measurements was held in Vancouver, Canada, in 2001. This is a method for identifying statistical data on employment, consumer spending, capital investment, government revenues and expenditures, foreign trade, and business expenditures. It has been adopted by many nations.

Tourism economists often refer to a *multiplier factor.* This term describes the several times the initial expenditures by travelers turn over again and again within an area. No common multiplier exists because of differences in how money stays in the community or leaves due to products and services that are imported from outside. Initial expenditures with direct impact on the community are usually on lodging, food service, entertainment, and local taxes. But even at this first round of spending, some travelers will spend money on travel agents and airline tickets purchased before leaving home, called leakages because they were not spent in the destination. Successive rounds of spending after the initial expenditures in a community may go to salaries, insurance, upkeep of physical plant, and further turnover by spending of local individuals for their housing, food, shopping, and other expenses.

An example of a search for better methodology, Frechtling and Horvath (1999) have made an in-depth study of tourism multipliers for Washington, D.C., as compared to other destinations. Their study concludes that economic multipliers vary greatly from area to area. The key difference lies in the number of suppliers located within the jurisdiction of tourism attractions. For example, Washington, D.C. showed a comparatively low economic multiplier as compared to such destinations as Miami, Door County, Wisconsin, and Sullivan County, Pennsylvania, because fewer jobs and earnings were generated by suppliers of goods and services within the jurisdiction. In other words, each case needs to be evaluated on its own to determine the secondary effects of tourist expenditures within a jurisdiction.

The relevance of economic multipliers to planning is especially significant because recommendations for growth may have a wide range of overall economic value to a community. Moreover, careful planning can help expand this value through the multiplier process. If planners consider the type of establishments needed to service new lodging, recreation, or other planned visitor facilities and seek to have such suppliers locate within the local area, they can reduce the leakage of income from the area and retain it to generate additional economic benefits.

Experience in the last few decades however has demonstrated that along with positive economic input comes many costs. By and large, it is not the individual tourism development that does the greatest environmental damage. Rather, it is the *collective development of mass tourism* that creates the major sum impact. In order to accrue maximum economic gain, a large collection of facilities and services compound the environmental impacts, especially upon natural resources. Expanded tourism development has often eroded basic resource foundations of vegetation, soils, wildlife, and waters. Equally significant has been negative social impacts of cultural clash and upset of local traditions and life styles. Other growth issues include stress on transportation systems, urban sprawl, deterioration of place distinctiveness, and stress on local infrastructure. Certainly, not all of these concerns can be resolved by planning standards but any views toward the future of tourism must include potential problems of unplanned growth.

The negative consequences of massive tourism growth have been documented by many, as cited by Mowforth and Munt (1998). Huge development worldwide has shown environmental, social, and cultural degradation; unequal economic benefits; spread of disease; and promotion of paternalistic influences. Deforestation, soil erosion, litter along mountain trails, and wildlife disturbance are among the negative impacts. In recent years, this clear evidence of the dark side of tourism has stimulated new plans and remedial action by both the public and private sectors. (See "Tourism: Positive, Negative," *Vacationscape,* 1997, Gunn, pp. 1–11.)

Perceptions of tourism growth vary. Murphy (1983, 10), in his study of some communities in England, found that local perspectives on tourism growth varied among three groups. The administrative group—political, professional, planning, and official—was most positive, believing that greater development would enhance employment and improve local facilities. While the business sector was positively inclined, it was somewhat skeptical that all the benefits from tourism would materialize as promised. Perhaps this was influenced by their concern over new competition and allegiance to local resident markets. Opinions of the third sector, the residents, ranged widely from very favorably inclined to negative. The author concludes that if leaders favoring tourism wish support from all three sectors, greater accountability and better understanding of tourism are required at an early stage of planning and development.

However, many localities make remarkable adaptation to invasions of tourists even though these outsiders are known to disrupt usual community life. Rothman (1978) found that resort cities realized that visitors required extra local effort to cope during the peak season: church schedules were changed, residents tended to avoid popular visitor places and the pace of activity and congestion increased. But for them, the tradeoff was worth it.

Throughout the world, many destinations have adapted to great volumes of visitors. Mega-attractions, such as Walt Disney World, were specifically designed to handle masses of visitors that have produced great economic growth throughout Orlando and the entire state of Florida. These experiences support the belief that growth is desirable and workable. But not all agree that growth is always positive. Molotch (1976, 328) has questioned land development growth as a prime political goal. Growth in numbers and land development can exact costs of environmental degradation, social problems, increased costs of infrastructure and public taxes, and may be perceived as benefiting only a few. Many wonder that "left to our own devices, are we destined to overdevelop all of our principal tourism attractions in a frenetic effort to maximize the influx of tourist dollars?" (Okrant 1991, 32). A single policy of growth may not be compatible with the realities of capacity limits—social, environmental, and economic.

In response to the damaging impact of mass tourism, the concept of *demarketing* is an action principle. Although it was coined in the 1970s as a means of discouraging customers in general, it is increasingly being viewed as a planning solution (Kotler and Levy 1971).

Taylor (1991, 29) observes that some destinations may not be able to accept more visitors no matter how well designed and managed. He states that "a concept of demarketing may have to be developed as it becomes necessary to reduce rather than increase the number of visitors to an area." For example, a panel on the integrity of Canada's national parks recommended that Parks Canada cease product marketing to increase overall use of parks and concentrate instead on social policy marketing and demarketing when appropriate" (*Unimpaired for Future Generations?* 2000, Vol. 1, 21). Although this has not yet become policy, it indicates the growing concern over tourism impacts on the environment.

As local resource managers and environmentalists are observing excessive negative impacts of tourism, the term has been applied to tourism. (Benfield 2000, 2001; Beeton 2001) In spite of its merit, it is difficult for proponents of tourism demarketing to accomplish their objectives of visitor reduction against the power of all the public and private marketing forces. Because the negative impacts are more apparent at the local level, it would appear that communities and their surrounding areas are best able to implement demarketing when needed (Bosselman et al. 1999). Benfield (2000) has identified several demarketing strategies, as paraphrased:

- increasing prices
- increasing warnings of capacities in attraction advertising
- reducing expenditures on advertising
- eliminating trade discounts
- reducing distribution outlets
- reducing quality of product
- providing vicarious visitor experience (interpretation)

In spite of resistance by promoters, the excesses of tourism growth may require demarketing in order to reach toward the goal of sustainability.

Another approach to economic growth is to regenerate existing physical development rather than develop new land. Many destinations contain sites with obsolescent or obsolete uses that, with creative design and planning, could be converted to tourism. The wave of growth of bed-and-breakfast facilities in the U.S. is evidence of this opportunity. If the emphasis is more upon improved *quality* than *quantity,* there may be fewer negative impacts.

Researchers Williams and Gill (1999) have reviewed past capacity theories and concluded that setting precise arithmetical limitations on visitor numbers is not feasible. Instead, a systems approach that relates incremental growth impacts on goals and objectives is more realistic. On a case by case basis, indicators of change can be established for a community's tourism objectives such as for social, economic, and environmental impacts. These indicators are monitored, triggering needed changes in management. The results may or may not indicate limits to growth.

Because of the finite quantity of some critical resource areas, a policy decision to manage visitor capacities may be necessary. Many now believe that the number of visitors to the Galapagos Islands has reached maximum capacity to be supportable without environmental damage. In response to increased scuba-diving tourism in the Cayman Islands, the government passed a marine conservation law and established a marine park system (Long 1990, 49) and Bermuda has reduced the number of cruise ship visitors. Perhaps the most fragile of all sites, demanding controls on numbers of visitors, are historic buildings. Mass use must be restricted to walkways and viewpoints that are designed for certain maximum capacities. Frequently, this planning issue can be resolved by establishing museums and visitor centers nearby that are designed for mass tourist use, providing an acceptable vicarious experience without damage to the resource.

SUSTAINABILITY

The Concept of Sustainability

The Brundtland Commission (World Commission on Environment and Development 1987) and the World Conservation Strategy (1980) are credited with initiating the term "sustainability," as a goal for all society, including tourism.

Sustainable tourism, just like ecotourism, has become a popular and yet ambiguous term. It is now used with meanings all the way from recycling waste and reduction of energy consumption to the prevention of human impact on natural and cultural resources. Many observers and writers have attempted to define sustainability, often adding more to confusion than clarity. One writer, John Pezzey (1989), found over 60 definitions.

Generally, most include the concept of fostering development that is least destructive in the long run of the resource upon which it depends. One definition calls it change that does not undermine ecological and social systems and requires new planning and policies in order to implement (Rees 1989, 3). Such a definition implies that wanton land use of past tourism that produced negative impacts—environmental, social, economic—was wrong and that such erosive development should be avoided, requiring new planning and advocacy for balancing tourism with resource protection.

Sustainability, for tourism as well as other development, grows out of the principle that all animal life, including mankind, is shaped by the environment, and conversely, organisms shape the environment. But, as Sargent (1974) explains, mankind is different from other organisms because of its dominance over all land and other organisms rather than restricted to ecological niches. Because of this significant difference, mankind can and must guide and control decisions on how land is used if long-range sustainability is to be accomplished. Planners, designers, developers, and managers can exercise judgment and new understandings of how to utilize land and yet not destroy it (Gunn 1991).

The concept of tourism sustainability points to the need for better spatial, environmental, and economic balance of tourism development, requiring new integrative public-private approaches and policies in the future (Godrey 1996, Coccossis 1996, Manning and Dougherty 1999). When the principle of sustainability is applied to new tourism development, it would mean that coastal hotels would not pollute their beaches with raw sewage, that hillside resorts would not incite soil erosion, and that sites of fragile and rare vegetation or wildlife would not be used for tourism except as scenery and interpretation. Tourist businesses can benefit by land use decision making that offers long-range protection of resources. The public sector can ensure such sustainability with long-range regulation and legislation. Only by accepting such responsibility will tourism be assured a continuing quality future. Throughout Part II of this book are principles and cases where efforts toward sustainable design and management have been demonstrated.

Achieving Sustainability

Sustainable design may be accomplished more easily with new projects on raw land than developed areas. Within heavily developed land, past mistakes in land use are more difficult to overcome. Buildings, roads, and utilities are already in place and not easily changed even though they may now appear chaotic, inefficient, and even wrong. But, even here, the new search for sustainability is stimulating modifications. Rather than destroy obsolete structures, many are being restored and modified for contemporary tourist use. Overlay historic protection ordinances are renovating entire sections of towns, for use by residents as well as visitors. Streets are being modified to either increase traffic flow or, where possible, convert them to pedestrian malls. In areas of scenic vistas, new building codes that restrict height are protecting important views. In both developed and undeveloped lands, the initiation, planning, and maintenance of sustainability depend upon traditional concerns over land use; political interests, goals, and values; and constant adaptation to change (Edwards and Priestley 1996, 188).

The best solution to sustainable development is likely to occur not from advocacy of environmentalists or governments but from voluntary action from developers of tourism. When the fundamental of the dependency of virtually all tourism upon the resource base becomes more apparent to developers, they will see it in their best interests to sustain the quality of the natural and cultural resources. The process by which this is to be achieved is through codes of practice and agreements locally— "information, monitoring, communication, and adaptation among an array of groups and individuals with different and similar interest" (Nelson 1991, 40). A new ethic for sustainable places has been suggested by Beatley and Manning (1997) as shown in Table 3-1.

Increasingly, examples of sustainable tourism development are appearing. An excellent case is the tourism adaptation on the island of Yap, 870 kilometers southwest of Guam (Mansperger 1992). Instead of typical enclave, large scale, and externally-owned development that often creates shock and conflict in a local society, tourism here has been slow, small-scale, and mainly locally owned, resulting in strong resource protection. Hotels employ mostly Yapese and the foreign exchange from over 3,000 tourists annually is adding to the economy. Controlled tours of historically important Bechyal give tourists insight into a native community house, a chief's house, shell money, a sailing canoe, and fish traps. The new locally-owned Manta Ray Bay Hotel caters to divers who enjoy the unusually close contact with manta rays and pristine coral formations. Every Saturday, a Cultural Show is held in the village of Maaq on Tamil-Gagil Island. This features native Yapese dances portraying the traditional search for stone that, in ancient times, was quarried and used for money. Actually, in this location, tourism is fostering the preservation of a culture. The Yapese have been very selective in accepting only those forms of tourism they feel are most appropriate. They want the benefits but are unwilling to sacrifice their culture to obtain them.

There continue to be many locations that because of good design and good management can accept great volumes of travelers yet at the same

TABLE 3-1

ETHIC FOR SUSTAINABLE PLACES

Current Ethic	Ethic of Sustainable Place
Individualism, selfishness	Interdependence, community
Shortsightedness, present-oriented ethic	Farsightedness, future-oriented ethic
Greed, commodity-based	Altruism
Parochialism, atomistic	Regionalism, extra-local
Material, consumption-based	Nonmaterial, community-based
Arrogance	Humility, caution
Anthropocentricism	Kinship

time protect the environment, accomplishing sustainable development objectives. National parks have increasingly demonstrated this principle. When visitors are guided only into areas where they do no damage to the environment, design has been successful. When management gives visitors descriptive information by means of literature, guidance, lectures, exhibits, and demonstrations, they gain rich experiences without damaging the setting.

Bosselman (1999) cites the case of South Pembrokeshire, Wales as a case of good tourism sustainable planning. A pilot project sponsored by the South Pembrokeshire Partnership for Action with Rural Communities (SPARC) provided funding in 1992 for a participatory planning program. The purpose was to develop plans for tourism growth based on local natural and cultural resources but also on limited growth for resource protection. The partnership and integrated approach stimulated new low-key attraction development of resources that added materially to the local economy through new lodging, food services, and shops.

An example is the planning for an interpretive resort complex, Wilpena Station, within Flinders Range National Park, Australia (Williams and Brake 1990). This is planned with two major objectives: (1) a satisfied park visitor, and (2) a managed outcome of the consequences of park visitation. The basic principle underlying all development here is: "Well planned and thoughtfully implemented site modification to accommodate an increase in the number and range of visitor groups to a particular site does not work against the conservation goals of a park." (More detailed description of this case can be found in chapter 10, "Site Planning Cases.")

A strong land ethic is traditional for Canada. In 1992, a conference entitled "The Tourism Stream of the Globe '92 Conference" was held in Vancouver. Resulting from presentations and discussions were the following elements of a Challenge Statement pertaining to sustainable development of tourism (Manning 1992).

1. Policy, Legislation and Regulation

   Building the institutions and the foundations for sustainable tourism.

   Protecting the resources base.

   Mobilizing industry action for sustainable tourism.

2. Technology and Research

   Understanding the natural resource base.

   Understanding cultural values.

   Measuring tourist demands and expectations.

   Measuring tourism impact.

   Information for better decisions.

   Mobilizing appropriate technology.

   Visitor management techniques and practices.

3. Economics and Finance

   Incorporating environmental costs.

   Modifying reporting procedures.

   Using market influences at home and abroad.

   Benefiting from the environmental market.

4. Communication and outreach

   Mobilizing the firms and employees.

   Self-regulation.

   Modifying decreased tourist expectations and actions.

   Becoming proactive.

Stanley (1991, 116), in his summary of a conference on sustainable development and tourism, concluded that no one should expect rigid standards for its achievement. Instead he identifies seven different threads of importance in research and policy if sustainable development is to be achieved. First, sustainable development is determined largely by *what the stakeholders want it to be*—a wilderness or a developed resort. Second, it can be accomplished only when people have found mechanisms for *working together.* Third, environmental impact results from *many forms of tourism* other than only visiting natural resources. Visiting friends and relatives, business travel, and visiting urban historic sites require special planning if sustainable development is to be achieved. Fourth, because most tourist establishments are small businesses, unable to obtain research and professional studies, *much education is needed.*

Guidelines and computer models may assist. Fifth, research can demonstrate that *sustainable development pays.* Sixth, *economic measures*, such as willingness to pay and contingent value, can demonstrate the real value of sustainable development. And seventh, *a review of cases* where sustainable development of tourism is being achieved can help communities and rural areas plan to reach their own sustainable objectives.

Planners and designers can now benefit from applying criteria for the evaluation of a potential tourism project's sustainability. For a creation of a checklist, the nonprofit organization, Groupe Developpment (Vellas and Barioulet 2000), sponsored a consortium of several international tourism organizations, universities, and countries (Indonesia, Madagascar, Dominican Republic, Seychelles). Using three types of indicators (economic, environmental, social), a 16-point checklist directed toward the goal of sustainability was prepared: physical impact, liquid waste disposal, solid waste disposal, water consumption, visual impact, job creation, continuing training of staff, local frequentation, safety, health, culture, information on child protection, control over development, and use of technologies.

For application of this checklist, a series of ratios were prepared for the criteria. The preparers of this checklist believe that it is flexible enough to be used in a variety of tourism projects and sites.

Development Adaptation

Increasingly, experience is demonstrating that there are differences between low-impact and high-impact development. Low-impact is tourism development that is characterized as small scale and slow progress whereas high-impact refers to large scale and rapid development.

There are arguments for high-impact development. In areas that are not yet known as destinations, some large-scale attractions and accommodations may be appropriate. High-impact development may be needed to attract support services such as airline access and promotion. This type of development may be needed to provide the noticeable change to attract local support for other tourism development. However, feasibility needs to be carefully examined to make sure that environmental, social, and economic change will not be detrimental.

Low-impact development, as exemplified by ecotourism, however, may be integrated more readily into the existing social and economic life of a community. Because developers are residents, they are more likely to be well acquainted with resource limitations. Low-impact development has a greater opportunity for feedback from each increment of growth.

However, this type requires long-range planning so that each new development is a logical part of the whole of tourism and not dispersed so widely that it is inefficient to service and confusing for visitors. Many regions are finding that both kinds may be needed, provided that they are kept in balance.

Much of the criticism by environmentalists of tourism development stems from the *quality of fit.* How well will this new development fit into the dominant qualities of the land and existing development? Sometimes it may be in conflict with the setting. Wight (1988, 15) describes this as extrinsic or alien to the existing environment and cites the proposal of a monorail from Ayers Rock to the remote Olga Mountains in Australia or a waterslide just outside Waterton National Park, Canada, as inappropriate tourism development.

This issue of appropriateness is founded in the two major factors influencing all tourism development. First is the factor of market segment preference. Aesthetically, it may be quite distasteful to visitors to see some development that just does not fit. For example, a garish food service structure beside a prominent historic site may actually reduce the quality of the visitor's appreciation of history. A more appropriate location for restaurants may be some distance away from the historic site and in a cluster of other commercial development. Or, an alternative, as is being done by McDonald's in many historic areas, is to break from the standard architectural style and adapt the facade to the site and architectural character of the surrounding historic district.

The second factor relates to the resource base. If the area has qualities that contribute to the attractiveness and user activity interests, it should be protected rather than eroded by inappropriate development. This suggests that major tourist services are most appropriate when located at communities rather than dispersed throughout natural areas. Tourists are able to select from a greater diversity of services and the criticism of "ugly commercialism" is avoided in natural areas.

Gayle Jennings and Daniela Stehlik (2001) studied farm vacation patterns in northeastern Australia as examples of authenticity. By means of participant observation, in-depth interviews, recordings and photographs, the study revealed three classifications of authenticity. First, several farms demonstrated "tourism experience as family life," whereby visitors are treated as part of the family. Host-guest relations show no distinction and visitors participate in family life. A second category is creation of the "ideal farm environment." This model placed visitors outside the farm quarters. Instead of sharing the actual farm activities, they were given staged demonstrations. Finally, a mixture of these two patterns was called "impressionistic farm life." Visitors were exposed to some farm life but guests were separated from hosts. The study suggests that market preference for authenticity can be offered several ways but that involvement offers stronger social impact.

ECOTOURISM

A recent addition to the lexicon of tourism that is sweeping the world is *ecotourism*. Definitions and interpretations are as broad and vague as for sustainability and also seem to overlap (Honey 1999). Like sustainable tourism, ecotourism is often referred to as "green," "conservation," and "sustainable." Even so, it has exploded as a concept that is different enough to have captured the interest of marketers, promoters, tour managers, and scholars. The "eco-" part of the word implies a tie with ecology—relations between living organisms and their environments, thus linkage between tourism development and the environment. While having roots similar to conservation, the interpretations do not necessarily imply protection or wise use of resources. The concept also has market appeal and is being directed to travelers interested in nature. Another market application is promoting tourists to facilities and services that have incorporated new low-energy consumption and recycling of waste, implying that the owners are environmentally sensitive.

Although the term, ecotourism, appears to be recent, the concept of balancing tourist use with resource protection was put forward years ago. The original national park act of the United States in 1916 mandated a dual policy of resource protection and public use. Smardon (1991, 704) points out that Dickert and Sorensen (1974) and Gunn (1978) called for application of ecological principles of tourism planning decades ago.

Increasingly, ecotourism is being defined as nature-oriented travel that promotes and finances conservation and resource protection and also adds to the local economy (Ziffer 1989, 24). Visitor revenues in protected areas can be derived from entrance fees, donations, ancillary services or products, and private investment. But, even though these offer potential support, each local situation must be evaluated on its own merits. Sometimes entrance fees, even though substantial, go directly into the nation's coffers and are not earmarked for protection of the areas visited. Because local people feel they already support protected areas with their taxes, a policy of variable entrance fees (larger for foreign visitors) may be initiated. Some tour operators make direct donations to conservation areas out of their profits. Gift shops, food services, and hotels may be able to donate funds to the protected areas nearby that bring them volumes of customers.

Another element in ecotourism operation is the role of tour companies. Here it is not so much the salient facts as the moral and judgmental aspects. In recent years, there has been increased recognition that tourism planning and management that protects natural and cultural assets is not mere altruism but good business (Malloy and Fennell 1998). For eco-tourism, tour operators are critical in the extent of their management of visitors in special settings and many have accepted responsibility for implementing ecotourism principles. Sirakaya and McLellan (1998) researched nature tour operators of Canada, the U.S., and Ecuador regarding their compliance with guidelines projected by The Ecotourism Society (1993). Important items of compliance included pre-departure information to tourists, education during trip (etiquette, waste procedures, fire, collecting), tour operator's personal support of conservation and cooperation locally, and relation to local accommodations. Among the results was the importance of voluntary application of ecotourism principles as compared to rules and coercion. Those operators who truly believed in environmental protection were best able to comply. Furthermore, they understood that compliance was good for their business. A concluding suggestion was new training for tour operators that emphasized the value of compliance for both environment and business success.

From the travel market perspective, some have attributed the new interest in ecotourism to a new class of intellectual and ecological traveler (Mowforth and Munt 1998). They have both the motivation and financial ability to pay for special interpretive programs and guided tours. It is a market segment that believes it was omitted in previous supply development. This esoteric new wave has spurred growth of new interpretive and tour professionals.

Colvin (1991, 578) describes one type of ecotourist as "scientific," having the following characteristics:

Wants an in-depth, "authentic" experience.

Considers the experience worthwhile, personally and socially.

Abhors large tour groups on strict itinerary.

Seeks physical and mental challenge.

Wishes interaction with locals, cultural learning.

Adaptable, often prefers rustic accommodations.

Tolerates discomfort.

Seeks involvement, not passive behavior.

Prefers to pay for experience rather than for comfort.

Managing such tourists is being accomplished through programs of the University of California Research Expeditions Program (UREP). Follow-up of presentations and exchanges of information is an important by-product.

Wight (1992) cites several cases in which ecotourism is stimulating greater collaborative interaction. For example, the Association of Independent Tour Operators (AITD) has taken a joint initiative with Green Flag International to promote sustainable development. Wight describes the case of Sobek Expeditions as providing a remarkably high amount of its proceeds (6.7–10 percent) to local conservation groups for their use in protecting resources, especially those related to Sobek's "Environmental Adventures." Green Flag is a nonprofit organization dedicated to working in partnership with tourism decisionmakers to improve the environment.

Wight has identified important principles that should underlie the concept of ecotourism:

- It should not degrade the resource and should be developed in an environmentally sensitive manner.

- It should provide first-hand, participatory and enlightening experience.

- It should involve education among all parties—local communities, government, nongovernmental organizations, industry, and tourists (before, during and after the trip).

- It should incorporate all party recognition of core values related to the intrinsic values of the resource.

- It should involve acceptance of the resource on its own terms, and in recognition of its limits, which involves supply-oriented management.

- It should promote understanding and involve partnerships between many players, which could include government, nongovernmental organizations, industry scientists, and locals (both before development and during operation).

- It should promote moral and ethical responsibilities and behavior by all players.

- It should provide long-term benefits: to the resource, the local community and industry (benefits may be conservation, scientific, social, cultural, or economic).

Planning Concerns

If ecotourism is to be planned and developed, consideration must be given to several issues in the earliest planning stages if they are to be avoided or overcome. Those identified by Collins (1991, 106) are paraphrased as follows:

- Displacement of local people.
- Direct and indirect erosion of resources.
- Loss of access to resources by local residents.
- Misleading promotion.
- Adequate funding for planning.
- Costs of establishing basic infrastructure.
- Cultural conflict with natives.
- Better road access may exacerbate resource destruction.

Although everyone, developers and visitors alike, would prefer to avoid controls, society and the environment cannot live together without them. For sustainable tourism development, certain controls are being accepted by both developers and visitors as necessary, even for their own welfare.. For example Wendt, of the U.S. National Park Service (1991, 537) lists several workable controls that have met with success in balancing tourist use with resource protection:

> *Entrance Stations.* These imply authority of management and provide information to visitors.
>
> *Visitor Centers.* Such centers reduce violations because the public is better informed. Environmental education begins here.
>
> *Effective and Courteous Law Enforcement.* Volumes of visitors require the same exercise of police protection as in other locations, perhaps even more so because tourists are in a more vulnerable attitude.
>
> *Resource Management.* The flora and fauna and land resources cannot be left to "nature" in order to sustain these valuable reserves.
>
> *Environmental Interpretation and Education.* Guided trails, evening programs, environmental education, extension programs in surrounding communities, living history interpretation, self-guided automobile tours, rock climbing school, and exhibits can provide rich visitor experiences without environmental damage.

In many ways, ecotourism development is similar to the land use issues of national parks. For most major resource areas, it is now being understood that major tourist services are better located at the edge rather than within the park. The diagram of Figure 3-1 (Gunn 1988) is a concept of how tourism functions could be planned and integrated with ecotourism and parks and protected preserves. A similar concept of park zoning was advocated by landscape architect Richard Forster in 1973 and endorsed by the International Union for Conservation of Nature and Natural Resources. He describes patterns of concentric use zones. A three-zone configuration includes a protected wild land core surrounded by an outdoor recreation buffer zone. An outer zone includes intensive use and visi-

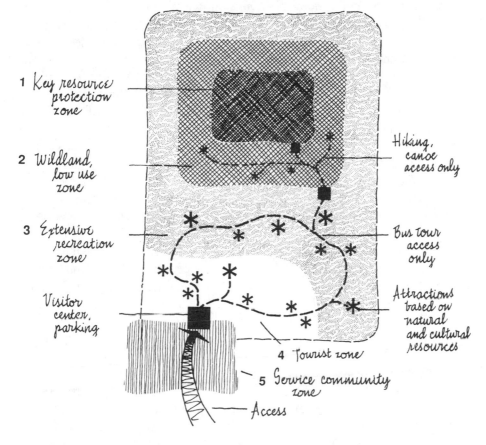

1 Key resource protection zone

2 Wildland, low use zone

3 Extensive recreation zone

Visitor center, parking

Hiking, canoe access only

Bus tour access only

Attractions based on natural and cultural resources

4 Tourist zone

5 Service community zone

Access

Figure 3-1. National Park Tourism Model. This planning concept recommends that most tourist services be located outside the park allowing protection of the park's major features. This concept encompasses five zones: resource protection zone, low use zone, recreation zone, tourist zone, and service community (Gunn 1988, 85).

tor services, such as lodging and food. A five-zone pattern was used by the Canadian National and Historic Parks Branch and included:

1. *Special Areas*—protected natural and cultural resource areas.

2. *Wilderness Recreation Areas*—protected resources but minor and controlled access.

3. *Natural Environment Areas*—protected, but greater visitor use.

4. *General Outdoor Recreation Areas*—planned use, such as campgrounds, trails.

5. *Intensive Use Areas*—major visitor services (Forster 1973, 63).

Fundamentally, these require not only a broader geographical context but a more comprehensive administrative action. By allowing nearby communities to function in their normal ways, park managers can be relieved of providing these visitor services internally. For example, most travelers seek communities for the majority of their services—food, lodging, entertainment, car service, shopping, and emergency health services. Communities also serve residents in many ways. Communities are able to provide basic infrastructure that is more costly to install and operate within remote areas of parks. By means of their own biological and ecosystem analysis of the park's resources, park administration can determine the extent to which visitors should be allowed within. Their control of visitor use within can prevent environmental erosion. Through reciprocal agreements, planning and technical park assistance could be provided to communities, and communities could return portions of tourism revenues to support conservation measures in the parks. This is sustainable tourism development.

The balance between service communities at park entrances has floundered primarily from poor cooperation between these two management jurisdictions. For political turf protection it has been difficult for park planners to reach out with assistance to adjacent communities. The community business sector has been reluctant to seek planning advice from the park bureaucracy. Wherever these barriers have been broken down, both sides have gained. Cooperative actions, such as between the gateway community of Red Lodge, Montana, and nearby national forest managers (Howe et al. 1997). Through a participatory process, a citizen-designed master plan providing for a nature-development balance was created in 1995. This was done with the catalytic assistance of professional planner Lee Nellis.

Figure 3-2 illustrates five ecolodge spatial options (Gunn 1995). Much preferred from both a business and environmental point of view is lodging outside the resource area, as illustrated in options A, B, and C. Option C, the Sabi Sabi Resort (figure 3-3), provides photography safaris into nearby Kruger National Park as well as in their own wildlife preserve. Visitor use of the resource area can be controlled and managed within capacity limits by the resource owner-manager. Option D showing ecolodges within prime resource areas poses serious environmental threats—to native animals, plants, and cultures. In a few instances, depending on the characteristics of the resource area, a private sector owner may be able to install a properly designed ecolodge within, as shown in option E. The responsibility of analyzing the resource features and determining suitable land use planning fall upon the owner. The underlying theme for all options is to balance resource protection with ecotourism development.

Many observers and researchers today are finding evidence contrary to many beliefs that tourism is always destructive of resource foundations. There are numerous examples demonstrating *increased resource protection* because of tourism—ecotourism practiced in a responsible manner.

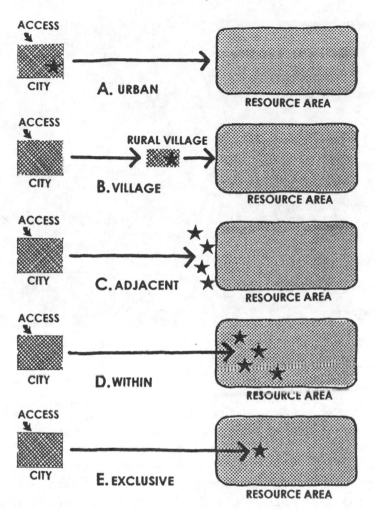

Figure 3-2. Ecolodge Spatial Relationship. Illustrated are five options for ecolodge locations. Preferred options are A, B, and C, outside the major resource protection area. Resource protection and use conflict occurs with option D and rarely is option E feasible.

Boo (1991, 518) defines this very simply as "nature tourism that contributes to conservation." She identifies key elements of planning, design, and management needed in natural resource areas such as national parks:

Entrance fee collection system.

Tourism training for park personnel.

Trail systems with interpretation.

Visitor center with interpretation.

System of monitoring environmental impacts.

Services: snack bar, lodging.

Figure 3-3. Sabi Sabi Ecolodge. The location of this private ecolodge in South Africa allows visitor access to photographic and viewing safaris in nearby Kruger National Park as well as their own resource reserve. Such a location favors business success as well as resource protection.

Although park managers and policymakers have accepted visitors in varying degrees for many years, their role in tourism has not been understood or incorporated into policy. Parks and conservation areas have been established, planned, and managed as isolated oases of special natural and cultural resources. Commendable as has been their protection of these resources, handling visitors has been treated mostly as nonconforming and only to be tolerated. Needed is new understanding of the role of parks and conservation areas in tourism. The new concept of reorganizational ecotourism is finally forcing this issue of providing a new awareness of the linkage between resource protection and tourism values—economic, personal, community enhancement.

Stewart et al. (1990) summarizes it thus: "comprehensive ecosystem management implies some degree of mutual understanding and reciprocity across multiple sociopolitical systems." He states further that ecotourism planning and development requires: "(1) integration of nonfinancial objectives; and (2) a planning process which encourages participatory decision-making encompassing entrepreneurs, land managers, host community, and interested tourists or visitors."

Perhaps the greatest premise of ecotourism for better planning is the concept of stronger linkage between the public and private sectors. The public sector role, as managers of national parks and wildlife reserves in Costa Rica (11.2 percent of the land), is protecting the assets of interest to ecotourists: deciduous forests, mangrove swamps, rain forests, marshes, paramos, cloud forests, Raphia swamps, oak forests, coral reefs, riparian forests, and a great diversity of bird and animal life (Fennell and Eagles 1990). The private sector entrepreneurs are developing the tours, tour guidance interpretation, lodging, food service, and transportation. There is a strong local understanding that these two forces are working together to a high degree, producing not only economic betterment but also increased public support for resource protection.

A combination of new planning and management of the Manu Biosphere Reserve, Peru, has effectively proven a balance between resource protection and visitor use (Groom 1991). The area encompasses 2 million hectares, including prime nesting and tourist viewing areas for black skimmer, large- and yellow-billed terns, pied lapwing, collared plover and sand-colored nighthawk. Unplanned and uncontrolled birdwatching tours were exacting severe disturbance upon this important wildlife. In 1987, several planning and management practices were implemented. Beaches used by birds were marked with red flags, and blue flags identified beaches suitable for tourists. Tourists are still able to observe the birds but nesting areas have been protected. Total visitor use has been limited to 500 persons per year. The success of this program has resulted from close cooperation among tour company managers, local and national governments, and park and reserve administrators as well as the Ministry of Tourism. Further planning and management concepts have been applied to visitors arriving by boat: motors equipped with silencers, prevention of animal harassment by tour groups, restricting some areas, and regular monitoring of visitor impacts. The State of Mississippi similarly cordons off a portion of the beach for terns' hatching season!

However, experience in the search for the noble goal of assisting the economy of small towns and protecting the environment through ecotourism is demonstrating some unfulfilled promises. The concept is flawed on two counts.

The introduction of visitors to any natural setting is bound to change the setting. A dominantly natural area contains plants and animals that have adapted to their habitat. Intervention by humans upsets this balance, increasing threats to the extinction of species. The proliferation of ecolodge development (accommodations within the ecotourism concept) is especially invasive. Often it is touted as environmentally innocent for its use of voltaic and wind power production and recycling of waste. While this technology is an improvement, it encourages rampant building development in rare and fragile environments. It introduces tourists, with their noise, trampling, and landscape erosion, into animal and plant ecosystems, resulting in conflict and decimation. This is exemplified by some ecolodge development where native plants and wildlife have been severely eroded. Indigenous landscape aesthetics on some arid sites have

Figure 3-4. Ecolodge-Site Conflict. Although new technologies of wind or solar electricity and recycling are desirable conservation techniques, they sometimes encourage ecolodge locations in remote and rare or fragile landscapes. Generally, tourist facilities and services are best located at the edge of resource protected areas rather than within them.

been destroyed and replaced with views of structures, wood walks, and steps (Figure 3-4, Gunn 1995). This issue requires judgment concerning how and where ecotourism projects are to be developed, an ethical and moral issue. Malloy and Fennel (1998) have elaborated on models for resolution of this issue.

Resource areas being considered for ecotourism development already have their own economic value as nonmarket goods. Most likely this is larger than tourist expenditures (Lee et al. 1998). This means that an ecolodge may destroy these intrinsic values unless extreme caution and environmental design and management are employed. Planners and managers have a moral obligation to respect economic values and landscape significance with site and building design as well as visitor education regarding behavior in such settings.

A second disappointment from ecotourism is the provision of economic impact and social enrichment in small villages, especially for native populations. Honey (1999, 311) reports instances where instead of local residents benefiting from tourism economics, returns are fraudulently diverted to unscrupulous governmental officials. Establishing ecotourism projects is faced with acculturation difficulties.

The case of Capirona, a community of 24 Quicha Indian families located in Equador's Amazon Basin, faces the challenge of creating an ecotourism project adapted to the rain forest and yet protecting their culture (Colvin 2000). Access is only by a strenuous two-hour walk from the trailhead at the nearest community of Tena. Led by a local Indian, Tarquino Tapuy, a visitor program includes forest walks, basketmaking demonstration, and participating in *mingha,* a local day of vegetable planting and clearing forest as do the natives. Women rotate the responsibility of preparing meals.

The first year of operation in 1992 raised more questions than profits. The total of 50 visitors were reasonably satisfied with enduring great hardship and primitive living. But the hosts soon realized that introducing outsiders to their culture and preserving it was a major task. Personally they were engaged in acculturation—beginning to use generators for electricity, wear western clothing, go to school, and purchase food from Tena. Yet they place visitors in candlelit huts and eat homegrown foods with native preparation. They face the dilemma of how far they must go to remain an authentic native village for visitors and yet move themselves rapidly into a modern world. And if their ecotourism project expands, new accommodations, food services, trails, and access may destroy the very concept they are trying to protect. They also face pressure from timber and oil companies who want their land. Cultural ecotourism requires special planning, management, and operational policies.

The concept of ecotourism is not a desirable option everywhere. Any planning of ecotourism must consider the future and predictable consequences. This is necessary to avoid problems of negative economic impact, resource denigration, and cultural conflict. Writer McLaren (1998) calls for new research, education, business objectives, local action, and media ethics if ecotourism is to prevent destroying itself. An extensive list of organizations and other resources on ecotourism is included in this reference.

CONCLUSIONS

Tourism's growth is showing increasing negative impacts.

The worldwide promotion of tourism development for economic purposes, a desirable and worthwhile goal, is now being questioned. In areas of mass tourism development, negative social, environmental, economic, and community impacts are causing rethinking of growth as a singular goal. Instead, new programs of better planning for lesser environmental impact are being considered.

The concept of fostering sustainability is creating new long-range planning.

Put forth today are new spatial, environmental, and economic planning concepts that are directed toward retention of basic resource values alongside tourism development. The natural and cultural resources can be eroded severely if only short-range planning and development are employed. Because these resources are the foundation for the majority of tourism, it is important that they be protected for the future.

Improved transportation facility is creating mass tourism glut at attractions and communities.

Better road, automobile, and air travel have produced exponential growth of travel volumes. This success is now showing congestion and eroded visitor satisfaction at travel termini. New planning concepts are needed at the community and attraction levels in order to cope with these hordes of new visitors.

Tourist saturation requires measures of control.

A new balance between promotion and development is needed in order to reduce excessive erosion of the key objectives of travel. In the future, quotas and other mechanisms may be required to keep volumes of visitors within the limits of area capacity. In some instances, demarketing, the reduction or elimination of promotion, is being applied to this issue.

The concept of ecotourism requires special planning and development.

Although new technologies of energy, water, and waste control are desirable, the promises of economic enhancement from low-scale nature tourism development in rare and fragile environments are not always being realized. The purpose of economic return to nearby communities is seldom fulfilled. And, the proliferation of ecotourism development worldwide is causing environmental and social stress in native societies. The goal of better visitor understanding of resources requires very special design and management of ecotourism sites.

DISCUSSION

1. What are the major consequences of too many visitors? Where are these pressures most detrimental?

2. In a community of your choice that is being overrun by visitors, discuss the major obstacles to controlling growth.

3. How can planners (professional, political, tourism leaders) stimulate the creation of solutions to excessive numbers of visitors and in what way?

4. Who can accept the responsibility of balancing tourism growth with protection of the resource foundations?

5. Because any development anywhere utilizes land resources, how can the ideal of sustainability be accomplished?

6. If an outside investor has the opportunity to build a resort in a foreign destination, what are the advantages and pitfalls of such action?

7. Discuss the role of designers/planners in evaluating tourism's impacts and recommendations for change.

8. What governmental or private agency is best able to guide sustainable development? What is the role of local citizens and professional designers and planners?

9. Discuss the spatial relationships between tourist facilities and services and major parks and resource reserves.

10. In the desire to satisfy the growing travel demand for nature tourism, how can ecotourism projects meet this need without damage to society and environment?

11. If interpretive visitor centers are a great aid to better tourism information and resource protection, what location and design principles are required?

12. What major issue does ecolodge development raise and how can it be resolved?

13. Who is best able to plan optimal locations for service businesses with regard to resource areas such as parks and preserves? What conflicts must be avoided?

REFERENCES

Beatley, Timothy and Kristy Manning (1997). *The Ecology of Place.* Washington, DC: Island Press.

Beeton, Sue (2001). "Cyclops and Sirens—Demarketing as Proactive Response to Negative Consequences of One-Eyed Competitive Marketing." *TTRA Annual Conference Proceedings,* June, pp. 125–136.

Benfield, Richard W. (2000). "Good Things Come to Those Who Wait—Demarketing Sissinghurst Castle Garden, Kent, for Sustainable Mass Tourism." *TTRA Annual Conference Proceedings*, June, pp. 226–234.

Benfield, Richard W. (2001). "Turning Back the Hordes—Demarketing as a Means of Managing Mass Tourism." *TTRA Annual Conference Proceedings,* June, pp. 137–150.

Boo, Liz (1991). "Ecotourism: A Tool for Conservation." In *Ecotourism and Resources Conservation,* J.A. Keesler, ed. Berne, NY: Association of Wetland Managers, pp. 517–519.

Bosselman, Fred P., Craig A. Peterson, and Claire McCarthy (1999). *Managing Tourism Growth.* Washington. DC: Island Press.

Coccossis, Harry (1996). "Tourism and Sustainability: Perspectives and Implications." *Sustainable Tourism? European Experiences.* G. K. Priestley, J. A. Edwards, And H. Coccossis, eds. Oxon, UK: CAB International.

Collins, Michael (1991). "Ecotourism in the Yucatan Peninsula of Mexico." Dissertation, No. 91–1, Faculty of Landscape Architecture. Syracuse, NY: State University of New York.

Colvin, Jean G. (1991). "The Scientist and Ecotourism: Bridging the Gap." In *Ecotourism and Resource Conservation*, J. A. Keesler, ed. Berne, NY: Association of Wetland Managers, pp. 575–581.

Colvin, Jean G. (2000). "Indigenous Ecotourism: A New Trend?" Unpublished paper.

Dickert, T. and J. Sorensen (1974). "Social Equity in Coastal Zone Planning." *Coastal Zone Management Journal,* 1: 141–150.

The Ecotourism Society (1993). *Ecotourism Guidelines for Nature Based Tour Operators.* North Bennington, VT: The Ecotourism Society.

Edwards, J. Arwell and Gerda K. Priestley (1996). "European Perspectives on Sustainable Tourism," pp. 189–198. *Sustainable Tourism? European Experiences.* Priestley, Edwards and Coccossis, eds. Oxon, UK: CAB International.

Fennell, D. A. and P. F. J. Eagles (1990). "Ecotourism in Costa Rica: A Conceptual Framework." *Journal of Park and Recreation Administration*, 8 (1): Spring.

Forster, Richard R. (1973). *Planning for Man and Nature in National Parks.* Morges, Switzerland: IUCNNR.

Frechtling, Douglas C. and Endre Horvath (1999). "Estimating the Multiplier Effects of Tourism Expenditures in a Local Economy through a Regional Input-Output Model." *Journal of Travel Research* (37) May, pp. 324–332.

Godrey, Kerry B. (1996). "Towards Sustainability? Tourism in the Republic of Cyprus." *Practicing Responsible Tourism.* Lynn C. Harrison and Winston Husbands, eds. New York: John Wiley & Sons.

Groom, Martha J. (1991). "Management of Ecotourism in Manu Biosphere Reserve,

Peru." In *Ecotourism and Resource Conservation,* J.A. Keesler, ed. Berne, NY: Association of Wetland Managers, pp. 532–540.

Gunn, Clare A. (1978). "Needed: an International Alliance for Tourism, Recreation, Conservation." *Travel Research Journal,* 2: pp. 3–10.

Gunn, Clare A. (1988). *Vacationscape: Designing Tourist Regions,* 2nd ed. New York: Van Nostrand Reinhold.

Gunn, Clare A. (1991). "Sustainable Development: A Reachable Tourism Objective." *Conference Proceedings, Tourism-Environment-Sustainable Development, An Agenda for Research,* Hull, Quebec, TTRA Canadian Chapter, October 27–29, pp. 15–20.

Gunn, Clare A. (1995). "How to Select an Ecolodge Site." Seminar on Ecolodge Planning and Sustainable Design. U.S. Virgin Islands, July 9–12.

Gunn, Clare A. (1997). *Vacationscape: Developing Tourist Areas,* 3rd. ed. Washington, DC: Taylor & Francis.

Honey, Martha (1999). *Ecotourism and Sustainable Development: Who Owns Paradise?* Washington DC: Island Press.

Howe, Jim, Ed McMahon and Luther Propst (1997). *Balancing Nature and Commerce in Gateway Communities.* Washington, DC: Island Press.

Jennings, Gayle R. and Daniela Stehlik (2001). "Mediated Authenticity: The Perspectives of Farm Tourist Providers." *TTRA Annual Conference Proceedings,* June 10 13, pp. 84–92.

Kotler, P. and S. J. Levy (1971). "Demarketing, yes, Demarketing." *Harvard Business Review,* 49 (6), pp. 74–80.

Lee, Choong-Ki, Ju-Heo Lee, and Sang-Yoel Han (1998). "Measuring the Economic Value of Ecotourism Resources: the Case of South Korea," pp. 40–47. *Journal of Travel Research,* (36) Spring.

Long, D. (1990). "Resource Management: A Global Perspective." *Tour & Travel News,* 179 (May 7): pp. 1ff.

Malloy, David Cruise and David A. Fennell (1998). "Ecotourism and Ethics: Moral Development and Organizational Culture." *Journal of Travel Research* (36) Spring, pp. 47–56.

Manning, Edward W. ed. (1992). "Challenges to the Tourism Sector for the Coming Decade." Based on presentations, Tourism Stream of the Globe '92 Conference, Vancouver, Canada, March.

Manning, Edward W. and T. David Dougherty (1999). "Planning Tourism in Sensitive Ecosystems." *Tourism Developments in Critical Environments.* T. V. Singh and Shalini Singh, eds. New York: Cognizant Communication Corp.

Mansperger, M.C. (1992). "Yap: A Case of Benevolent Tourism." *Practicing Anthropology,* 14 (2) (Spring): pp. 10–13.

McLaren, Deborah (1998). *Rethinking Tourism and Ecotravel: The Paving of Paradise and What You can do to Stop It.* Hartford, CT: Kumarian Press.

Molotch, H. (1976). "The City as a Growth Machine: Toward a political Economy of Place." *American Journal of Sociology,* 82 (2) (September).

Mowforth, Martin and Ian Munt (1998). *Tourism and Sustainability: New Tourism in the Third World.* London: Routledge.

Murphy, P. E. (1983). "Perceptions and Attitudes of Decisionmaking Groups in Tourism Centers." *Journal of Travel Research,* 21 (3) (Winter): pp. 8–12.

Nelson, J. G. (1991). "Are Tourism Growth and Sustainability Objectives Compatible? Civics, Assessment, Informed Choice." In *Tourism-Environment-Sustainable Development: An Agenda for Research*, proceedings of the Travel and Tourism Research Association Canada Conference, Hull, Quebec: October 17–19, pp. 38–42.

Okrant, M. J. (1991). "A Skeptics View of Sustainability and Tourism." *Tourism-Environment-Sustainable Development: An Agenda for Research*, proceedings of the Travel and Tourism Research Association Conference Canada. Hull, Quebec: October 17–19, pp. 31–33.

Pezzey, J. (1989). *Definitions of Sustainability.* Discussion Paper No. 9, University of Colorado, Institute of Behavior Sciences.

Rees, W. E. (1989). "Defining Sustainable Development." *CHS Research Bulletin*, University of British Columbia, May: 3.

Rothman, R. A. (1978). "Residents and Transients: Community Reaction to Seasonal Visitors." *Journal of Travel Research,* 16 (3): pp. 8–13.

Sargent, Frederick (1974). *Human Ecology.* New York: American Elsevier.

Sirakaya, Ercan and Robert W. McLellan (1998). "Modeling Tour Operator's Voluntary Compliance with Ecotourism Principles: A Behavioral Approach," pp. 42–55. *Journal of Travel Research* (36), Winter.

Smardon, Richard (1991). "Ecotourism and Landscape Planning, Design and Management." In *Ecotourism and Resource Conservation,* J. S. Keesler, ed. Berne, NY: Association of Wetland Managers, pp. 517–519.

Stanley, Dick (1991). "Synthesis of Workshop Sessions." *Tourism-Environment-Sustainable Development: An Agenda for Research,* proceedings of the Travel and Tourism Research Association Canada Conference. Hull, Quebec: October 17–19, pp. 116–118.

Stewart, W. P. et al. (1990). "Sustainable Tourism Development: A Conceptual Framework." Presentation, Third Symposium on Social Sciences in Resource Management, Texas A&M University.

Taylor, G. (1991). "Tourism and Sustainability: Impossible Dream or Essential Objective." *Tourism-Environment-Sustainable Development: An Agenda for Research,* proceedings of the Travel and Tourism Research Association Canada Conference. Hull, Quebec: October 17–19, pp. 27–29.

TSAs (2001). News Release: "TSAs: Revolutionizing the View of the Tourism Industry." Ottawa, Canada: Canadian Tourism Commission.

Unimpaired for Future Generations? (2000). Protecting Ecological Integrity with Canada's National Parks, Vol. II, "Setting a New Direction for Canada's National Parks," pp. 10–21. Ottawa, Ontario: Minister of Public Works and Government Services.

Vellas, Francois and Herve Barioulet (2000). *Checklist for Tourist Projects Based on Indicators of Sustainable Tourism.* France: Groupe Developpment.

Wendt, Charles W. (1991). "Providing the Human and Physical Infrastructure for Regulation Ecotourism Use of Protected Areas." In *Ecotourism and Resource Conservation,* J. S. Keesler, ed. Berne, NY: Association of Wetland Managers, 2: pp. 520–528.

Wight, Pamela (1988). *Tourism in Alberta.* Edmonton: Environment Council of Alberta.

Wight, Pamela (1992). "Ecotourism: Ethics or Eco-Sell?" *Journal of Travel Research*, Winter, pp. 3–9.

Williams, P. W. and Alison Gill (1999). "A Workable Alternative to the Concept of Carrying Capacity: Growth Management Planning," pp. 51–64. *Tourism Development in Critical Environments*. T. V. Singh and S. Singh, eds. New York: Cognizant Communications Corp.

Williams, M. and Lynn Brake (1990). "Wilpena Station, Flinders Ranges National Park: Planning for Cultural Tourism." *Historic Environment,* (7): pp. 3–4, 61–71.

World Commission on Environment and Development (1987). (The Brundtland Commission) *Our Common Future*. New York: Oxford University Press.

Ziffer, Karen A. (1989). *Ecotourism: The Uneasy Alliance*. Washington, DC: Conservation Foundation.

Chapter 4

Policy

INTRODUCTION

Around the world, the dominant tourism development policy continues to be *economic enhancement*, no matter the several impacts. Many nations have already developed a very strong tourism economy. Others, especially those that have a low or flat economy, now seek tourism as a solution to economic improvement. In order to strive toward this aim, new investment in facilities, services, attractions, transportation, information, and promotion are expended. The intent is that through these developments, more people will be employed, obtain incomes, new tax revenues will be received, and new wealth will accumulate. Such policies have had undeniable success as tourism has grown exponentially, and from the national to the local level.

In recent years, however, it has become clear that this singular policy direction is too narrow. *Visitor satisfactions* have been impaired by huge numbers of tourists, especially from transportation delays, congestion, and increased hazards. Host communities are not always hospitable to visitors, especially with their different cultural standards creating *cultural and social conflict*. Many nature-oriented travelers now are sophisticated enough to disapprove of many local land use policies and management, such as abusive use of wild animals from mass tours. Frequently, rare and fragile resources, such as marine parks, are being threatened by too many divers (Davis and Harriott 1996). Many *negative environmental impacts* are now well known. And many communities have been overwhelmed by so many mass tourists that their life styles and quality of life are threatened, demonstrating *poor integration between tourism and communities*.

If tourism is to be successful in the future, public and management must strive for all four goals: *enhanced visitor satisfactions, improved economy and business success, sustainable resource use,* and *community and area integration*.

Although desirable, this is very difficult to accomplish because of tourism's many complexities and other influences. Such action is thwarted by the tremendous volume of players who have their own goals and objectives. No matter how difficult this may be, the development of tourism policy must incorporate all levels and all participants involved in every public and private facet of tourism.

Tourism policy-making and administration is generally considered as the prerogative only of government but actually is done by all three sectors—governments, nonprofit organizations, and the commercial sector. Every tourist business has its own set of policies on building design, location, target markets, service quality level, and level of acceptable profits. Generally, the commercial sector can adapt to market changes more rapidly than governments. Nonprofit organizations tend toward social and environmental policies regarding tourism development. For example, policies of historic protection organizations focus primarily on the cultural importance of preserving buildings and heritage. Adjustment to tourist use requires many modifications of policies such as providing parking and access, special facilities for visitor use, and interpretation. Often policies of nonprofit interest groups exercise political influence on tourism matters in their roles.

Interest groups and their missions and policies toward tourism may be focused at all levels, national to local. They could be classified as either pro or con tourism. Chambers of commerce, made up of business members, generally are protective of their interests and favor greater promotion and more tourist business success and opportunity. Trade associations, such as for lodging and food service, sports events, special tours, air transport, and better roads, often lobby governments for legislation favorable to their causes and fewer restrictive policies. Interest groups opposing tourism include park agency leaders who believe tourist services are outside their responsibility, environmental groups who accuse tourism of damaging the environment, ethnic and cultural groups who resent tourist corruption of their values, and local residents who do not wish their settings and lifestyles upset by nearby tourism development, the NIMBY (not in my back yard) syndrome. But in recent years, these separate attitudes and policies have started to change, to reflect the growing need for broadened policies and better cooperation and collaboration. Needed is a better understanding of the multifaceted roles of policies in tourism.

PUBLIC POLICY

Generally the tourism policy scope of national tourism offices in the past has overtly been promotion—all means of enticing more visitors.

Omitted has been the influence of many other governmental agencies that also bear upon tourism, such as those for transportation, health, safety, environment, heritage, agriculture, forestry, wildlife, industry, and parks. Seldom is tourism implicit in their mandates and policies. Fortunately, in recent years, increased cooperation among the agencies and with the private sector has begun.

New Cooperation

An example in the United States was the creation of the Western States Tourism Planning Council (WSTPC 2000). Established in 1990 and facilitated by the consulting firm of Hunt and Hunt, this was the first collaborative policy effort in the nation, bringing together the public and private sectors of tourism. Its Memorandum of Understanding was endorsed by 13 federal agencies and 7 western state government tourist offices (Alaska, California, Nevada, New Mexico, Oregon, Utah, and Washington).

This effort was stimulated by many conflicts between these sectors principally because the public sector owns and manages such vast areas in the West. Local communities and tourism interests were increasingly opposing governmental land use policies. Tourism was often in conflict with agriculture, ranching, mining, forestry, national parks, and water use. Even though the state tourism offices were largely supported by governmental funds they were an action arm of the private commercial tourism sector.

The stated purposes of the WSTPC were:

• to enhance the experience of visitors,

• to support the long-term economic viability of the travel and tourism industry and communities that serve visitors,

• to protect and where appropriate, restore the natural, environmental, cultural, and historic resources that are the foundation for tourism, and

• to respect the needs and values of these people who live in the West.

This unprecedented and innovative consortium took action toward these goals with a series of summit meetings and special white papers. For instance, one summit addressed five major concerns: improved public access, infrastructure, fees and management, the environment, and information and marketing. The White Paper on Environment (Harvey and Kelsay 1996) included documentation of negative trends, such as air pollution, water shortage, threats to resources from mining, excessive timber harvest, impacts of oil production, erosive grazing, urban sprawl, and environmental deterioration from agricultural practices.

Unfortunately, this important effort has floundered due to lack of continued interest and support from the private sector and state tourism offices. When governments become compartmentalized and bureaucratized, it seems very difficult to create mechanisms for cooperation and collaboration, either internally or with others.

In the United States, some states, such as Texas, have begun to overcome some of the governmental fractioning of tourism policy (*Strategic Travel* 1994). Formed in 1988, the Texas State Agency Tourism Council (TSATC) was formed and includes representatives from key state agencies: Tourism Division, Division of Travel and Information, Public Lands Division, Texas Parks and Wildlife Department, Texas Historical Commission, Texas Commission on the Arts, Texas A&M University System, Texas Department of Public Safety, Texas Department of Agriculture, Texas Department of Transportation, State Preservation Board, and Office of Music, Film, Television and Multimedia Industries. In addition, a great many private sector groups work closely with this Council.

The TSATC meets four times a year with the purpose of integrating policies and actions of the various agencies as they relate to tourism. The Texas A&M University, in keeping with its research and educational role, acts as an objective catalyst for the deliberations and actions of TSATC. In the report of 1994, several accomplishments are cited: establishing a new park reservation system, providing tourism development assistance to communities holding beach cleanup events, tourism development, upgrading public facilities in park and recreation areas, providing hospitality training programs, and finalizing a Memorandum of Understanding with four prominent federal resource development agencies. This effort demonstrates a sincere dedication to cooperation among state agencies. However, as yet there are no policy statements on major tourism development issues, such as transportation glut, erosive land use, or impacts on the environment.

Policy Perception

Policymaking is often seen as based on facts of public need. Although factual foundations are helpful, some researchers emphasize that policies are heavily value-laden, that policy really grows out of values held by those establishing them (Hall and Jenkins 1995). What governmental officials and publics perceive—want, feel, believe in, and hold as tradition—are more fundamental sources of policymaking than facts. Several cases described by Hall and Jenkins illustrate the importance of values in tourism policies. For example, in the political wave of governmental downsizing in Australia in 1989, the Industries Assistance Commission Inquiry into Travel and Tourism recommended the elimination of the national tourism office, leaving tourism promotion to the private sector. Because of the differing political values between the United States and Cuba, travel to Cuba from the United States was restricted for many years. Tourism policies of East Germany before the merger with West Germany were focused on trade union support of domestic travel to sites for recreation combined with political indoctrination. Since the merger, tourism policies have shifted to the economic value of promoting international as well as domestic tourism from all markets. One must conclude that in the planning of tourism development, concerns over policy values must be added to those of research facts.

What nations value highly for social and economic reasons influences greatly their national policies toward tourism. For example, in some countries tourism has purposely been directed toward increased immigration in the hope that visitors will become residents, enhancement of national image (that may have been tainted in the past), stimulation of investment to increase national wealth, and expansion of exports (Richter 1994). These purposes, when implemented, will influence plans for land use, kinds of development favored, and markets to be promoted. Professional building and site designers will need to condition their traditional planning processes in order to meet these policy guidelines.

Policy Foundations

As governments accept roles of tourism planning and development at any scale, national-to-local, they must obtain the training and understanding of how tourism functions. They hold the responsibility for research and guideline development, especially in three areas: travel markets, existing and potential; tourism physical plant; and linkages between market needs and physical plant development (Taylor 1994, 150). The chain that provides this linkage between policy and practice is significant in guiding growth that respects cultural and natural resource protection, social adjustment, and adaptation to community continuity.

When viewed in this broad perspective, it is easier to understand the influence of the diversity of tourism policy around the globe. Nations and communities vary greatly in their political structures and traditions and views toward tourism. In capitalistic countries where laissez-faire tourism development by the private sector has dominated, there was little intervention by government in policy making. However, in recent years the aggregate impact of fragmented development on the environment and society has shown need for some overall governmental intervention.

Tourism policy development has had a checkered evolution as compared to other economic development (Williams and Shaw 1998). In the 1950s it was evident that mass tourism was threatened by constraints at the borders of nations, especially across Europe. Varying customs and their administration caused delays and confusion on the part of travelers. Among the barriers were limitations on traveler spending in other countries, exit visas and fees, restrictions on numbers of trips per year, and prohibition of travel to countries of contrasting religious or political systems (Edgell 1990). As nations became aware of the growing power of tourism, attempts were made to remedy these barriers.

Among those who addressed the issue were the International Civil Aviation Organization (ICAO), the Organization for Economic Cooperation and Development (OECD), the International Monetary Fund (IMF), and the Customs Cooperation Council (CCC). A leading organization for change was the European Union and Commission of the European Communities (Williams and Shaw 1998, 318). Among the changes recommended were: foster freedom of movement across borders, improve tourism working conditions, a common transport policy, protection of European heritage, and governmental assistance in the development of tourism supply. An effective tool for reducing travel barriers is the General Agreement on Trade Services (GATT), agreed to in 1993 by over 100 countries (Edgell 1999). When incorporated into policy, this holds promise of improving customs regulations, formalities at borders, and regulations on travel services. In recent years, many of the restrictions have been removed with the exception of increased inspections for drug and terrorist related travel. This aspect of policy is cited here because these variations in policy will influence how designers and planners create their project plans.

Because tourism policy is influenced greatly by tradition, a major variable in tourism planning and development is entrepreneurship, or lack thereof. In nations without an entrepreneurial tradition, a vacuum exists in the freedom to create supply side development. Individuals have neither the understanding nor desire to visualize a market demand, obtain financing, seek and purchase a site, establish facilities, and obtain operational management. When this does not exist locally, the vacuum is often filled by an outside international source, frequently encouraged by national policy.

Investment by outsiders such as transnational organizations (TNCs) has the advantage of experience and financial backing for tourism development. However, this advantage is often offset by problems. The promise of hiring local people, thereby making new economic input with jobs, may not happen because of the lack of local training and experience. Even when locals are hired, the jobs are likely to be low-pay and unskilled. Because financing came from outside, little economic return may accrue to the local community (Hall 1998). In addition, because planning and design are likely to be performed by outside specialists, local cultural and natural resource sensitivity may be lacking, causing societal and environmental conflict. This issue of the source of development is raised here because it is related to planning and policy. A national policy could be initiated to sponsor education in tourism planning and development most appropriate to the special conditions of the country and area. Such a policy could be a valuable instrument for screening outside and local investment that is best for the area. McNulty and Wafer (1990) have recommended that host nations have the power to guide and control the actions of TNCs in several ways: prohibiting foreign ownership of land, recommending local construction materials, limiting role of foreign management, and guaranteeing local investment and financial return.

As demonstrated by the September 11, 2001 terrorist attack on the United States, maintaining a policy on passenger transportation is extremely difficult. Before this event that caused a severe drop in travel demand, a travel glut had become an issue. Communities and tourist attractions were unable to cope with burgeoning mass travel. A high economy combined with new automobile, highway, and air technology had stimulated an enormous increase in tourist travel. It was estimated that airport delays and difficulties of air traffic management cost travelers and related tourism businesses over $5 billion in 1998 (Progress and Priorities 2001).

Then, after September 11, an economic recession combined with fear of travel devastated the airlines. Huge layoffs and threats of bankruptcy impacted tourism as never before. No public or private policy on transportation was in place to contend with this unprecedented change. This event suggests the need for entirely new collaboration between the public and private sector for better travel policy and management.

Tourism policy, as with other policies of the state, is formed by the interaction of the many institutions of the state and varies greatly from nation to nation. For some, tourism is virtually ignored whereas others are heavily involved in the planning, development, and management of tourism. And public policy for tourism does not derive from only one agency. Hall and Jenkins (1995, 20) have identified the following as groups at the national level that often are involved in direct or indirect participation in tourism: law enforcement, executive branch, legislative branch, judiciary and regulatory, public service, quasi-public agencies, statutory authorities, and government enterprises.

Policy and Environment

When examining environmental damage related to tourism, a policy maker must distinguish between true causes. Whereas some erosion and pollution of resources is caused by great numbers of visitors and tourism land use, most environmental damage is caused by lack of plans, policies, and action to prepare for any economic growth. Most of the ills cited by environmentalists are the result of the failure by governments and private sector leaders to cope with any economic growth, not just tourism. With growth comes need for decisions on natural areas to be protected and designed for park usage; location and design of water supply and waste disposal systems; land use locations that are compatible; and on protection of cultural resources, such as historic and archeological sites. Tourism cannot be blamed for environmental deterioration caused by bad policies and decisions rather than real visitor impacts. Some examples may illustrate this point.

For many year Santa Catalina, a small island off the California coast, had maintained a reasonable balance between environment protection and tourism development. Most of the island had been under a resource preserve policy with the exception of the little town of Avon, the harbor tourist village. A major limitation that kept tourism in balance with the environment was water supply, dependent wholly on rainfall catchment. However, this balance has been broken by the installation of a desalinization plant allowing a major new resort condominium development to deface some of the natural landscape. It is not tourism per se that is now destroying some of the natural environment but rather the political and private policy that allowed further development based on a new water plant. Because so many areas around the world have insufficient infrastructure for tourism, the World Travel and Tourism Council, a leading tourism organization, has established an Infrastructure Task Force whose mission is directed toward overcoming this deficiency (*Progress and Priorities* 2001).

Another case of misguided development policy is Walden Pond. Worldwide, Henry David Thoreau is known and revered for his insightful writings about man and nature. Many of these writings were inspired while living alongside a small lake near Concord, Massachusetts. For years, the lake and forested setting were much the same as Thoreau saw them in 1845. By the 1980s, recreational congestion on the lake and destruction of shoreline vegetation by visitor overuse became evident. But the true cause of environmental deterioration from overuse by visitors was the policy of the state agency that fostered use of the area not as an historic shrine but as a recreational area. In recent years, this policy has been reversed and honors the setting as the natural resource inspiration for many of Thoreau's concepts and writings. The area is now named the Walden Pond State Reservation. New planning and use policies have reduced erosion from visitor use, added interpretive programs, and established rigid controls including restricted hours of use, elimination of recreational uses except limited swimming, picnicking, and non-motorized boats (Rules and Regulations 2001). This illustrates the planning and design environmental improvements that can take place when policies of land use are changed.

In another instance, the island of Moorea, near Tahiti, was studied in 1990 by a Task Force, sponsored by the Pacific Area Travel Association and the French Polynesian Government (PATA 1991). Market surveys had documented visitors' rewards as the scenery, unspoiled settings, and the sea—outstanding resource assets of the island. However, the findings of the Task Force showed that the comparatively low occupancy of resort hotels was due largely to lack of planning and management of the very resources cited as important to visitors. There was evidence of mountainside erosion, waterfront sewage pollution, reduction of wildlife habitat, and threats to rare archeological sites. Virtually no planned visitor access and interpretation had been developed to meet market needs of enjoying the rich and abundant natural and cultural resources. It was not excessive tourism but the lack of planning and policy by the French government that caused environmental problems. The failure to build central sewage treatment, to place the mountain and several beach areas under public park control and planned uses, to establish new attractions and visitor centers, and the failure of the resort owner-managers to utilize off-site resources for enrichment and enjoyment were the causes of poor business and resource threat. If public and private policies and planning were improved, the environment could be enhanced and the tourism economy increased.

Especially difficult and controversial are policies regarding wildlife management in and around protected areas, such as national parks. The tourist appeal of viewing and photographing wildlife has grown exponentially, stimulating increased economic value to communities and nations not only in Asia and Africa but elsewhere. A major interference with tourism has come from wildlife practices that slaughter animals for the sale of products, especially elephant tusk ivory and rhinoceros horns for aphrodisiacs. And, even if management practices within park boundaries balance wildlife with habitat maintenance, animal ecosystems frequently go beyond the boundaries, creating conflict with land owners, particular ranchers and residents of gateway communities. Demanded are new policies and cooperation among all parties for resolution of these issues.

Finally, if growth and sustainability are to be reconciled, new politics and practices need to be implemented, and at all levels. At the national

level, federal governments usually have the power to establish many policies relating to tourism, such as entrance customs, transportation, health,
and visitor controls. Although most countries have placed priority on
marketing, they also can initiate demarketing where necessary. Many
countries have policies for land use, especially federal jurisdiction. In
recent years, concerns over desertification of forests, air and water pollution, wildlife extinction, and global warming have reached federal levels.
When nations are struck with natural disasters or are engaged in political
upheaval or war, major changes in national tourism policies take place.

Even more critical are growth and sustainability policies, backed up by
legislation, at the area and local levels. It is here that policy decisions are
made on infrastructure, such as roads, water supply, waste disposal, fire
protection, health, and power. Excessive tourism growth may demand
revision of policies at this level. The natural and cultural amenities
important to tourism can be protected an controlled by communities and
surrounding areas. Demanding conservation policies are local parks,
beaches, historic sites, shrines, and landmarks. Today it has become
apparent that local jurisdictions must guide tourism so that it does not
impair the social welfare and quality of life. The following are essential
in the formulation of policies for tourism:

- a clear definition of issues and purposes that is understood by all.
- a consensus on vision and goals for tourism development.
- an amalgam of all sectors affected by policy preparation.
- utilizing the best and most recent research and technical information.
- directed toward specific objectives, not mere platitudes.
- elements of a new tourism organizational structure.

(Adapted from Lamb and Davidson 1996)

*Governments of nations, provinces, states, and communities have the
choice of doing nothing or doing something constructive about public
tourism policy.* They have not only the opportunity but also the obligation
of creating and implementing policies that plan tourism and maintain it in
a positive manner. Especially important is coordination with the private
sector and concern over the public weal.

PRIVATE SECTOR POLICY

Policies and practices of the private sector are equally important and now
in a state of flux. Businesses are adapting to market segmentation,
accepting the great diversity among tourist interests. Hoteliers, restaura-
teurs, and tour operators now recognize the need for targeting specific
travel market groups (Cleverdon 1992). Factors influencing diversifica-

tion include demographic and social change, economic and financial developments, political and regulatory changes, and advances in technology. In turn, these factors influence business plans, decisions, and policies—location, site selection, building and site design, product offerings, as well as level of service and profit-loss relationship. Because these changes are dynamic, business must constantly monitor them, even though this process is time-consuming. Information sources of assistance are marketing firms and worldwide sources such as the World Tourism Organization.

Tourist businesses, as compared to those in shopping malls, are especially susceptible to crises, such as war, terrorism, abrupt transport changes, violent acts of nature, or quick shifts in markets. Crises make planning difficult. But specialists in crisis management recommend steps that can assist business in responding to this problem. Cassedy (1992) suggests a four-step process. In the "pre-crisis" stage, many catastrophes can be avoided by being alert to warning signs of impending crisis. The "turning point" is the actual point of crisis impact when management assesses the damage. This is then followed by "ongoing crisis" when every effort is made for remedy. Finally, at the stage of "crisis resolution," the event is over and business renovation and adaptation can possibly renew service. A business policy that increases awareness of potential crises is a great aid for continued success. But the most difficult threat to predict is terrorism, such as the surprise attack on New York and Washington, DC in 2001.

Many studies of travel behavior have shown that previous visitor experience at a destination is a major factor to influence continued visits in spite of a local catastrophe (Sonmez and Graefe 1998). As travelers become familiar with an area their degree of risk perception of safety and satisfaction increases. They are more likely to travel back to such a destination as compared to a traveler inexperienced at this location, even though a major disaster has occurred. Publicity on a tragedy will generally have less discouragement on the repeat traveler than on the novice tourist. For example, in spite of negative press following the Ixtok I oil spill in the Gulf of Mexico in 1979, repeat visitors quickly returned to the Texas Gulf Coast (Restrepo 1981). They soon learned that their experience on the Texas coast was not impaired and that the only area to receive beach damage was along a very small portion of the Mexican coast.

A relatively new tourist business policy among the private sector is environmental sustainability and ethics. Although sporadic, private sector hotels and tour companies are implementing new rules of business operation and visitor behavior that are protective of the environment. Nelson (1991, 40) advocates the creation of codes of practice and other agreements locally to monitor tourism on the ground. This has been accomplished by consensus of tour operators of the Queen Charlotte Islands and Southwest Alaska (Falconer 1991, 21). All of the 40 tour operators have agreed to a code of conduct "primarily to regulate our own activities in Gwaii Hannas/South Moresley, British Columbia." The content is extensive and includes the following key issues: etiquette; wildlife; archeological, cultural, and historical sites; food gathering; garbage; waste; camping; and local cooperation.

After examining a great number and diversity of implementation issues related to sustainability, Dubois (2001) has concluded that this is an ethical problem and therefore subject to many different interpretations. He offers three scenarios. First are "win-win" strategies by tourist businesses that recognize economic and marketing values that voluntarily reduce environmental erosion. Energy use efficiencies, protection of waterfront values, and location of hotels near access are among these sustainable actions. Second, new collaboration between the tourist business sector and environmental advocates may provide more effective sustainability of the environment. Finally, sustainability may require an underpinning of public policy and regulation. All these scenarios should receive a great amount of new promotion, education, awareness, discussion, and debate in order to make real progress.

Policymaking is equally important for nonprofit organizations. An example is the lobbying efforts of environmental groups, tourism groups, and community organizations regarding public policies affecting tourism. For example, the many federal agencies that dominate lands in the western United States are of concern to tourism businesses there. Local communities often believe that policies of agencies such as the National Park Service, Forest Service, Bureau of Land Management, Fish and Wildlife Service, and the U.S. Army Corps of Engineers, deny them the opportunity to determine their own destinies regarding land use, especially for tourism. Therefore, many nonprofit organizations regularly challenge these institutions.

An example of private sector collaboration on tourism policy is the Ontario Tourism Council (OTC), based on the recommendations of the Tourism Advisory Committee in 1993 (Hawkins 1994). Over 500 tourism business representatives held over 50 meetings to guide the formation of the OTC as an independent nonprofit organization on August 23, 1994. The stated purposes included: better communication within tourism business, overseeing a tourism strategy, assisting in the formation of a marketing and advocacy organization, ensuring sound business principles, and lobbying for better tourism. A seven-member Board of Directors administers the program that for the first time is separate from government and provides a business forum for Ontario's private sector.

No design or planning of tourism anywhere can be accomplished without the influence of public and private policies. *The private sector of tourism (business, nonprofit) has the choice of doing nothing or doing something constructive about better planning policies.* Increasingly, they are beginning to recognize this is not mere altruism, it is the best way to reach their own tourism goals and objectives.

CONCLUSIONS

The formulation and administration of tourism policy must take place by all sectors.

For best tourism planning and development, policies need to be developed and implemented by the private commercial sector as well as the nonprofit and public sector. Too often in the past, these sectors have not communicated nor integrated their policies for tourism, a challenge for the future.

Tourism policy development is the prerogative of all levels, national to local.

Because tourism planning and development crosses all jurisdictional boundaries, policies require integration. Federal to local cooperation is essential for decisions within each supply component—attractions, services, transportation, information, and promotion. In many instances, barriers of tradition, mandates, and turf protection hamper such cooperation.

Tourism policy is hollow without support by action.

Many nations have discovered that mere passage of legislation or presidential declarations are impotent if not followed up by implementation. The creation of action processes and their organization are essential elements of policy development in a real world of tourism objectives.

Needed is better integration of tourism policies directed toward all goals.

Too frequently, promotion has been the sole policy of all levels of public and private tourism agencies. Such a narrow approach does not provide for integration of policies, research and education, and guidance for better tourism planning and development. All levels must include resource protection, visitor satisfactions, and community integration as well as an economic goal.

Tourism policies must be flexible because tourism development is dynamic, not static.

Although the creation and documentation of tourism policies is a first step, they require constant evaluation and revision because markets and supply development change. Such action sometimes requires legislative change in order to keep up with new trends.

Tourism policies of the components of the tourism supply side are interrelated and interdependent.

Too often, new attraction enclaves are created without reference to policies pertaining to services, transportation, information, and promotion. Every development within each supply side component is related to and often dependent upon policies in other components. Essential is a mechanism for integrating policies in all components.

DISCUSSION

1. For any nation, what are the most needed tourism policies at the federal level and how can they be carried out?
2. Discuss the pros and cons of governmental intervention with investment and control of tourist services such as hotels, restaurants, tour companies.
3. Discuss the process by which residents can be involved in participatory development of tourism policies and at all levels.
4. Explore Web sites for nations and destinations for the status, details, and problems related to the development and application of tourism policies.
5. Construct a framework for new and effective private sector tourism policies that encompass social, environmental, and economic impacts.
6. Discuss public and private tourism policy changes needed in order to cope with terrorism.
7. How can the private business sector have a greater influence on protecting the natural and cultural resources that their business depends upon?

8. Discuss why public policy on tourism has virtually ignored the need for supporting more and better research and education in the many facets of tourism.

REFERENCES

Cassedy, Kathleen (1992). "Preparedness in the Face of Crisis," pp. 169–174. *World Travel and Tourism Review,* Vol. 2. J. R. B. Ritchie and D. E. Hawkins, eds. Oxon, UK: CAB International.

Cleverdon, Robert (1992). "Global Tourism Trends," pp. 87–92. *World Travel and Tourism Review,* Vol. 2. J. R. B. Ritchie and D. E. Hawkins, eds. Oxon, UK: CAB International.

Davis, Derrin and Vicki J. Harriott (1996). "Sustainable Tourism Development or a Case of Loving a Place to Death?" pp. 422–444. *Practicing Responsible Tourism.* L. C. Harrison and W. Husbands, eds. New York: John Wiley & Sons.

Dubois, Ghislain (2001). "Codes of Conduct, Charters of Ethics, and International Declarations for a Sustainable Development of Tourism," pp. 61–83. *TTRA Annual Conference Proceedings.* Travel and Tourism Research Association, June 10–13.

Edgell, David L., Sr. (1990). *International Tourism Policy.* New York: Van Nostrand Reinhold.

Edgell, David L., Sr. (1999). *Tourism Policy: The Next Millennium.* Champaign, IL: Sagamore Publishing.

Falconer, Brian (1991). "Tourism and Sustainability: The Dream Realized." *Tourism-Environment-Sustainable Development: An Agenda for Research.* Proceedings of the Travel and Tourism Research Association Canada, Hull, Quebec: pp. 21–26.

Hall, Colin Michael and John M. Jenkins (1995). *Tourism and Public Policy.* New York: Routledge.

Hall, Derek R. (1998). "Central and Eastern Europe: Tourism, Development, and Transformation," pp. 345–373. *Tourism and Economic Development,* 3rd ed. A. M. Williams and G. Shaw, eds. New York: John Wiley & Sons.

Harvey, Bob and Diane Kelsay (1996). *White Paper on Environment.* Western Summit on Tourism and Public Lands. From www.wstpc.org/publications/whitepapers as of July 14, 2000.

Hawkins, Ann E. (1994). "The Ontario Tourism Council: A Common Ground for Industry," pp. 116–123. *Proceedings, TTRA Canada.* Niagara-on-the-Lake: Ontario, Nov. 13–15.

Lamb, Barbara and Sandy Davidson (1996). "Tourism and Transportation in Ontario, Canada." Pp. 261–276. *Practicing Responsible Tourism.* L. C. Harrison and W. Husbands, eds. New York: John Wiley & Sons.

McNulty, Robert and Patricia Wafer (1990). "Transnational Corporations and Tourism." *Tourism Management,* December, pp. 291–295.

Nelson, J. G. (1991). "Are Tourism Growth and Sustainability Objectives Compatible? Civics Assessment, Informed Choice." *Tourism-Environment-Sustainable Development An Agenda for Research.* Proceedings of the Travel and Tourism Research Association Canada Conference, Hull, Quebec, October 17–19, pp. 38–42.

PATA (1991). *Moorea Tourism,* Report of Task Force sponsored by Pacific Asia Travel Association and French Polynesia Government. Sydney, Australia.

Progress and Priorities (2001). London: World Travel and Tourism Council.

Restrepo, Carlos, F. Charles Lamphear, Clare A. Gunn, Robert Ditton, and John P. Nichols (1981). *Ixtoc I Oil Spill Economic Impact Study.* Washington, DC: Bureau of Land Management.

Richter, Linda K. (1994). "The Political Dimensions of Tourism," pp. 219–231. *Travel, Tourism, and Hospitality Research*, 2nd ed. J. R. B. Ritchie and C. R. Goeldner, eds. New York: John Wiley & Sons.

Rules and Regulations, Walden Pond State Reservation, from www.nanosft.com/walden/wprules.html. as of May 11, 2001.

Sonmez, Sevil and Alan R. Graefe (1998). "Determining Future Travel Behavior from Past Travel Experience and Perceptions of Risk and Safety." *Journal of Travel Research*, 37, November, pp. 171–177.

Strategic Travel and Tourism Plan (1994). Austin, TX: Tourism Division, Texas Department of Commerce.

Taylor, Gordon D. (1994). "Research in National Tourist Organizations," pp. 147–154. *Travel, Tourism, and Hospitality Research,* 2nd ed. J. R. B. Ritchie and C. R. Goeldner, eds. New York: John Wiley & Sons.

Williams, Allan M. and Gareth Shaw (1998). "Tourism Policies in a Changing Economic Environment," pp. 375–391. *Tourism and Economic Development,* Williams and Shaw, eds. New York: John Wiley & Sons.

WSTPC (Excerpts from www.wstpc.org. as of August 20,2000 and contacts with state tourism offices, 2000 and 2001.)

Part II
Concepts and Examples
of Tourism Planning

Finally, after many decades of apathy and even distrust, the reality of tourism planning can be documented, albeit erratic and inconsistent. Finally, it is being recognized that the many complexities and great growth of tourism cannot be left to muddling along. Too much is at stake—in terms of the economy, society, and the environment—for this to be the dominant mode of development.

Because the purpose of this book is toward a better balance between the environment and tourism development, the focus is mostly toward spatial and physical planning. Based on the fundamentals described in Part I, Part II encompasses planning concepts and principles illustrated by a selection of applications. Because there are differences between macro and micro planning, this part is organized on three scales—regional, destination, and site.

As used here the term *regional* applies to the larger scale, such as for states, provinces, and nations. At this scale most physical planning for tourism is at the level of policy and guidelines rather than land development. Therefore governments play a larger role than does the private sector. Concepts and principles for tourism planning at the regional scale are presented in chapter 5. A variety of cases are offered in chapter 6.

Although the term *destination* is popularly applied in many ways, in this book it refers to a subdivision of a region, encompassing one or more communities and their surrounding areas. At this level developers, consultants, and local governments exercise important roles in physical planning for tourism. Participatory planning (involvement of local people) is becoming an important part of the planning process at this level. Chapter 7 describes concepts and principles of tourism planning at this scale, followed by examples of application in chapter 8.

It is at the *site* scale that tourism development is actually built. Design and construction of structures and landscape modification result in attractions, services, roads, and other physical features. Although governments and nonprofit organizations often influence planning at this scale, the major role is carried out between professional designers (architects, landscape architects, urban designers, planners) and developers. In chapter 9, concepts and principles of tourism site design are described followed by cases in chapter 10.

Although the organization of Part II lends itself best for book presentation, it must be emphasized that the totality of tourism planning will be the result of integration of all three scales. Planning policy, and practice at each scale influences and is influenced by the others. Working toward the overall goal of integrated tourism planning is the major challenge today for all tourism leaders, designers, developers, and local citizens. A major conclusion is that tourism planning must be dynamic, not static, because that is the true nature of tourism. The aim is toward the goals of better visitor satisfactions, increased economic benefits, adaptation to local society, and protection of basic cultural and natural resources, and on a continuing basis.

The book concludes with a brief Epilogue. Rather than repeat the discussion questions and conclusions presented after each chapter, the authors conclude with some pertinent comments. These relate key revelations from the experience of revising the book. In spite of the travel depression following the terrorist attack on the United States in September 2001, the authors are convinced that the information presented in this book continues to be of value for planning and developing a better tourism world.

Chapter 5

Regional Planning Concepts

INTRODUCTION

Even after gaining new understandings about the many characteristics of tourism—its functioning system and components—the question remains, how can tourism plans and planning be made and implemented? Are there techniques and processes that nations, provinces, and states can follow that will assist them in reaching their tourism development objectives? Some enlightenment on these questions is offered in this chapter.

Experience is demonstrating that planning is taking place and producing results. World-wide there is an increasing awareness that tourism can no longer rely only on heavy doses of hucksterism—that greater planning and care must be exercised to avoid negative social, environmental, and economic impacts and reach the positive objectives desired.

Tourism planning approaches at the regional scale (national, provincial, state) are so extremely diverse that compressing them into a uniform set of principles becomes difficult, almost impossible. The main reason for such diversity is the difference among the many political ideologies and traditions around the world. Some nations have no tourism plans whereas others have sophisticated plans that have been in place for some time. Many have engaged in annual revisions of marketing plans. Others have invested in airlines and hotels. In addition, planners and scholars have put forth concepts and principles for regional tourism planning.

Although research is needed, it is now known that planning for tourism is more art than science. Certainly the results from scientific inquiry should be utilized. Logical sequences—the *discursive* aspects—should be incorporated into planning whenever possible. However, equally important, and perhaps in greater need, is *intuition*. Conscious reasoning needs to be tempered by intuitive perception and judgment. Leaders of tourism planning and development as well as local constituencies must allow feelings of rightness as well as facts and logic to influence plans and decisions.

However, experience is also showing that there is no right way to plan, especially at so large a scale. And, probably no one should look for such an ideal model. Rather, any process that can respond to regional needs should be applied. The purpose of this chapter is to offer some planning models, ways of discovering zones of potential and planning processes for regions in their desire to create and implement better tourism. Planning concepts for destinations and sites follow in chapters 7 and 9.

BACKGROUND

Years ago a few scholars, planners, and geographers expressed interest in how tourism was being developed and responded with new models and processes for planning tourism. Formica (2000) analyzed four tourism journals (*Annals of Tourism Research, Journal of Travel Research, Tourism Management,* and *Tourist Review*) for content on tourism planning. Because the study was not confined to regional planning, the papers covered topics of urban, rural, and coastal planning of tourism. From the 1930s to the '60s, the focus of most articles was toward geographic distributions of development and transportation. From the 1960s to the '80s, planning articles varied but many were econometric models involving employment, income, and multiplier effects. From the 1980s, most included a systems analysis approach directed toward a continuous planning process. In the last two decades, the negative environmental and social impacts have influenced planning concepts. Earlier an emphasis was on planning for growth whereas more recently the consequences of growth have been considered.

Researchers and planners have put forth a variety of models of regional tourism planning based on their response to existing tourism development and the general lack of integration of all the parts. For example, Getz (1986) reviewed 150 models of tourism planning and classified them into several categories. *Whole system models* viewed tourism as many integrated components that were dependent upon environmental foundations. *Spatial/temporal models* grew from the geographer's interest in locational reference to host-guest relationships. Included were distribution of resorts and attractions and an area's dependence upon tourism. *Motivational and behavioral models* delved into the socio-psychological aspects of travel demand, particularly as related to attractions. *Impact models* were oriented dominantly to economic impacts including multipliers. Few of these dealt with social or environmental impacts. *Forecasting models* were intended to project future trip generation but revealed the difficulty of assessing a great variety of influences upon such projections. Finally, *planning/management models* (the main focus of this book) were directed toward problem-solving and guiding rational and optimal decisions for tourism development. An integrated systems approach identified goals and the needed planning processes in support of these goals. A first requirement was an understanding of the tourism system.

REGIONAL DEVELOPMENT HIERARCHY

Since planning involves prediction of consequences of the manipulation of a number of development factors, it is necessary to identify these factors. Based upon past tourism experience, a dependency hierarchy, such as that diagrammed in Figure 5-1, can provide a useful foundation for planning. This hierarchy summarizes many of the topics and principles stated earlier in this book, such as the tourism system and influential factors, chapter 2.

Regional development of tourism (A in Figure 5-1) generally must have an increase in the volume of participation (B in Figure 5-1). More people must go to a region and spend money on tourism activities in order to generate new jobs, new incomes, and new tax revenues. However, increased participation depends upon two very important factors.

First, there needs to be a heightened demand to visit the given region. In this context it means that more people, at their home origins, must be able to exhibit both the *desire* and *ability* to travel to the region and participate in its offerings. If prospective visitors do not have a desire to visit the given region, it is doubtful if they will. In addition to desire, they must have the time, money, transportation, and equipment necessary to make the visit.

Second, if more people are to do this, changes in present levels of offerings—the supply—must take place (C2). Either the capacity of the present physical plant or the total number of offerings must be increased. In other words, either more people or shifts of markets must be accommodated at more attractions, lodging, food service, transportation, and retail sales and services. Furthermore, if the region has a reputation of low attractiveness or poor service, this image must be reversed. And, from an economic point of view, the local system servicing tourism should have the highest "export" ability; that is, it should import the fewest services and goods. Finally, whatever changes are made must be appropriate to both national and local political and social goals. For these

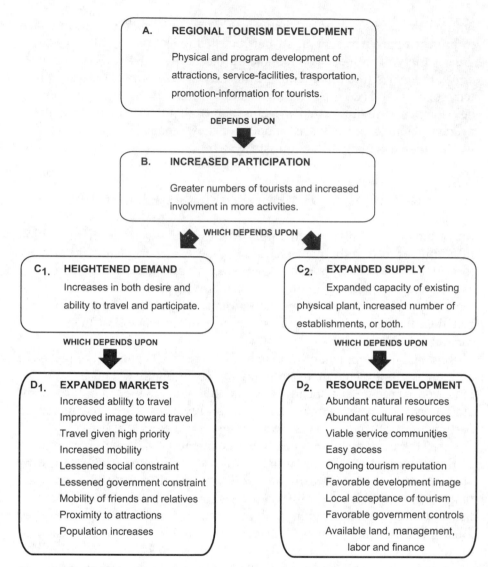

Figure 5-1. Tourism Development Dependence Hierarchy. Those who seek to increase tourism will need to heighten travel demand and especially expand supply development. This diagram models some of the factors important for such action.

to happen, there need to be changes in both markets (D1) and resource development (D2). Some of these changes can be manipulated from the standing of the region; others cannot.

Expanded Markets

Changes in D1 are often influenced by overall cultural and economic trends of the nation (for domestic tourism) and the world (for foreign tourism) and are not subject to easy manipulation by a tourism region. Some, however, can be influenced. A review of some of the important factors within markets (D1) may provide some insight into the opportunities and limits of market manipulation.

Ability. Shifts in the ability of people to travel and to spend money on travel objectives can have a very important impact on a region. Factors such as increased incomes, greater job security, and discretionary income can also have great impact. These, however are not easily manipulated by the tourism region as they are general conditions of a total society and economy at the source of markets. Sometimes the lowered pricing of transportation and destination services can increase a market's ability to travel to the destination.

Image. Markets have either negative, neutral, or positive image values toward regions. These image values can have a good or bad influence upon the popularity of a region. Their creation is extremely complicated and not easily manipulated, as has been discovered by tourist agencies who have been disappointed with massive image-changing programs. In the long run, better supply development at destinations can enhance the region's image, producing increased word-of-mouth recommendation.

Priority. People must place high priority on travel in their expenditure of time and money if tourism is to thrive. If society shifts its priorities into other consumer targets—homes, automobiles, education—tourism will lose. Although advertising and promotion may have some impact, it is difficult to change the priority ratings of families and individuals. While some modification of priorities on personal expenditures can be determined from the standing of a region, it is very limited.

Mobility. Obviously, mobility of the market—opportunity to travel from home to the destination—is very important. The numbers of people able to travel to tourist destinations may be drastically affected by continued changes in costs of automobile, bus, and air travel. To some degree, mobility both within and relating to a region can be changed by influence from a destination area. However, increased ownership of automobiles and controls of gasoline and plane ticket costs are difficult to influence from the standing of a region. A receiving region can take action on relieving travel congestion.

Social Constraint. Over history, social controls have been increasingly liberalized. Changes in dress, conduct, and an increase in the variety of recreational activities have influenced travel land use in recent decades. However, such cultural shifts come from market sources and are not easily manipulated by the destination region. A local society may even have values in conflict with those of visitors. Better information and educational programs can be used to lessen social conflict between hosts and guests.

Governmental Constraint. Federal governments often control border entrance through custom regulations. Taxes on gasoline, air travel, and properties exert constraints on certain recreational activities. Governmental distribution of subsidies to highways, airports, reservoirs, and recreation areas may favor some activities over others. In some areas "blue laws" place constraints on purchases of recreational goods on Sunday. Land use regulations and environmental controls are of increasing importance. Political agreements (or disagreements) between nations, such as for control of disease or terrorism, can foster (or restrict) travel. To some extent tourism regions can exert control over governmental constraint.

Friends and Relatives. A major market segment is the visiting of friends and relatives. The locations of these people make a difference in the destinations of this segment of travelers. Knowing the locations and changes in locale of friends and relatives is important but it is difficult for a tourism region to have any influence over this factor.

Proximity. Generally, the shorter the distance between home and tourism region the better. Markets nearby or within a region are usually the ones most productive. One way a region can affect this proximity is by favoring destination development at locations most accessible from markets. Another way is to tap new market segments within the market area. Improving the time-distance relationship, such as discount air travel between home and destination, can sometimes improve the effect of proximity.

Volume. Given today's burgeoning environmental issues, it is fanciful to suggest increasing the population in market areas in order to increase tourist business. However, for certain destinations, it may be within a region's power to tap increasing volumes of new market segments in cities of origin within their market range, especially those that exhibit elements of economic growth.

Resource Development

Although market manipulation is not the subject of this book, it has been introduced to more clearly identify the relationship to the role of resource development (D2) in changing economic and social impact of tourism. In other words, if anything major is to take place within a tourism region the changes in the supply—providing more for people to see and do— become very important. If economic-impact generators, primarily attractions, are to increase, upon what factors do they depend?

Natural Resources. The given natural resource assets that lend themselves to development are important if growth is to take place. If a region has an abundance of usable surface water, aesthetic and game-laden forests, interesting topography, buildable soils, and favorable climate, it has greater potential for tourism development than one without these assets. Of course reservoirs can be built, forests can be planted, wildlife can be increased with controlled management, and hostile climates can be ameliorated (enclosures with heat in winter and with air conditioning in summer). However, these come at a cost, a cost which often can be lessened by avoiding areas needing environmental modification and by selecting areas with most suitable natural resource assets. Certain rare natural resource assets cannot be replicated and therefore demand special protection. Many developed attractions, such as natural resource parks, scenic overlooks, resorts, cruises, scenic tours, hunting and fishing areas, depend greatly upon natural resource assets.

Cultural Resources. Many participants seek their travel objectives less from developed natural resource assets than from those that are the result of man's cultural imprint. Religious, scientific, and educational institutions, trade centers, national shrines, ethnic customs and crafts, engineering feats, and manufacturing processes as well as historic and archeological sites, buildings, and artifacts are examples of the wide range of cultural resources important to tourism development. These often already exist within a region but have not been developed for tourism.

Viable Service Communities. Generally, the cities that lie within a region being considered for tourism development serve two functions. First, they frequently contain attractions or resources with potential for attraction development. Second, they provide many needed services, facilities, and products. The primary ones, of course, are lodging and food service, but equally important for many tourists are other services, such as police, communications, medical aid, shopping, and banking. In addition, cities offer the infrastructure—water, waste, fuel, police, electrical power, governance—necessary for tourism development. Tourism expansion, therefore, depends upon the distribution and viability of cities. Generally, the city that is larger and possesses a strong economy is better able to provide these supporting functions.

Easy Access. Tourism expansion depends heavily on access and not all regions are equally served by transportation and access. Within the continental United States, the highway system is very important because automobile travel dominates. However, for certain localities and activities, air travel is an important access factor for tourism. Even when expansion is considered, existing routes generally are favored over new ones.

Existing Development. An area that already possesses an ongoing tourism development has a stronger factor in its favor for future expansion than does raw and undeveloped land. The existing development may have established a reputation that is well known in the market place. Existing development, such as public parks, theme parks, historic sites, and beaches, can provide many clues to the relative importance of resource potential of an area.

Favorable Development Image. Image is a product of both the supply and user attitude and therefore cannot be dealt with only on a resource development basis. However, if an area, no matter the reason, now has a reputation of poor (or excellent) quality of tourism participation experience, it can deter (or favor) further expansion greatly. Changes of development can alter this image but not without massive change.

Local Acceptance of Tourism. Expansion of tourism depends greatly upon the local attitude toward expansion. If the local electorate and leadership fully understand the implications of tourism and favor its development, further expansion has support. However, if attitudes and cultural norms are antagonistic or hostile, it will be difficult to develop tourism.

Favorable Government Controls. Tourism development can be accomplished best with the least governmental constraints. If too many legislative controls are enacted against it, certainly development is restricted. But modern tourism saturation and negative impacts may require controlled management. Jurisdictional problems can sometimes limit full opportunity of developing legislation that gives tourism greater chances of growth. Care needs to be exercised in evaluating controls. Many of the recent environmental controls may appear to work a hardship on some development but, upon deeper examination, may be protecting and perpetuating tourism attraction assets.

Available Land for Development. Tourism development certainly depends upon space for development. Some segments use more land than others. Beach use may be very intensive, whereas hunting and wild land recreation are probably the most extensive. Vacation home development uses relatively large areas of land. Probably a greater constraint is not being able to purchase land that is properly located. Since modern concepts of ecology do not allow any land to be classified as "waste," new tourism development must be a tradeoff from existing or other potential land use. Land price and purchasability are important. Therefore, tourism development depends upon the availability of suitable land for expansion—suitable in both quantity and quality.

Availability of Entrepreneurs, Managers. The tourism development of a region will depend upon the availability of entrepreneurs and managers. If a region does not have these resources, they will have to be imported. However, this can cause conflict with local aspirants who thought they could qualify for the positions. The greater the supply of the several types of developers and managers—for attractions, transportation, lodging—the more favorable a region is disposed to development.

Availability of Labor Pool. Tourism employs a wide range of job categories from highly skilled to unskilled. The nature of development will determine the labor needed and whether the region will have to import labor or supply its own. If new labor is imported, local underemployed or unemployed people may resent their coming and resist tourism expansion. Therefore, source of labor is an important factor in tourism development.

Availability of Finance. Tourism development demands great amounts of capital investment. Perhaps it would appear that the means of finance would not matter as long as development takes place. This may be true in the economic sense that some additional development is better than none. However, investment sources often carry with them contingencies that may or may not be compatible with local interests. For example, local residents may realize the value of certain land use controls to protect resource assets. An outside investor may refuse to invest under such conditions. Availability of finance is important from many standpoints.

This hierarchy of dependency identifies a number of variables, any one of which can make considerable difference in opportunities for tourism development. Some are geographic; others are not. Some are slow to change, others change very rapidly. Some are subject to legislative control; others are determined—sometimes permanently—by given environmental conditions. Some are subject to the caprice of society—either that population who lives in and controls land use of a region or that coming in as visitors. Some can be maneuvered from the standing of the region; others cannot.

Study of this hierarchy can reveal a number of factors that are within the realm of regional planning possibility.

REGIONAL TOURISM PLANNING CONCEPTS

Following are a few conceptual approaches that have been put forward for tourism planning at the regional scale. Their diversity reflects the complexity of tourism and the difficulty of creating a standard planning effort that will apply to all situations. Each national plan is greatly influenced by tradition and policy. Even from such differences, some common threads can be observed.

A BASIC TOURISM PLANNING CONCEPT

As early as 1950, Gunn identified factors important for tourism business location in the state of Michigan. As part of the Cooperative Extension Service program for tourism development, his planning guide, *Planning Better Vacation Accommodations* (1952), advised investors to consider locations with favorable natural resource factors, such as water, wildlife, land forms, forests, and climate and man-made factors of markets, transportation, competition, history, market access, and characteristics of neighborhood. These factors appeared to be necessary for business success.

In the early 1960s, as a doctoral dissertation, a regional concept for tourism planning and development, was prepared (Gunn 1965). This concept had grown out of twenty years of critical observation and technical assistance to tourist and resort businesses through the Cooperative Extension Service of Michigan State University. The purpose of this study was to create a planning model and concepts whereby tourism could be developed in an orderly manner and toward desirable goals. ("Application to the Upper Peninsula of Michigan" is described in chapter 6.) Following is a digest of this regional concept and recommendations.

Research

Because tourism development takes place on the land, an important research phase is to describe land characteristics pertinent to tourism development. The final product of user (visitor) satisfaction depends greatly on land development for attractions, services, facilities, communities, access, and circulation. These developments are made by public agencies, commercial enterprise, and private individuals. For these action agents, it is important to have information on three land characteristics: geographic position, geographic content, and landscape expression.

Geographic Position

Figure 5-2 illustrates key elements of a region's geographic position on the earth's surface. This is a fixed relationship that places a region in its tourism development context. Relationship of traveler access to regions is an important factor in regional tourism development. Air travel increases the ability of many tourists to reach remote destinations whereas automobile travel is restricted to highway access.

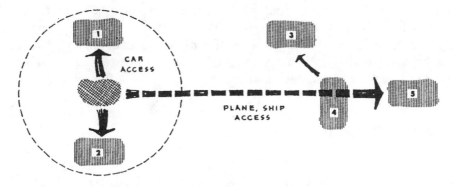

Figure 5-2. Tourist Region Geographic Position. Regions vary in their geographic relationship to travel market demand. Both market segments and destination development will be influenced by distance, time, and ease of access (Gunn 1965, 7).

Geographic Content

Internal resource characteristics influence the kinds of development and their locations. The number, quality, and location of several physical characteristics of a region need to be documented. The surface *geology* of a region varies and may or may not support certain types of tourism development. *Soils* vary and influence buildability and erosive qualities. *Land relief* (topographic change) may be suitable for a variety of activities—scenic appeal, mountain climbing, ease of travel. *Vegetative cover* influences viewing scenery, nature interpretation, and habitat for wildlife. *Surface water* supports a great many tourist activities and its quality and location are significant factors for planning. *History* in certain areas may have potential for development of events, cultural interpretation, and artifact preservation. Finally, the existing *quality and distribution of cities and villages* is a fixed characteristic and influences many planning and development factors—infrastructure, travel termini, governance, local population.

Landscape Expression

Describing and mapping of the above characteristics can suggest qualities of landscape expression. For example, is the land characterized by a domesticated and urbanized landscape or does the wild and undeveloped character give it its distinction? Is the land dominated by agriculture, industry, or city development? From an examination of landscape expression, it is helpful to deduce three major characteristics. A conclusion of *universality* demonstrates similarity of characteristics throughout the region. On the other hand, the region may be *divided* into subregions of similar characteristics. Finally, it is useful to deduce those areas of unusual *distinction*. This is a systemic approach to research that provides clues to potential tourism development.

Evaluation of Potential

Demand

The next step is to synthesize the above findings with travel demand. At the time of this study, market surveys and resulting characteristics were not plentiful. One travel market surveyor identified eleven categories of total traveler visitor: in transit, personal pleasure, convention, business, governmental, military, educational, personal affair, and combined purpose (Crampon 1961). Suggested to planners were these criteria concluded by the author of this concept:

- representative of the demand
- current, not outdated
- indicative of trends in competing regions

Owners' Rewards

For evaluation of potential, the purposes of action agents must be added to resource analysis and traveler demand. Owners and developers have different objectives whether in government, commercial enterprise, or private individuals. Most public agencies are focused more on the social weal than profitmaking objectives. Location, design, and management however are directed toward profitmaking by the commercial sector. Relationship to attractions is very important. Personal rewards from vacation home ownership are primarily aesthetics and private recreation.

Synthesis

Just as analysis separated many factors for examination, the results demand synthesizing into a whole. The mix of resource factors, travel demand, and owner rewards will vary greatly in its meaning from region to region.

Some key questions at this point include:

- What is the land character at points of entrance?
- How well has development reflected potential opportunities?
- What types of activities are not offered?
- What are the seasonal opportunities and limitations?
- How well are communities adapted to tourism?
- How does the distribution of communities relate to tourism potential?

The Planning Units of a Region

From this study, it was concluded that a region contained several basic geographic and tourism functional parts. Understanding these parts can stimulate the regional tourism leaders and designers/planners to guide development in desirable directions. Reduced to their most succinct level, a region can be described as having three major functional units: *community-attraction complexes, circulation corridors,* and *a non-attraction hinterland,* as illustrated in Figure 5-3. (A later revision redefined community-attraction complex as destination, the term used in this book.)

The Community-Attraction Complex

Conceived here is a geographic unit that encompasses a community, its surrounding area, its attractions, and access linkage. If a region is to be planned, the beginning step would be to start with the discovery of these units (also called destination zones). By means of the earlier phases of synthesizing resource amenities with travel demand, these generalized subregions can be identified. Communities serve several functions. All transportation modes terminate at communities. They contain many attraction features and potentials. They provide the basic infrastructure important to visitor services as well as those for residents. They offer governance.

The term attraction complexes implies a grouping of attractions rather than a scattering. By clustering, they gain strength and are more efficiently served by access linkage. Attractions may be diagrammed as illustrated in Figure 5-4. The "nucleus" is the main feature based on a natural or cultural foundation. It is the focal point of appeal. Because these sites may be eroded in importance and aesthetic appeal with incongruous development nearby, recommended is an "inviolate belt" surrounding the

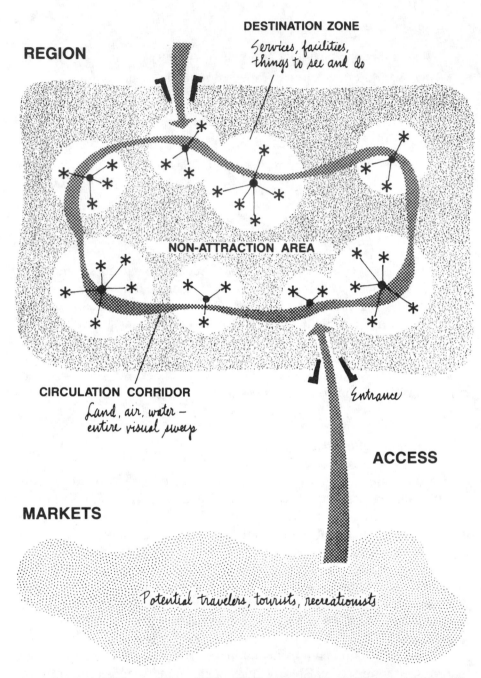

Figure 5-3. Regional Planning Concept. This model illustrates the three major geographical parts of a tourist region: circulation corridor, community-attraction complex (destination), and non-attraction hinterland. Such a concept helps the planner identify potential destination zones and guide future development (Gunn 1965, 26).

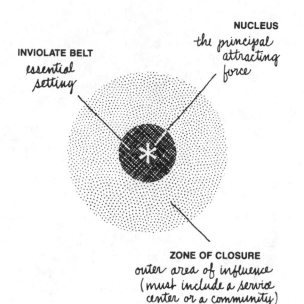

NUCLEUS
the principal attracting force

INVIOLATE BELT
essential setting

ZONE OF CLOSURE
outer area of influence (must include a service center or a community)

Figure 5-4. Model of Attraction. Tourist attractions could be modeled to include three major spatial zones. The main attracting feature for visitor interest is the *nucleus*. An important planning zone around the nucleus is an *inviolate belt* where land use functions and aesthetics can provide a compatible introduction to the attraction. Essential is the surrounding *zone of closure,* an area of overall interest, especially community services important to tourists (Gunn 1965, 26).

nucleus that provides a complementary setting. Completing the planning unit is a "zone of closure," a surrounding area that provides access and includes one or more service communities.

A design aspect for consideration is the relationship between attractions, as illustrated in Figure 5-5. If compatible, it may be desirable to interconnect several attractions into a cluster. If they are incompatible it may be necessary to keep them separate, by design and management. In some cases, such as historic sites at different locations, they may gain by relating them to each other with tours.

Circulation Corridor

A critical planning factor that provides visitor access to community-attraction complexes and linkage with market origins is transportation. Although most systems historically were developed for commerce and general mobility, their planning and design is critical for visitor pleasure as well as mobility. Where corridors enter the region, special gateway design opportunities should be given consideration. Such points give visitors first impressions of the region and its amenities.

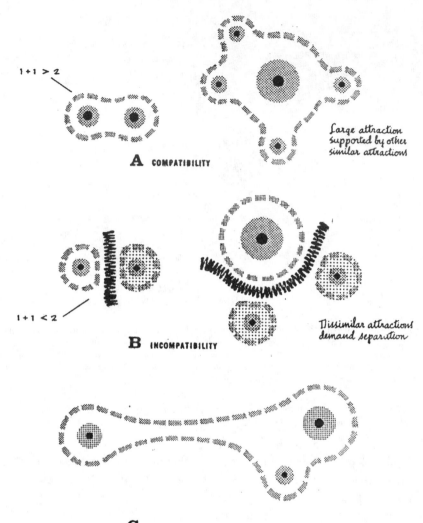

Figure 5-5. Attraction Spatial Relationships. Illustrated are three planning considerations. A, *Compatibility*. Efficiency and greater visitor use are often fostered by grouping. B, *Incompatibility*. Dissimilar attractions may need to be separated by physical planning and management. C, *Distant complementarity*. Some attractions, such as historic sites, even though some distance apart, gain by tour and management linkage (Gunn 1965,38).

The Non-Attraction Hinterland

When circulation corridors and community-attraction complexes are subtracted from the entire region's geography, the remaining land area could be identified as a non-attraction hinterland. It may have potential for tourism in the future but only if new communities and attractions are discovered, especially as influenced by changes in market trends.

Conclusions

This study concluded that the several goals of tourism development could be fostered by planning a region for its main functions. The major functional unit is the community-attraction complex. By means of resource analysis and synthesis with travel demand and developer objectives, these important zones can be identified. Future implementation will then depend on the principal actors for their response to these guidelines.

An Appendix

An Appendix to this report was a special investigation of travel imagery. Based on reference to the psychology of perception and a special survey in Michigan's Upper Peninsula, three phases of imagery were identified (Bruner 1951). Before traveling to an area an individual already has a set of preconceived images of the travel destination. This set is strengthened by previous experience and other influences, such as from promotion. Then, when the traveler actually participates in an attraction, this pre-image may or may not be confirmed. If promotion has exaggerated the appeal or if other conditions of crowding or weather reduce the quality of the experience, it is disappointing. Finally, the traveler comes to a conclusion regarding a new image, derived from this sequence of imagery.

This report is cited here because it was an early regional tourism planning concept that developed basic planning directions for tourism development.

A PLANNING PROCESS

Regional plans and planning for tourism take on many forms. Some are skewed heavily toward *promotional plans.* Often, these are improperly labeled marketing plans because they focus mostly on how more travel can be sold. Some nations develop *marketing plans* that include both improved supply side development as well as how to merchandise it.

Supply side planning, the primary focus of this book, can range from policies and guidelines to specific action strategies for physical and program development. For this, today's planning suggests there is value in a nation or region performing two types in concert—*a supply side plan,* updated regularly (perhaps every five years), and *continuous planning action.* Following is description of these two approaches.

Supply Side Plan Project

Put forward here are key steps in a process for developing a regional tourism plan. It must be emphasized that this process assumes that it has been preceded by several important antecedents. First, a well-represented public-private commission or organization should be in place to sponsor the process. Second, the several goals (visitor satisfaction, economic improvement, resource protection, and local integration) are equally balanced motivating forces for planning. Third, all parties have agreed to involve local constituencies throughout the process. And fourth, recommendations will be directed toward all three sectors involved in tourism development in the region—governments, nonprofit organizations, commercial enterprise.

Step 1. *Setting Objectives.* The key objectives of a regional plan would be to provide planning action on the following topics:

Solutions to constraints and issues.

Identification of destination zones with greatest potential.

Action objectives and strategies.

Optional objectives might include concepts for projects and new policy statements. Upon completion, these objectives should be expressed in two major ways—a document and public forums.

Step 2. *Research.* This step can usually be accomplished through use of secondary data—existing reports, maps, literature. However, the planning team will benefit from reconnaissance and public workshops throughout the region.

Two sets of factors need to be studied. An understanding of the *physical factors* is necessary for five reasons. First, this is essential in determining *potential destination zones.* Second, as these resources are studied, the planning team and regional leaders should be stimulated to *identify concepts, projects, and solutions to issues.* Third, this study helps to place the region in proper *geographical and competitive context.* Fourth, information thus derived for physical resources is essential to the establishment of *new and improved attractions,* usually the first order of tourism growth and expansion. Finally, such an examination of resources identifies existing *threats to the environment* and guidance for future expansion capacity limits.

A prevalent fallacy of tourism development is the popular notion that all areas and communities have equal potential. The strong desire to improve local economies tends to overlook the many differences among areas. The assumption is made that the sole factor for tourism growth is promotion, that with greater expenditures on advertising, any area can enjoy the fruits of tourism. This fallacious belief has resulted in many disappointments when communities discovered that even increased advertising could not make up for deficiencies of basic foundations. Sometimes governmental policy has provided financial incentives for tourism development in sterile geographical locations.

The examples described in chapter 6 contain application of processes that seek to identify zones that contain the best and most abundant factors that support tourism development. Of the many physical factors, experimentation has resulted in using a series of factors that can be generalized and aggregated to produce composite maps illustrating areas where the several factors occur in greatest mass. This process is based on the assumption that travel market interests can be generalized into two overall categories—those activities based on development utilizing natural resources or those utilizing cultural resources.

Several *program factors* also will need study. Essential is basic information on market preferences and other characteristics to determine gaps in present developed supply. National and international trends are also part of this study. An evaluation needs to be made of the present *promotional* and *informational* systems. An identification of *constraints* and *issues* is part of this study. During this process, the planning team should obtain a sense of socio-environmental concerns in the region as well as special influences of government, labor, finance, and available management and policies.

Step 3. *Synthesis-Conclusions*. Many consultant projects omit an essential step at this point in the planning process. Instead of going directly into the stage of recommendations, it is wise to evaluate what the research step has revealed. For example, the nine statements of "conclusions" for the Upper Peninsula study (chapter 6) provided a basic foundation for the creation of recommendations and concepts. The main purpose of this step is to derive meaning from mixing together the several findings from the research stage. Conclusions from both program and physical data are then derived.

Step 4. *Concepts.* It is at this step that creativity and ideation have full sway. Because tourism planning is more art than science, the first three steps are not determinants—they suggest and stimulate propositions. It is at this step that local citizens, public and private developers, and professionals review the findings and conclusions. This requires many meetings and workshops throughout the region. The final recommendations depend on how well all parties can visualize change in order to produce desired results.

A major aid to the development of concepts is the discovery of destination zones with greatest potential. Local areas will benefit from a regional assessment that will place them in proper perspective. This process utilizes the following steps:

1. Mapping natural and cultural resources.

2. Weighting and aggregating these maps by computer.

3. Interpreting zones with highest quality and quantity of resource factors.

Step 5. *Recommendations.* At the regional scale all recommendations must be generalized but not necessarily vague or weak. To the contrary, the focus is on planning toward making the tourism system function more smoothly and more productively. Actors in all parts of the region will then see how their efforts can be helped by contributing to the success of the whole.

Suggested here are recommendations on four aspects of tourism development.

a. *Physical development.* At this step, the kinds of development needed in order to improve the supply side match with demand will be described. Because this is at the regional scale, these would be generalized but would suggest types in the following categories: attractions, transportation, and services. By identifying types of development and management changes needed, actors within all three sectors would discover opportunities for their following through with development plans at the destination and site scales. In addition there would have to be recommendations on the infrastructure (water supply, waste disposal, police, fire protection). Finally, the issues of capacities, environmental impacts, and sustainability would be addressed.

b. *Program Development.* Recommendations would be provided on the need for improvement in tourist information systems—information and interpretive centers, descriptive literature, videos, Internet, tape narratives, maps, and directional systems. An assessment of the promotional program would suggest recommendations for improving advertising, publicity, public relations, and incentives. This final step would also include recommendations to resolve issues and eliminate or ameliorate constraints on tourism development.

c. *Policies.* Most regions lack a policy statement that provides a framework for tourism development throughout the region. This policy would be the result of joint agreement by public and private sectors. All destinations and individual enterprises can then have a sense that they are part of a whole that has agreed upon overall dimensions of tourism development. This policy should set forth the importance of the four goals addressed in chapter 1—enhanced visitor satisfactions, resource protection, economic expansion, and integration into the local environment and economy. It should also clarify the roles of government and the private sector and especially within each destination zone. How will the roles of promotion, research and education be allocated? What public agencies are involved and how will their policies support tourism? How do regional taxation and governance issues relate to tourism? Are new public and private institutions needed? These and other policy questions may need to be addressed.

d. *Priority.* A major concern of tourism planning, especially at the regional scale is that the volume of recommendations may be so great that the task seems formidable. Probably this is the greatest reason for many well-thought-out plans to remain unimplemented. A solution is to review all recommendations with the purpose of assigning priority. Some will take much greater funding and a longer time span than others. Top priority should be given to those that can be accomplished most readily in order to demonstrate improved tourism at the regional level. High priority must be allotted to organizing destination zones for their own planning and development.

These five steps can produce a region's blueprint for improved tourism. Major emphasis is upon action strategies that regional organizations and agencies can and should implement. If the planning antecedents were properly considered at the start, implementation of the recommendations should be well under way.

But one of the major outcomes will be identification of destination zones. Having accomplished this at the regional scale, the next question to be dealt with is getting these zones organized so they can initiate their own planning, as described in the next chapter. This final step requires close collaboration between those responsible for planning at all levels, regional, destination, and site.

The final report resulting from the regional planning effort may best be presented in two kinds of publications. First, there may be need for a full report that describes completely all steps taken and the documentation to support them. This is not publicly distributed but is placed on file with all principals for future action and reference. From this a second report—condensed, clearly written, action oriented—would be prepared for widespread distribution. Emphasis would be placed on the action strategies needed at the regional level with followup recommended for the proposed destination areas. This publication can be a useful tool to stimulate involvement by both regional and local levels.

It must be emphasized that such a plan, although of great value, is bound to the time period during which it was prepared. For this reason, such a plan should be revised at some future date and be accompanied by "continuous planning action."

Continuous Planning Action

Increasingly, in recent years, planners and scholars of planning theory have been giving attention to planning as a continuous process (Figure 5-6). Much of this activity is reaction to the inadequacies of the project or

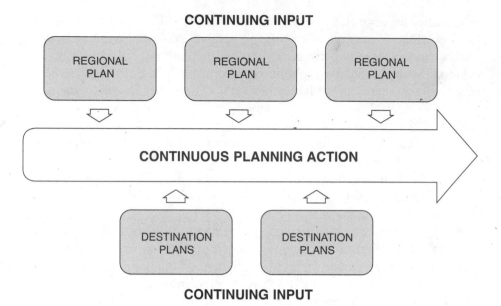

Figure 5-6. Continuous Planning Action. Because of the dynamics of both development of the supply side and market trends, all tourism plans must be revised regularly. Especially important is integrating plans at all levels as conditions change.

plan approach (sometimes master plan) "which gave a detailed picture of some desired future and state to be achieved in a certain number of years" (Hall 1975, 269). The project approach grew out of architecture and landscape architecture which dealt with specific buildable site development. Today, it is increasingly recognized that in addition to this approach, planning as an *ongoing process* has great merit.

The process that checks back upon itself has grown out of the science of cybernetics, coined by the mathematician, Norbert Wiener, in 1948. This process was presented as a means of controlling complicated mechanisms by interrelating important information. It was applied not only to internal control exercised by the nervous system in an animal but also to the engineering control of equipment such as guided missiles. An important aspect of cybernetics is feedback, in which corrections are made as necessary in the functioning of a system such as the path of a missile. From this, the concept of systems planning developed, meaning the integrated and operational planning of the entire system as a whole composed of interrelated parts. The concept of continuous planning is an application of systems planning to existing agencies, organizations, and the private sector.

Within each component of the tourism functioning system is massive involvement by public agencies. In addition, many organizations outside government exercise great influence on functions of each component.

Seldom, however, are these ever integrated. In fact, they are frequently counterproductive. Certainly, for the sake of diversity and countercheck, it is desirable to have such an array of agencies and organizations. But there are instances, particularly at the planning stages of tourism development, when even a small amount of collaboration and cooperation would be constructive.

Not only must the tourism system receive better continuing planning but also tourism must be integrated with all other planning for social and economic development. Many governmental agencies at the federal-to-local level are engaged in programs fostering new jobs, housing, and general social welfare. However, seldom do these programs include tourism. A review of structure plans (official planning documents) in England (White 1981, 40) revealed that tourism was seldom mentioned. In recent years, many nations have begun to integrate tourism planning with other federal plans.

A continuous tourism planning function could be modeled as an interactive system whereby each sector is not subjected to a superior level of planning. Instead, each sector, *on its own initiative* interacts with all others in its own decision making. Since this is not a legislated planning model, it does not depend on a planning bureaucracy or hierarchy. It capitalizes on its own self interest to benefit from communication and interchange with other sectors.

For example, the accommodations sector (entrepreneurs, managers, organizations), in order to stimulate greater business, would on its own initiative interface with the historical restorers, festival backers, park and recreation managers, and entertainment sectors because they are the ones to stimulate more visitors. Because accommodations can be affected by the policies and decisions of the other sectors, the leaders would open up communications with them. While some overall committee, council, and other structure may be needed to integrate action, it would seem that the greater each sector increases its own sophistication regarding tourism integration, the more it will contribute to overall integrated planning for tourism.

Such an approach may be getting closer to Lang's call for better integrated planning—" ... to create a new sense of commonality which may then motivate actors to seek new forms of collaborative action; and to build capability to respond effectively to change as well as to generate change when that becomes necessary" (Lang 1988, 98). He utilizes the concepts of "domain" (Trist 1985), used to describe a set of interdependencies among stakeholders in a transactional (shared) environment. Collaboration begins with stakeholders recognizing that they have mutual interests and that their problems are too complex and too extensive for organizations to go it alone. Each one may be willing to engage in a give-and-take exchange on the basis of mutual gain. Only when each sector sees the advantages of interactive functioning will it reach out beyond its traditional and mandated turf.

In order to activate such interagency and intersectoral cooperation a detailed review of existing practices, policies, and legal mandates may be necessary. If, for example, a federal agency's enabling act makes no mention of tourism in spite of actions impinging greatly on tourism development, it may be necessary to amend the legislation. To illustrate further, a highway department may have great competence in engineering construction but lack planning data on human tourism trends that influence future highway planning. Until the department's official mandate is amended to include greater responsibility for traveler needs, it may be expecting too much of officials to reach out and cooperate with tourism developers on a voluntary basis.

Even though a continuous planning process ideally would encompass great integration of all actors it may be necessary to empower a central tourism agency at the highest level of government to be the catalyst for continuous planning. Most tourism agencies today at the federal level are promotion-oriented and do not have powers of coordination and integration of overall tourism. When the key tourism agency is given responsibility for more than promotion, more effective planning and decisions can be made by both public and private sectors. One way of accomplishing this is for a regional public-private tourism consortium or council to sponsor annual reports of planning opportunities.

PASOLP

In 1977, Lawson and Baud-Bovy published their approach to planning outdoor recreation and tourism development called Product's Analysis Sequence for Outdoor Leisure Planning (PASOLP). Baud-Bovy (1980, 1982) then elaborated on his earlier experiences of applying some of his concepts and principles of tourism planning in several countries. He stresses integrated planning—planning that breaks from the traditional technical planner's approach. By integration, he asks that tourism planning be integrated with the nation's policies, with the physical environment, with the related sectors of the economy, into the public budget, into the international tourism market, and with the structure of the tourist industry.

His experimentation with an updated PASOLP approach in Niger resulted in the concept shown in Figure 5-7 (Baud-Bovy 1982, 312).

Baud-Bovy emphasized that planning should be a continuous process because of the vagaries of tourism over time—economy, politics, fashion. Required is a regular monitoring system.

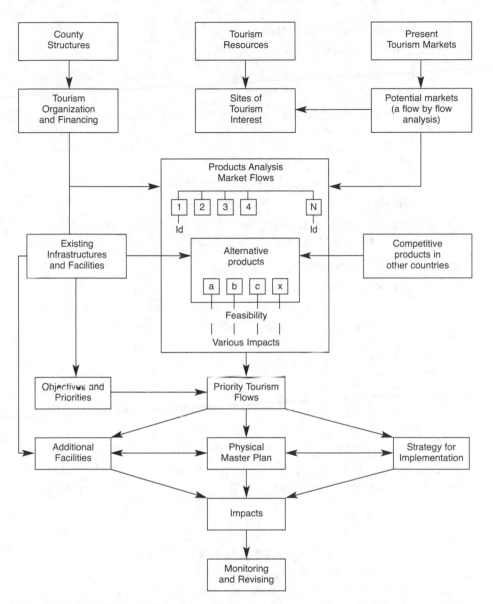

Figure 5-7. PASOLP Model. One of the earliest approaches to tourism regional planning was that put forth by Lawson and Baud-Bovy in 1977 (Product's Analysis Sequence for Outdoor Leisure Planning). Essential features within the planning process were market flows, resource analysis, monitoring, and revising (Baud-Bovy 1982, 313).

He conceived a four-phase planning process:

A. Scientific investigation and analysis

—principal tourism flows (existing and potential) are compared with attractions and resources.

—the nation's structures, politics and priorities are analyzed.

B. Identification of development objectives

—each market segment is examined.

—existing tourism products are compared to market segments.

—destination attractions are examined.

—feasibility, as well as socioeconomic and environmental impacts of new development are examined.

—priority development is identified.

C. Creation of physical plan

—three preliminary studies are made: needed new facilities, estimated impact on sites, and preferred destinations.

—based on results of Phase B, conclusions and recommendations for needed planning are stated.

D. Impacts

—socio-economic and environmental impacts of proposed development are assessed.

—resource issues and problems are given particular attention.

Although this concept was put forward several years ago, its basic elements are often found in other tourism planning approaches in more recent years.

MARKET-PLANT MATCH

An important process step for planning tourism in a region is to match development with needs and desires of travel markets, as described in chapter 2. As an experiment, Taylor (1980, 96) applied market analysis in several countries to the Canadian tourism plant. It was found that of the six segments of Swedish travelers, the Canadian plant could satisfy only one segment. West German markets were matched by only two. All American market segments, however, matched the available Canadian plants.

A special contract to develop a plant-market model for Canada was offered by the Canadian Government Office of Tourism (CGOT). The objectives were to:

Assemble and store key data on tourism market and plant.

Find where markets should be directed to meet expectations.

Show gaps in tourism plant.

Show Canada's position in world plan.

Relate CGOT's program to market/plant.

(DPA 1981, 1)

Running parallel with this investigation was a second study directed toward elaborating a federal perspective on tourism destinations throughout Canada. Particular reference was to be made to four markets: the United States, overseas, interregional Canada, and intraprovincial. The investigation was limited to review of existing documents on planning and zones. The study (Gunn 1982) resulted in two main conclusions. First, several functioning destinations have developed even though they are not necessarily the result of plans. Market data for the four market thrusts of the project were applied to known Canadian resources, access and developed destinations. The "ideal" criteria used for new delineation of zones were: access, service centers, natural resource base, cultural resource base, other resource factors, and program-management factors.

Second, in the effort to delineate zones based on these criteria, several observations suggested that a new approach to national planning was needed. The existing basis for destination delineation seemed to have several weaknesses:

The provinces used different criteria.

The definitions of destination zones were not uniform.

The dates of basic data of provinces and federal government varied considerably.

It was concluded that a new methodology for identifying destination zones within the entire nation was needed. This resulted in a three-phase planning approach to discover zones of potential: (1) research of physical factors, (2) research of program factors, and (3) conclusions for destination potential (Gunn 1982). These recommendations were distributed to all provincial tourism agencies, many of whom implemented all or parts of the concept. In recent years, the provinces have taken on much greater responsibility for tourism planning and development.

SPATIAL PATTERNS

At the same time that planning consultants and governments awakened to the need for tourism planning, several geographers began to take interest.

Pearce (1981) built upon earlier work of Gilbert in England, Miege in France, and Poser in Germany to identify structures and processes of tourist development, evaluation of resources, and analysis of impact of tourist development.

Pearce (1981, 6) identified five elements of tourism supply: attractions, transport, accommodations, supporting facilities, and infrastructure. He divides attractions into three types: natural features, man-made objects, and cultural, such as music, folklore, cuisine, and others. Planning issues are identified: areas with greatest developmental potential, the need to foster growth in dispersed areas, and the need to ameliorate local cultural disruption.

Because tourism involves elements of great interest to geographers—spatial differentiation and regularities of occurrence (Pearce 1979, 247)—few disciplines have contributed as much to the literature of tourism development. Pearce identifies interest topics such as spatial patterns of supply (Thompson 1971; Wolfe 1951; Piperoglou 1966; Pearce 1979), spatial patterns of demand (Wolfe 1951; Deasy and Griess 1966; Boyer 1962, 1972), geography of resorts (Pearce 1978; Pigram 1977; Relph 1976), tourist movement and flows (Williams and Zelinsky 1970; Guthrie 1961; Archer and Shea 1973; Wolfe 1970; Campbell 1966; Mariot 1976), impacts of tourism (Christaller 1954, 1964; Coppock and Duffield 1975; Archer 1977; Pearce 1978; Odouard 1973; White 1974; Smith 1977), and models of tourism space (Miossec 1976, 1977; Yokeno 1977). Van Doren and Gustke (1982, 543) analyzed shifts over time (1963–1977) of the growth of hotel development with particular reference to the "sun belt" of the United States. This sampling of scholars and topics emphasized the fact that economics and promotion, while dominating political interest in tourism, are not the only topics of study important to planning.

Other geographers have studied special aspects of tourism. Demars (1979, 285) traced the development and distribution patterns of resorts in North America and Britain. Murphy (1979, 294) investigated the spatial imbalance in travel patterns and the place of camping in market development strategies in Canada. He concluded that market plans must be much more aware of the specialized behavioral preferences for destinations. Britton (1979, 326) in his study of Third World tourism concludes that, "Host governments must convince an arrogant and powerful industry that local citizens and tourists would benefit from the honest representation of places." Geographical studies in developing countries have been made by many geographers including Helleiner (1979, 330), Hyma and Wall (1979, 338), and Collins (1979, 351).

Pearce (1981, 83) concludes that there are many constraints on planning for tourism. One problem, particularly in developing countries, is the temptation to use models from elsewhere due to lack of data and expertise. This may lead to highly inappropriate development. Implementation is also an issue, especially when roles of the several sectors are not clearly understood. Coordination between agencies is often difficult. Planning, as a process, is preferred over a plan due to the dynamics of tourism. Feedback, monitoring, and flexibility to meet changing conditions are needed.

Transportation for travelers has shifted, causing spatial changes in tourism development. Accommodation patterns have diversified and a wider range of other services are in demand. While infrastructure is not revenue-producing, it is essential to tourism. "Successful tourist development depends in large part on maintaining an adequate mix, both within and between these sectors." Developmental factors of interest to geographers include location, land tenure and use, carrying capacity, and analysis of tourism impacts.

Fortunately, the early works of geographers and planners have gradually been accepted and incorporated into recent national plans for tourism. Insight on the spatial aspects of tourism planning has been brought forward by Fagence (1991). His focus has been on frameworks that help identify those "certain areas" (destination zones) that have special suitability for tourism. His basic working premise is:

> that locations, regions, resources, amenities and infrastructures have an unequal potential and capacity for particular forms, types and scales of development. (Fagence 1991, 10)

Based on this premise, he advocates a coordinated and collaborative approach among all public and private interests and especially to stimulate entrepreneurial initiative. This geographically-referenced framework could provide national tourism policy makers, planners, and developers with a tool that would be able :

1. to express the spatial aspects of any national or regional policy—locations, concentrations, geographic linkages, travel routes and networks, areas of amenity, distributions, and so on;

2. to monitor the evolving patterns of geographic suitability—that is, beyond the status of mere physical capability for development as measured through impact assessments of various kinds—within such concepts and strategies as balance, diversity, complementarity;

3. to facilitate the geographical integration of the various types of tourism development, and of tourism development with other forms of economic activity—so as to avoid inharmonious relationships, creation of economic monocultures, incompatible forms, and quantities of development and servicing need;

4. to assist the integration of tourism development with other forms of economic and regional restructuring;

5. to more efficiently identify locations or "zones" for tourism development so as to maximize the utilization of indigenous spatial, economic, and environmental resources, and so as to pursue specialization and balance rather than duplication, replication, perhaps at the expense of sub-optimization of realistic opportunities; and

6. to formulate integrated strategies to accommodate local/regional/national government and entrepreneurial initiative, so as to provide a context of investment and development confidence, and so as to achieve a coordination of transport, communication, utility infrastructure, and other public capital works programmes.

(Fagence 1991, 11)

Perhaps the strongest endorsement of the Fagence approach is its practical applicability. It again emphasizes the need for tourism management tools that are not only spatially sound but also are directly related to business success, environmental protection, and local social integration. His lucid paper offers further framework concepts that include *points, lines, and areas* as essential physical elements of tourism planning, development, and management (Fagence 1991, 14).

Figure 5-8 illustrates results of study by Lue and Crompton (1992) that revealed at least five different spatial configurations:

1. *Single Destinations*—most activities within one destination.
2. *En Route*—several destinations visited en route to a main one.
3. *Base Camp*—others visited while at a primary destination.
4. *Regional Tour*—several destinations visited while in a target region.
5. *Trip Chaining*—a touring circuit of several destinations.

From a planning perspective, these patterns again endorse the need for the policy makers, planners, developers, and managers at all levels to work together. Spatial patterns are influenced both by distribution of resources within a region and market interest in travel to and within the region. Smaller and rural destination zones can benefit from cooperating with larger ones for development and promotion. Destinations visited for only a short time can benefit from food service, retail sales, and admissions whereas lodging may take place in larger destinations. These patterns are useful in evaluating the potential based on resource factors —some with abundant natural and cultural resources may be favored over others.

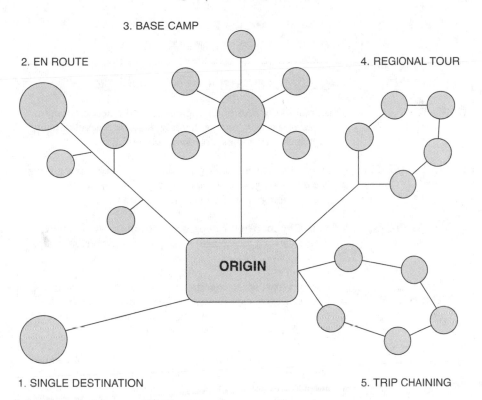

Figure 5-8. Geographic Pattern of Destinations. This diagram illustrates a variety of options for tourism planning and development. These suggest that close cooperation among marketers, planners, and developers is needed (Lue, Crompton and Fesenmaier 1993, 294).

DISCOVERY OF TOURISM POTENTIAL

An important direction for spatial planning of tourism has been the process of discovery of zones of potential. Based on the mix of travel demand and geographic factors, some areas reveal much greater potential than others. As was described earlier in this chapter, key resource foundations for tourism development could be identified as shown in Table 5-1 (Gunn and Larsen 1988). The significance of each factor for the provision of visitor activities, and hence the opportunity for development of the supply side, is also illustrated in this table.

In order to identify zones with greatest potential, the mapping of overlays of the several factors and their aggregation for a sum is a desirable technique. The premise can be stated thus: *wherever these factors occur in the greatest quality and quantity are areas of greatest potential for tourism development.* From the standpoint of a region, research of each factor could result in a generalized map showing areas of greater and

TABLE 5-1

DEVELOPMENT-RESOURCE RELATIONSHIP

RESOURCE	CHARACTERISTICS TO INVESTIGATE
1. WATER	suitability for resorts, campgrounds, parks, second homes, cruising, boating, fishing, hunting, historic redevelopment, organization camping; freedom from pollution
2. TOPOGRAPHY, SOILS, GEOLOGY	suitability for snow skiing, mountain climbing, hang gliding, scenic viewing, resorts, building construction, scenic roads, photography; freedom from erosion
3. VEGETATION, WILDLIFE	suitability for parks, campgrounds, hunting, photography, scenic viewing, organization camps, nature trails, second homes
4. CLIMATE, ATMOSPHERE	freedom from severe storms, excessive humidity, cold or heat, excessive cloudiness, precipitation or fog and pollution (sounds, odors, chemicals); impact of high altitude; suitability for outdoor recreation activities
5. ESTHETICS	suitability for nature appreciation, scenic beauty, photography; freedom from cluttered, ugly and abused landscape; adaptable to development of resorts, campgrounds, scenic drives; attractive streetscapes for urban activities
6. EXISTING ATTRACTIONS, INSTITUTIONS	extent of present tourism development and its image; extent of parks, marinas, resorts, campgrounds, urban attractions; abundance of industries, universities, cultural centers that may have potential; freedom from dangerous industry
7. HISTORY, ETHNICITY	suitability for developing historic and ethnic sites; abundance of customs, legends, foods, crafts, arts; places of prehistoric, historic, and ethnic significance
8. SERVICE CENTERS	distribution, size and qualities of cities; infrastructure; urban attraction potential; extent and quality of services; downtown potential; accessibility
9. TRANSPORTATION	location, modes, and excess capacity; need for new routes; frequency, convenience and market match; distance and access to attractions and services

lesser importance. The challenge has been practical implementation of this process.

At first, the method used by landscape architects entailed hand-prepared tracings of each factor and physically overlaying them to produce a composite. This cumbersome technique became very difficult and visually confusing when more than three overlays were used. Even when numerical values were assigned to each overlay, the process was tedious and prone to inaccuracy. A result of applying this to a region of Texas is shown in Figure 5-9. (Gunn 1974)

As computer mapping became available, the process was expedited greatly. A first application employed the SYMAP program, the Synagraphic Mapping System designed at the Laboratory for Computer Graphics at Harvard University. An application of this technique to a region in Texas is illustrated in Figure 6-14 (chapter 6) (Gunn and McMillen 1979). Although this technique was an improvement over the tracing method, it

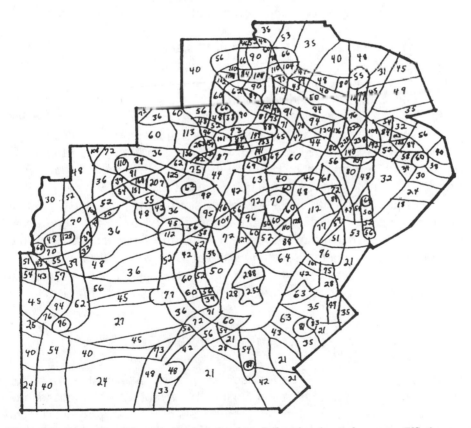

Figure 5-9. Composite of Resource Tracings Overlays. Before the advent of computer GIS, the several hand-drawn tracings of resource factors were aggregated to determine the combined spatial values of the sum of these factors. This was a difficult and very inaccurate process, resulting in this example for an area in Texas (Gunn 1974, 67)

was time consuming and sometimes inaccurate. It required conversion of each factor map into a grid that then could be digitized and assigned a punch card.

Over the 1980s, new generations of geographic information systems (GIS) appeared. They provided more rapid and more accurate composite maps when the several factors were aggregated. This technique was applied to the state of Oklahoma in 1987 to determine zones of greatest potential, as illustrated in Figure 6-17 (Price Waterhouse and Gunn 1987). A later application was made for the state of Illinois, based on study of natural and cultural resource foundations (Gunn and Larsen 1993). From this information the planner can then interpret zones of potential. For example, for Illinois, this resulted in the generalized map shown in Figure 6-22. (More detail for this case is presented in chapter 6.)

It must be emphasized that GIS and other techniques are merely tools to expedite a process of analysis. They still require the input of planners and implementing actors with their skill, experience, and collaborative judgment for planning concepts and recommendations for development. Many other influences, such as the social, environmental, and economic impact on local areas and the nation must be added.

Although this technique has well demonstrated its ability to identify potential destination zones within a region, several caveats must be stated. This approach provides a general regional perspective but depends on follow-up at destination zones and sites. Even though the process may have revealed potential, further examination is necessary. The destination may already have reached its visitor capacity. The high level of existing tourism development may be resulting in resource erosion and social conflict. A federal plan that suggests more growth may meet with opposition locally. This method of destination zone identification demands frequent updating because both demand and supply are dynamic, not static. And new research in the future may reveal revision of the factors used and capability of new planning techniques. The new wave of environmental concern and the ideal goal of sustainability are supplements that need to be added to this planning process for regional tourism development.

NATIONAL PLANNING POLICY

Comprehensive planning for tourism is a relatively recent concept for nations. Although planners and scholars have hypothesized concepts and processes, governments generally have taken interest only in recent years. Based on his world-wide study and experience, and the results of deliberation by representatives of many international tourism planning and development interests, Edward Inskeep (1991) put forward an action strategy for planning sustainable tourism development.

This proposal encompassed many policy implications. Paramount is the concern over all impacts on natural and cultural foundations that would threaten their availability and quality for future generations. This means the avoidance of all actions that have irreversible negative impacts. Guidelines should offer positive recommendations that do not impair social, environmental, and economic foundations. Governments can foster a process that includes research, create land use plans, develop standards, create educational awareness, collaborate with the private sector, and integrate plans of governmental agencies.

Paralleling the public role is that of the private sector. Non-government organizations and tourist businesses need to accept new responsibilities of tourism's many implications. Issues such as environmental impacts, increased visitor safety, and conservation standards are essential for future tourism business success. Finally, international tourism organizations can foster greater cooperation and dissemination of information on sustainable tourism planning.

This proposed planning at the national level supports the social and environmental concerns expressed in many other plans.

WTO GUIDE FOR PLANNERS

Based on reviews of many national approaches to tourism planning at the national level, the World Tourism Organization (WTO) (McIntyre 1993) proposed a generic planning process that includes five major phases.

1. *Study Preparation.* It is essential that a clear statement of Terms of Reference be prepared. Although government may take the lead, a multidisciplinary team of private and public specialists is best able to determine planning directions. This team should encompass representatives of physical planning, marketing, economics, environment, sociology, and infrastructure.

2. *Determine Goals and Objectives.* Several attempts at stating goals and objectives need to be explored and then refined. Feedback from all those impacted is the next step. These statements of goals and objectives should represent a balance of economic, environmental, and social issues.

3. *Make Surveys.* Planning should be based on research information on many factors that make up the foundation for the future. Among the factors are existing travel patterns, potential markets, existing facilities and services, land use, economic situation, social conditions, natural and cultural resources, policies, and availability of capital.

4. *Analysis and Synthesis*. Based on stated goals and objectives and a market-supply study, opportunities and constraints for future development may be identified. This phase provides conclusions on needs for new or revised attractions and services, and addresses capacity issues.

5. *Policy and Plan Formulation*. It is in this final phase that planning policies and recommendations are made. Essential is the integration of tourism plans into other plans at this level. It is at this stage that tradeoffs between potential positive and negative impacts of development are stated, especially social and environmental impacts of tourism development.

A very important outcome of such a tourism planning process is a new understanding of several elements of national tourism planning: the whole rather than only parts of tourism development, tourism as a functioning system, the need for protection of basic resources, revised policies on transportation, identified zones of potential, and stimulation of planning at the community and site levels that are compatible with the national plan.

A COMPETITIVE POSITION CONCEPT

An essential aspect of regional tourism planning is the identification of the area's competitive position. The purpose of this effort is to be sure that recommendations for future development take into account existing competition for travel attractions and services. Ritchie and Crouch (2000a and 2000b) have researched and conceptualized important factors that influence an area's competitiveness. Figure 5-10 is an adaptation of their model.

As a beginning point they use the metaphor of human environmental and heredity influences. For example, on the heredity side, regions have built-in characteristics, their DNA, that is already in place: human resources, physical resources, knowledge resources, capital resources, infrastructure, tourism superstructure, historical and cultural resources, and size of the economy. These are endowments that a planner can begin with as guidelines are developed for future tourism development. Much then depends upon what is done with and to these important given factors. If a region has a strong DNA for future development, it has more opportunities for a competitive edge than other regions.

Reference to the figure reveals a series of building blocks that provide several levels of analysis from the foundation to the top. First are "Supporting Factors & Resources" that include the infrastructure, accessibility, facilitating resources, hospitality, and enterprise. Based on this foundation, pillars then support the next level, "Core Resources &

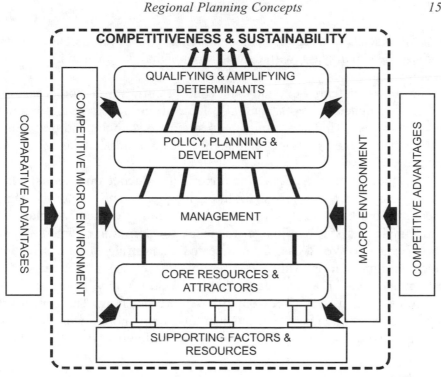

Figure 5-10. Model of Competitiveness and Sustainability. This model adds the dimensions of competitiveness and sustainability to the basic planning process. It is another case that demonstrates the great complexity of the planning and development of tourism (Adapted from Ritchie and Crouch 2000).

Attractions." Important characteristics include physiography and climate, culture and history, market ties, mix of activities, special events, entertainment, and superstructure.

The next two levels include policy and management. Just having the resource foundations does not automatically suggest that development has taken place. Within "Destination Management" are many factors: resource stewardship, marketing, finance and venture capital, organization, human resource development, information/research, quality of service, and visitor management. Generally these are tangible factors that lend themselves to identification and description. Less tangible but equally important are tradition, ethics, and judgment, factors within "Destination Policy, Planning & Development." These factors reflect what a region wants to become and is capable of becoming. Included are: system definition, philosophy, vision, audit, positioning, development, competitive/collaborative analysis, and monitoring and evaluating.

Then, rounding off these levels of building blocks is "Qualifying and Amplifying Determinants" for identifying "Destination Competitiveness and Sustainability." Within are factors such as location, interdependencies, safety/security, awareness/image/brand, and cost/value. Finally, in order to understand a region's competitiveness, the development of assets (resource endowments) will depend upon what is done with them— resource deployment.

The authors defend their study and concept but warn that other factors, not yet researched in depth, may also have an influence on the region's competitiveness. This approach should stimulate scholars and practition- ers to make further study of regional and destination analysis for future planning.

From the market perspective, an important influence on regional com- petitiveness is traveler image of these locations. This factor is being researched but because it is so dynamic it becomes difficult to plan for. An example of a study of image was made for domestic travelers within New Zealand (Pike 2001). This revealed that influences on selection were: accommodation, value, comfortable drive, scenic beauty, cafes/ restaurants, weather, lots to see/do, beaches, and friendly locals. Study of a nation's tourism image is essential for future planning. The concept of competitiveness cited here suggests that this is one more factor that needs to be cycled into national plans for tourism development.

CONCLUSIONS

From this review of regional (national, provincial, state) tourism planning concepts and processes, some pertinent conclusions may be reached. (Implementation cases are cited in chapter 6.)

Regional tourism planning concepts reveal much similarity.

A cursory review of concepts suggests great diversity. Researchers and planners approach planning from different perspectives. Consideration of applications to different nations reveal adaptation to separate cultures, traditions, and political forms of government. In spite of this, most con- cepts include regional planning fundamentals such as: research of natural and cultural resources, market study, synthesis of research information, variation in geographic potential, environmental sustainability, and potential impact on local societies.

Planning concepts now include more environmental concerns.

Early planning concepts focused primarily on growth of tourism, where and how it could be expanded, no matter the consequences. Recently there is a growing recognition of environmental issues. Included are concerns over negative impacts on natural and cultural resources and conflict with local societies. Oversaturation of tourism on communities has become a concern. New planning guidance is being put forward that can prevent or ameliorate these pitfalls.

Regional tourism plans must have a clear and acceptable vision.

At the outset of plan development, a nation must have consensus on a clear statement of purpose. Such a statement has several special dimensions. The scope of the plan needs definition—economic, social, geographic, environmental. Important is identification of plan sponsorship—federal government, other governmental agencies, private sector, influencing agents. A distinction between goals and objectives must be clear—goals are continuing aims, objectives are to be accomplished within a specific time frame. And, important is the extent to which the potentially impacted publics have been involved in the plan preparation.

Implementing agents and their responsibilities need early declaration.

Many well-researched and professionally prepared regional plans have been aborted because implementation was not addressed in the beginning of the planning process. Unless those who are best able to take action, public and private, have accepted their responsibilities, implementation is unlikely. Such implementation may require revision of legislation, new collaborative policies, and even a new national tourism planning and development organization.

Planning often requires new public-private cooperation and collaboration.

Because development of burgeoning mass tourism is extremely complex and relatively recent, nations are generally ill prepared with a suitable management structure. Because many governmental policies and agencies influence tourism but have their own focus—agriculture, environment, parks, health, social services, transportation—new interagency policies and actions may be needed for tourism planning and management. Because tourism is global as well as domestic, international cooperation is also required.

Techniques have improved the discovery of potential destination zones.

Modern computer GIS and other techniques aid greatly in the description and analysis of geographical resources for tourism development. These techniques can provide better information of value to policymakers, planners, and managers of tourism development. At the same time these techniques are tools and require the addition of other influences for planning, such as planning aims, tradition, professional experience, and national philosophy.

Destination zone identification requires local follow-up.

A regional plan that identifies potential destination zones is but one step toward the goal of development. This must be followed by concerted tourism planning at the zone and site scales. The zone designation includes communities and surrounding areas in a configuration quite new to local jurisdictions and populations. It will be their responsibility to take the next steps of creating their own tourism visions, cooperation, studies, policies, and guidelines for development.

DISCUSSION

1. Discuss the factors that would influence one national tourism plan to be different from that of other countries.
2. What are the similarities among national tourism planning concepts and processes?
3. Discuss the difference in the viability of implementation between national tourism plans and continuous planning.
4. What is the role of the private sector in the creation and implementation of regional tourism planning and how can it be more effective?
5. Discuss changes that must take place if a nation is to incorporate physical, social, and environmental factors in its tourism planning rather than only promotion.
6. Discuss how national tourism planning can maintain better competitiveness.
7. Name the reasons for continuous tourism planning in addition to periodic plans.
8. What should be the role of government in tourism research and education regarding tourism planning?

9. Assuming a nation of your choice, create the best concepts and processes for tourism regional planning.

REFERENCES

Baud-Bovy, Manuel (1980). "Integrated Planning for Tourism Development." Presentation, CAP/SCA Seminar, Colombo, Sri Lanka, May 8–18. Madrid: World Tourism Organization.

Baud-Bovy, Manuel (1982). "New Concepts in Planning for Tourism and Recreation." *Tourism Management*, 3 (4): pp. 308–313.

Britton, Robert (1979). "The Image of the Third World in Tourism Marketing." *Annals of Tourism Research*, 6 (3): pp. 318–329.

Bruner, Jerome (1951). "Personality Dynamics and the Process of Perceiving." *Perception*. R. R. Blake and G. V. Ramsey, eds. New York: Ronald Press.

Collins, Charles (1979). "Site and Situation Strategy in Tourism Planning: A Mexican Case." *Annals of Tourism Research,* 6 (3): pp. 352–366.

Crampon, L. J. (1961). *Tourist Travel Trade.* Boulder: Bureau of Business Research, University of Colorado.

Demars, Stanford (1979). "British Contributions to American Seaside Resorts." *Annals of Tourism Research,* 6 (3): pp. 285–293.

DPA Consulting (1981). "Plant/Market Match Model," unpublished report, November 24 to Canadian Government Office of Tourism. Ottawa.

Edgell, David L. (1990). *Charting a Course for International Travel.* Washington DC: U.S. Travel and Tourism Administration.

Fagence, Michael (1991). "Geographic Referencing of Public Policies in Tourism." *The Tourist Review,* March: pp. 8–19.

Formica, Sandro (2000). "Tourism Planning," pp. 235–242. *Annual Conference Proceedings,* Travel and Tourism Research Association, June 11–14.

Getz, Donald (1986). "Models in Tourism Planning: Towards Integration of Theory and Practice." *Tourism Management,* (7) pp. 21–32.

Gunn, Clare A. (1952). *Planning Better Vacation Accommodations,* Cir. R–304, Tourist and Resort Series, Cooperative Extension Service/Agricultural Experiment Station. East Lansing, MI: Michigan State University.

Gunn, Clare A. (1965). *A Concept for the Design of a Tourism-Recreation Region.* Mason, MI: BJ Press.

Gunn, Clare A. (1974). *Ranch, Hill, and Lake Country.* College Station: Texas A&M University.

Gunn, Clare A. and Jay Ben McMillen (1979). *Tourism Development—Assessment of Potential in Texas,* MP–1416. College Station: Texas Agricultural Experiment Station, Texas A&M University.

Gunn, Clare A. (1982). *A Proposed Methodology for Identifying Areas of Tourism Development Potential in Canada.* Ottawa: Canadian Government Office of Tourism.

Gunn, Clare A., and Terry R. Larsen (1988). *Tourism Potential—Aided by Computer Cartography.* Aix-en-Provence, France: Centre des hautes Etudes Touristiques.

Gunn, Clare A. and Terry R. Larsen (1993). *Illinois Zones of Tourism Potential.* Prepared for A. T. Kearney, Inc. and Illinois Bureau of Tourism.

Hall, Peter (1975). *Urban and Regional Planning.* New York: John Wiley & Sons.

Helleiner, Frederick (1979). "Applied Geography in a Third World Setting: A Research Challenge." *Annals of Tourism Research,* 6 (3): pp. 330–337.

Hyma, B., and G. Wall (1979). "Tourism in a Developing Area: The Case of Tamil Nadu, India." *Annals of Tourism Research,* 6 (3): pp. 338–350.

Inskeep, Edward (1991). *Tourism Planning.* New York: Van Nostrand Reinhold.

Lang, Reg (1988). "Planning for Integrated Development." *Integrated Rural Planning and Development*, F.W. Dykeman, ed. Sackville, NB: Rural and Small Town Research and Studies Programme, pp. 81–104.

Lawson F. and Manuel Baud-Bovy (1977). *Tourism and Recreation Development.* London: Architectural Press.

Lue, Chi-Chuan, John L. Crompton and Daniel R. Fesemaier (1993). "Conceptualization of Multi-Destination Pleasure Trips." *Annals of Tourism Research*, 20(2): pp. 289–301.

McIntyre, George (1993). *Sustainable Tourism Development: Guide for Local Planners.* Madrid: World Tourism Organization.

Murphy, Peter (1979). "Tourism in British Columbia: Metropolitan and Camping Visitors." *Annals of Tourism Research,* 6 (3): pp. 294–306.

Pearce, Douglas G. (1979). "Towards a Geography of Tourism." *Annals of Tourism Research*, 6 (3): pp. 245–272.

Pearce, Douglas G. (1981). *Tourism Development.* London: Longman.

Pike, Steve (2001). "Destination Positioning: Importance-Performance Analysis of Short Break Destinations." Poster presentation, *Annual Conference,* Travel and Tourism Research Association, June 11–14.

Price Waterhouse and Clare A. Gunn (1987). *Proposed Master Plan for Travel Marketing and Development.* (For the Oklahoma Department of Recreation and Tourism) Washington, DC: Price Waterhouse.

Ritchie, J. R. Brent and Geoffrey I. Crouch (2000a). "Are Destinations Born or Made—Must Competitive Destinations Have Star Genes?" pp. 306–315. *Annual Conference Proceedings,* Travel and Tourism Research Association, June 11–14.

Ritchie, J. R. Brent and Geoffrey I. Crouch (2000b). "Special Issues on the Competitive Destination." *Tourism Management* (21) 1.

Sandiford, John S. and John Ap (1998). "The Role of Ethnographic Techniques in Tourism Planning." *Journal of Travel Research,* (37) August, 3–11.

Taylor, Gordon D. (1980). "How to Match Plant with Demand: A Matrix for Marketing." *Tourism Management,* 1 (1): pp. 56–60.

Trist, Eric (1985). "Intervention Strategies for Interorganizational Domains." In R. Tannenbaum, et al. (eds.) *Human Systems Development: New Perspectives on People and Organizations.* San Francisco: Josey-Bass.

Van Doren, C. S. and Larry Gustke (1982). "Spatial Analysis of the U.S. Lodging Industry." *Annals of Tourism Research*, 9 (2): pp. 543–563.

White, Judy (1981). *A Review of Tourism in Structure plans in England.* Centre for Urban and Regional Studies. Birmingham, UK: University of Birmingham.

Chapter 6

Regional Planning Cases

INTRODUCTION

Today, many nations, states, and provinces are seeking "investment in tourism," just as they would investment in manufacturing or processing plants. But, as has been described in Part I, tourism cannot be accomplished the same way as obtaining investors in plants. All three sectors—governments, nonprofit organizations, and businesses—"invest" in tourism and have different objectives. But these investors need a great amount of information before they will be convinced that tourism development is feasible. Even though tourism leaders may have some ideas on what needs to be developed and where, this is usually based on opinion and a limited factual base. Experience has shown that some areas are far better suited to tourism development than others, but only after failures and successes have become evident. It would seem that the time is way overdue to have more sophisticated approaches that can at least provide guidelines for best development.

Experimenting continues in the search for approaches toward improved tourism planning and development as described in chapter 5. Individuals, governments, scholars, and consultants have entered into planning in many ways. But, as yet, even though there are many similarities of concepts, no universally accepted methodology has emerged. So, in order to provide a base for improved approaches in the future, this chapter includes older and newer cases of planning tourism at the regional level. Because of differences in policies, traditions, and philosophies at this level, it will be observed that these planning efforts vary but exhibit some common threads.

MICHIGAN'S UPPER PENINSULA

In 1945, Michigan established the Tourist and Resort Service, an extension advisory program based at Michigan State University. An outgrowth of study and counsel with tourist business people was a planning project for recreation and tourism growth in the Upper Peninsula (Blank et al. 1966). A loosely-knit organization, the Upper Peninsula Committee on Area Progress, and the regional Cooperative Extension Service, under the direction of Uel Blank, sponsored a project to analyze and make recommendations for future development.

Preceding this project was a special extension program directed toward stimulating interest in improved tourism. It was called the "It Pays to Know" campaign (Gunn 1964). This was in response to surveys that showed visitor hosting was less than desired. Although the focus was on improved hospitality (6,000 people attended host training seminars), it was the first program to break tourism lethargy in the region. It served well as a foundation for the guideline project.

The Region

The Upper Peninsula of Michigan is a region between Lakes Superior and Michigan, approximately 334 miles (538 km.) east-west and 215 miles (346 km.) north-south. First occupied by Chippewa and Menominee native tribes, the first white explorer was Etienne Brulé in 1620. Following French, British, and American control, it became a part of the State of Michigan in 1837. Its history has been dominated by forestry, mining, and agriculture with the addition of resort development in the late 1800s. Camping and low-scale tourist and resort development followed with road access in the 1930s. The only tourism policy for the region was that of promotion performed by the Upper Peninsula Development Bureau.

Process

The purpose of this project was to stimulate growth but in a manner that would conserve resources. The concepts of planning were adapted from the dissertation study by Clare A. Gunn of 1965, as reported in chapter 5.

A participatory process, innovative at this time, was used for the project. Figure 6-1 describes the general elements of this model. The three main steps were *research, conclusions,* and *recommendations.* The "Study Consultant Team" consisted of Dr. Uel Blank and Dr. Clare Gunn as principals. Other members of Michigan State University provided assis-

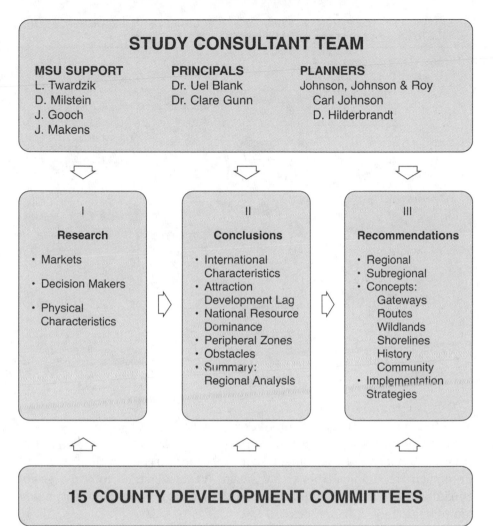

Figure 6-1. Framework for Upper Peninsula Tourism Plan. Innovative at this time in the planning process was public participation through the input of 15 county tourism development committees. Planning leadership and technical service was provided by a university and a professional landscape architectural firm.

tance—Louis Twardzik, David Milstein, Jim Gooch, and J. Makens. Special planning and design input was provided by the landscape architectural firm of Johnson, Johnson, and Roy (JJR) with input from Carl Johnson and Donald Hilderbrandt. The process of planning this region also followed the model shown in figure 6-2 (Gunn 1979, 240). It included extensive review of literature, interviews with university specialists, and interviews with pertinent state and federal agencies.

A key element of the process was to establish tourism planning committees in each of the fifteen counties. These committees were made up

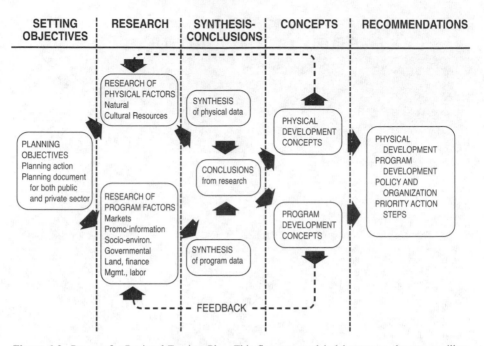

Figure 6-2. Process for Regional Tourism Plan. This five-step model elaborates on the process illustrated in Figure 6-1. Important is regular feedback that compares the results of new steps with earlier information (Gunn 1979, 240).

of public and private representatives of residents, tourism interests, and governmental entities. They were charged with several functions—aid in research, review and contribute to concepts for tourism development, and implement recommendations. By means of leadership from the study consultant team, a great many county meetings were held for many months throughout the process.

From earlier extension effort in the region, several local planning projects for tourism development had been suggested. In order to provide assistance for these initiatives, JJR created sketch plans for six of these proposed projects.

Report Guidelines

A comprehensive report of findings and recommendations was published in 1966, *Guidelines for Tourism-Recreation in Michigan's Upper Peninsula* (Blank et al. 1966). This report was presented in six major chapters and was given widespread distribution to local interests as well as public and private agencies that would be involved in implementation. Following is a digest of the report contents.

1. Demand

Review of secondary information on *travel markets* produced generalizations useful to planning. The geographical relationship is shown in figure 6-3. Sources indicated that recent market trends were shifting. Some of the Chicago-Detroit-Milwaukee-Minneapolis markets were becoming interested in travel to the Ozarks and Kentucky lakes regions because of two factors. First, new Interstate highways provided easier access, and second, more modern facilities, including air conditioning, were breaking down the tradition of "going north" and the resistance to "going south." These market findings suggested the need for new and better attractions and facilities in the Upper Peninsula.

Because of the region's abundance of natural and cultural resources, these were documented as strong market attractors. Figure 6-4 illustrates the result of a survey of travelers' interests in attractions. It revealed great diversity. Other attractions in the region included ski areas, Fort Mackinac, Soo Locks boat excursions, tours of copper and iron mines, youth camps, and fall color tours. Markets were quite diverse and came primarily from population concentrations such as Chicago, Detroit, Milwaukee, and surrounding states.

2. Supply

An inventory of existing development revealed a reasonable number of lodging and food services but many in need of upgrading. Older resort

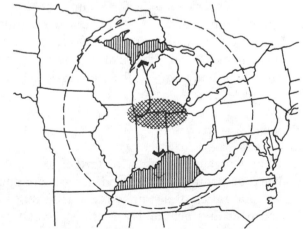

Market competition
   between Michigan's
   Upper Peninsula
   and Kentucky

Figure 6-3. Competitive Regions. In the 1960s there was evidence that new tourism development in the Kentucky region, due to the introduction of air conditioning and new Interstate highways, was beginning to compete with the attractiveness of Michigan's Upper Peninsula (Gunn 1965,7)

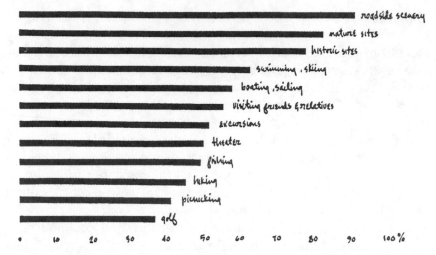

roadside scenery

nature sites

historic sites

swimming, skiing

boating, sailing

visiting friends & relatives

excursions

theater

fishing

hiking

picnicking

golf

•   10   20   30   40   50   60   70   80   90   100 %

Figure 6-4. Survey of Visitor Preferences. Although local opinion favored fishing as the most important attraction to visitors, a traveler survey in Michigan's Upper Peninsula proved eight other activities of greater significance to tourists (Blank et al. 1966, 16).

hotels, once catering to a steamboat trade, were having difficulty in meeting the need of an automobile travel market. The abundance of natural and cultural assets have yet to be developed with good access and interpretation. The success of some attractions, such as the Soo Locks, Grand Hotel, and Mackinac Fort, and interpretation at Ft. Wilkins State Park provide clues to the need for more such development. Governmental agencies own over 4.1 millions acres of land, primarily for forestry, and over two million acres are in large private holdings, such as for mining, timber harvest, and electricity generation. Federal, state, and local park and recreation areas provide for fishing, boating, winter sports, hiking, hunting, camping, and other outdoor recreation activities. Highway access is good from the Lower Peninsula but not as well developed through Minnesota and Wisconsin to the western portion of the region. Air access is provided by North Central Airlines.

3. Regional Characteristics

The region's greatest strength for further tourism development lies in its abundance of natural and cultural assets. It contains over 4,000 inland lakes, 11,000 miles of streams, over 1,700 miles of Great Lakes shoreline, and over 200 beautiful waterfalls. Over 90 percent of the land is forested and the western half is mountainous with plains and ancient seabeds to the east. The influence of prevailing westerly winds and the lakes produces huge amounts of snowfall in winter, sometimes exceeding

200 inches annually. The land provides habitat for deer, bear, moose, and other wildlife such as coyote, fox, bobcat, and raccoon.

Communities are located around the perimeter and depend on an economy supported by mining, lumber, and tourism. Yet to be exploited are the historic assets. From early exploration and Indian occupation through timber harvest and mining, many older life styles, entire ethnic settlements, and rich legends and lore offer great opportunity for increased attraction development.

For analysis of these resources and development of the region, base maps issued by the U.S. Geological Service were used. The several resource factors were graphically highlighted on these maps. From this visual assessment, review of literature on resources, and personal observation, the geographic content of the Upper Peninsula was prepared as illustrated in Figure 6-5. Further study by the planning team reduced

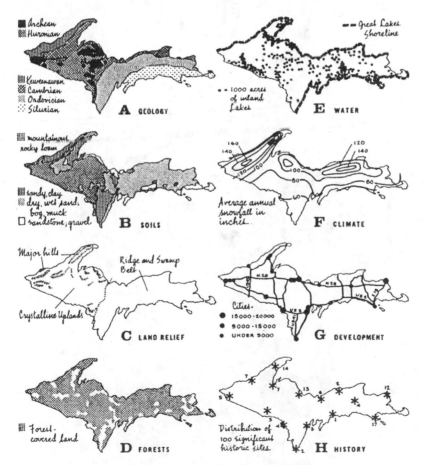

Figure 6-5. Geographic Content Factors. Research of the Upper Peninsula region revealed eight geographic factors of great significance for tourism development. These demonstrated the heterogeneity of resource foundations (Gunn 1965, 10).

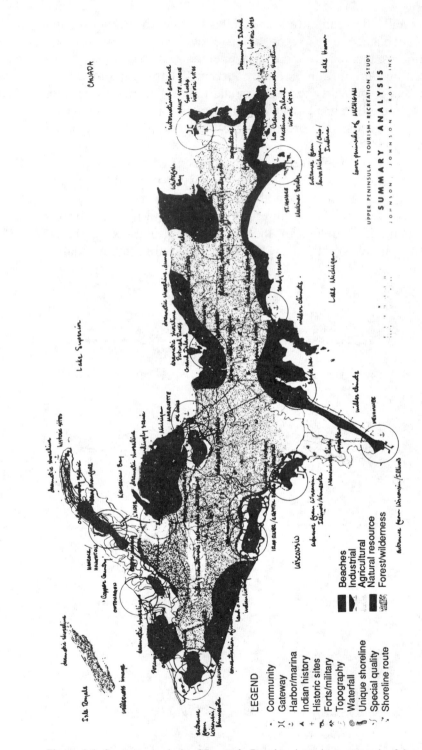

Figure 6-6. Summary Analysis of Research. Based on interviews, research of documents, and reconnaissance of the Upper Peninsula Region, the resources fundamental to tourism were identified and mapped. The legend lists major geographic factors (Blank et al. 1966, 42–43).

the foundation factors to the "Summary Analysis," as illustrated in figure 6-6. The circles indicate the community locations. Coastal and mountainous areas are indicated with the dark tone and the central shaded area represents a vast area of forests and undeveloped land. The few lands in white are generally in agriculture. Brackets show important entrance points. In addition, a great many detailed descriptions indicate locations of mining, historic sites, inland lakes, scenic areas, and beaches.

4. Interpretation for Growth

Following many meetings and discussions between county tourism committees, advisors, and the planning team, a long list of meanings of the first three planning steps was prepared. From this, consensus was reached on a reduced set of conclusions. These follow:

1. Visitation to the Upper Peninsula for pleasure already is great, comes primarily by automobile, and is mostly from heavily populated areas one day away. Visitors come for a variety of recreational reasons but primarily those related to the original natural (not man-made) resource base.

2. The region, as never before, is now caught up in interregional competition, especially with the new and massive man-made and nature-interpretive attractions equally available to its markets. This appears to be the major cause of the present arrested growth of tourism there.

3. Investigation showed many possible reasons for stagnation, chief of which are: a lag in development of powerful attractions, voids in certain services (especially those catering to higher socioeconomic brackets), and proliferation of natural resource use.

4. The bulk of recreation land development policies and practices (and therefore the dominant control of decision making) now lies in four main groups: small commercial tourist enterprises, private personal land holdings, large private non-tourist-oriented businesses and government (state, federal, local) forest and recreational agencies.

5. If there is any one dominant characteristic which sets the Upper Peninsula apart from its surrounding competitive regions, it is its wild vastness of undeveloped forests and water-blessed land.

6. Investigation revealed clues to promise for growth in spite of recent relative lag: its fortunate geographic position relative to markets, its ease of access, its abundance of natural resource assets, its unspoiled frontier, its unique and rich historic development, and a very desirable distribution of fine community service centers.

7. The greatest amount of existing development for tourism and the greatest potential for future expansion appear to lie most heavily in the peripheral zone, leaving the interior a predominantly undeveloped region.

Yet within this outer zone, a variety of resource bases suggests varied and interesting development potential.

8. The major obstacles to expansion, therefore, are not those imposed by nature, nor by gross errors in development to date. Rather, the growth of visitation and its economic corollaries appear to lie in major new development—both quantitative and qualitative.

9. The greatest opportunities shown by this search are in the design, location, investment, and management of new attractions clusters and their supporting services and facilities, but developed in a manner to enhance the valuable natural and man-made resource assets.

From the analysis of the region's foundation for tourism and the conclusions on development, the planners applied the concepts described in chapter 5. The following criteria were used to delineate subregions with greatest potential:

- a series of attractions, including existing ones, based on resource assets
- one or more viable communities as service centers
- access linkage with main travel corridors
- subregional unity derived from distinctive and compatible qualities.

From this examination, nine subregions (potential tourism development zones) and a central non-attraction hinterland were derived, as illustrated in figure 6-7.

The theme of subregion "A" could well be "Voyageurland" because of the dominance of historic fact and lore, perhaps in greater concentration than anywhere else in the Upper Peninsula. Major historic complexes at St. Ignace and Sault Ste. Marie could depict: early exploration, Indian occupation, French claims, missionary activity, fur trading activity, forts and battles, British Control, War of 1812, American settlement, and the heyday of the Victorian resort era. When land and waterway interpretive tours as well as pageantry and festivals are added these can be outstanding attractions. Additional attraction potential of this subregion includes:

Gateway complex at both St. Ignace and at Sault Ste. Marie.

A natural resource culture center.

Major marina development.

Second homes subdivisions.

A "Soo Locks" transportation park.

A year-around resort complex and convention center.

The other eight zones were described in the same manner. The central zone of the region was described as "deserving of restraint and conservation." Because it is farthest from established community centers, and

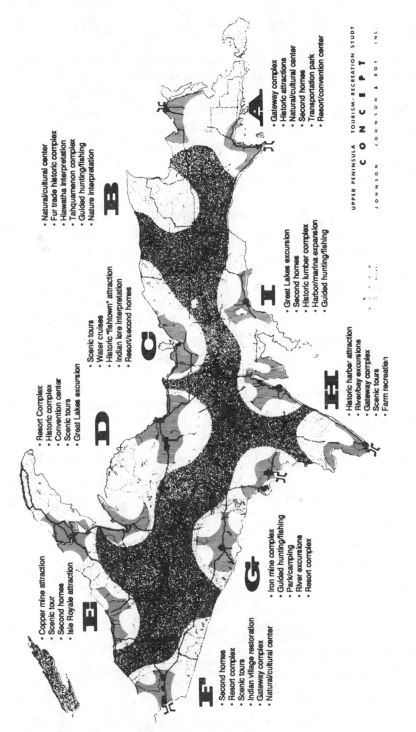

Figure 6-7. Destination Zones. Interpretation of the research resulted in identifying nine potential development zones. Less desirable for tourism development and best suited to conservation and other development was the central zone, J. For each zone the study identified a theme and specific project opportunities (Blank et al. 1966, 54–55).

because the other subregions contain the strongest resource potentials for tourism and recreation development, this area could be eroded by scattered and haphazard development. For the time being, it may be best suited to other economic uses, such as mining, forestry, and agriculture.

Many other opportunities grew out of this study. New urban-oriented campgrounds for transient tourists, greater development of local and ethnic crafts and events, and additions of evening entertainment were recommended. Tour packages, new vacation villages, and interpretive tours of historic mining sites and local industries were seen as opportunities. Especially needed throughout the region was upgrading tourist facilities and services.

5. Concept Development

A significant part of the project was to provide concepts and sketch plans for individual development opportunities, especially those already identified by County Tourism Committees. Illustrated in Figures 6-8 through 6-12 are a few of these projects, further elaborated by JJR. In all cases, these were designed as integrated within the regional context as well as based on local assets.

6. Implementation

Implementation was to be carried out locally and to include the following topics:

1. Attraction development.

2. Regional Committee.

3. Dissemination of report findings.

4. Specific development tasks.

5. Subregional action.

6. Training and education.

7. Marketing study and analysis.

8. Upgrading and analysis.

9. Promotion and advertising.

10. All-season activities.

The implementation phase actually began at the start of the project in 1963 and continues today. The fifteen County Extension Agents, by administrative request, were involved from the start in working with local constituencies and the planning team. As the project recommendations

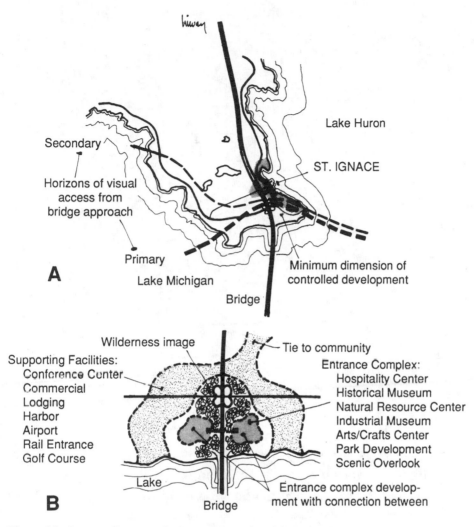

Figure 6-8. Gateway Concept at St. Ignace. The main tourist approach to the Upper Peninsula was from the Mackinac Bridge, suggesting the need for entrance planning concepts (Blank et al. 1966, 63). (Compare with the Straits plans of two decades later, figures 10-6 through 10-8).

were being formulated two types of air tours were held. One invited several travel writers from major cities to fly the region so that they could see firsthand and then write about investment needs and potential. The other flight included investors who were exposed to the region and local representatives at several stops throughout the Upper Peninsula.

After ten years from the start of the project an assessment was made of progress. The following are some of the major items that had been started or established as reported by Ray Gummerson, area development specialist of the Upper Peninsula Extension Service (Gunn 1973).

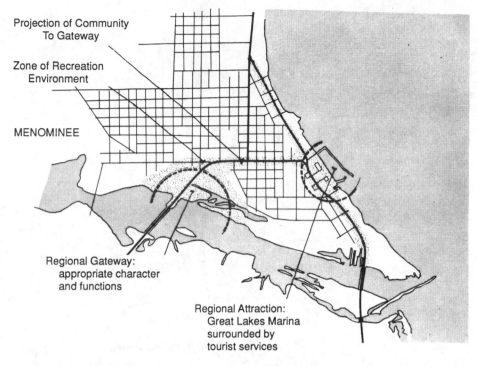

Figure 6-9. Gateway Concept at Menominee. Design of major entrance to the Upper Peninsula from Wisconsin. Emphasis is placed on the features and aesthetics of the entrance corridor leading to the downtown attractions (Blank et al. 1966, 91).

- A 1600-acre lake with canoe liveries and guides.
- Big Sea Development of the U.S. Forest Service.
- Pictured Rocks National Lakeshore established—new marinas, boat tours, renewed fishing.
- Memorial statue to Bishop Baraga as part of historic complex.
- New convention center complex at Marquette.
- Restoration of Carp River Forge.
- Quincy Hoist (copper mine) has been restored.
- New ski-flying facility was installed.
- Large forest-lake tract, Sylvania, established by Forest Service.
- Mystery Ship and Memorial Marina established at Menominee.
- Overlook and interpretive tower established at Sault Ste. Marie.
- Many new motels, food services, and other travel services.

Reflection on this project reveals several implications. Planning implementation takes time; it cannot be expected to produce immediate results. There is much value in bringing together local constituencies and

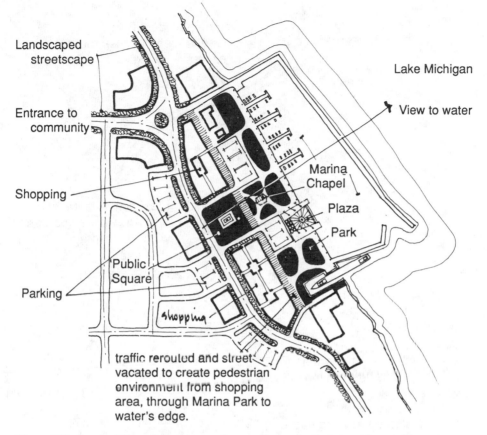

Figure 6-10.

Landscaped
streetscape

Entrance to
community

Shopping

Parking

Public
Square

shopping

traffic rerouted and street
vacated to create pedestrian
environment from shopping
area, through Marina Park to
water's edge.

Lake Michigan

View to water

Marina
Chapel

Plaza

Park

Figure 6-10. Detail of Menominee Gateway Concept. Design detail of planned community and waterfront assets and attractions. Note conversion of street into pedestrian mall (Blank et al. 1966, 93).

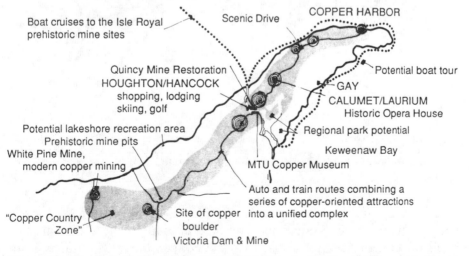

Boat cruises to the Isle Royal
prehistoric mine sites

Scenic Drive

COPPER HARBOR

Quincy Mine Restoration
HOUGHTON/HANCOCK
shopping, lodging
skiing, golf

Potential boat tour

GAY

CALUMET/LAURIUM
Historic Opera House

Potential lakeshore recreation area
Prehistoric mine pits
White Pine Mine,
modern copper mining

Regional park potential

Keweenaw Bay

MTU Copper Museum

Auto and train routes combining a
series of copper-oriented attractions
into a unified complex

"Copper Country
Zone"

Site of copper
boulder

Victoria Dam & Mine

Figure 6-11. Concept for Copper Country Zone. Design ideas for attraction development in this rich historic and scenic subregion. Examples of capturing the copper mining theme linked together with the main scenic travelway (Blank et al. 1966, 83).

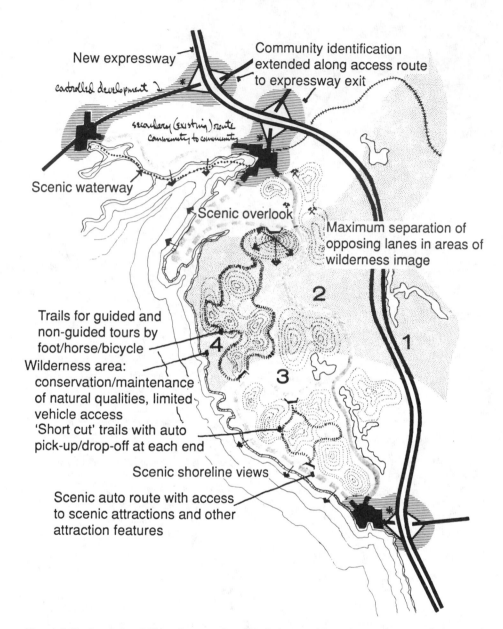

New expressway

controlled development ↓

secondary (existing) route
community to community

Community identification
extended along access route
to expressway exit

Scenic waterway

Scenic overlook

Maximum separation of
opposing lanes in areas of
wilderness image

2

Trails for guided and
non-guided tours by
foot/horse/bicycle

4

1

Wilderness area:
conservation/maintenance
of natural qualities, limited
vehicle access

3

'Short cut' trails with auto
pick-up/drop-off at each end

Scenic shoreline views

Scenic auto route with access
to scenic attractions and other
attraction features

Figure 6-12. Scenic Land-Water Concept. A generic design to relate main access to coastal resources, so important in the Upper Peninsula region: 1, highway; 2, scenic forest drive; 3, access to wilderness trails and overlooks; 4, protected conservation area (Blank et al. 1966, 69).

professional planners-designers. As Blank described it, "There is not a simple linear-research-education continuum. Rather, there is a multidimensional matrix, with study-action as one facet, the range of local-state-federal as another, public-private decision makers as yet another facet, plus many others" (Gunn 1973). This project allowed practical application of the theoretical concepts of planning developed by Gunn (1952, 1965). (See chapter 5.) It dramatized the value of analyzing several geographical factors to determine how they combined into potential destination zones—an aid to discovery of potential. Finally, there was great value in spreading the project over a three-year time span to allow maximum interchange and communication between all planners and stakeholders. (See "Great Lakes Crossroads Museum and Welcome Center," chapter 10, as further follow up of this project.)

CZECHOSLOVAKIA

Another early tourism plan for a region was developed for Czechoslovakia in 1968. (Prikryl 1968) After World War I, the Czechs and Slovaks formed the nation. Later, under Communist rule, the government took over all business, industries, churches, and schools. It is remarkable that under these conditions, there was sufficient interest in tourism to seek planning for its development. As part of the economic plan of the country, the intent of the tourism plan was to foster development appropriate to the various natural resources of the nation, especially for the domestic health and leisure travel markets. The major objective of the plan, therefore, was to make a thorough examination of the resource base in order to identify zones best suited to the overall purpose.

The Planning Process

The planning process involved six major steps.

1. *Evaluation of territorial qualities.* This step included study and mapping of natural and cultural resource conditions, human activities, facilities, transportation, and negative influences (air and water pollution, spoiled landscape, and insect pests).

2. *Criteria for identifying potential zones.* The planners identified the criteria for selecting zones based on market and geographic factors. For example, favored were zones with rich natural resource qualities such as mountains, forests, slopes, best micro-climate, and least damaged by existing development. Cultural resources, such as historic sites and archeological sites, were favorable assets.

3. *Urban centers.* An evaluation of 379 towns was made in order to determine their favorable touristic qualities. Included were factors of communication level, social importance, places of distinction, historic significance, therapeutic value, and spas.

4. *Determination of regions.* Using the criteria for selection, four categories of regions were identified: best conditions, good conditions, some

conditions, and least value. Primary market categories include: rest and relaxation, water sports, mountaineering, winter sports, games, therapeutic, folklore festivals. Figure 6-13 illustrates these regions (shaded areas indicate regions not desired for tourism development).

5. *Capacities*. By means of a specially designed formula, the tourist capacity (visitor use, accommodations, etc.) was evaluated. The conclusion was that there was ample room for growth of tourism development at that time.

6. *Transportation*. Recommended for expanded tourism were several changes: increase railroad capacity, improve railroad quality, improve existing roads, enhance road landscapes, multiply food and lodging, and modernize water transport.

The study and plan concluded that there are many opportunities for tourism development in the country, especially for domestic health and social markets.

Placing this plan in historical context, several similarities and contrasts with the Michigan plan can be made. It was similar in process in making a thorough study and analysis of key geographic factors, such as natural resources, cultural resources, urban centers, and transportation. Because Czechoslovakia was then under Communist rule, their tourism development plan was dictated by central government and did not involve local people, as did the Michigan plan. The purpose was primarily social welfare and private enterprise was ascribed only a token role. Only domestic markets were considered. The role of cities as providing major tourist facilities, services, and infrastructure were excluded from the plan. Perhaps the greatest contrast was the participatory process used in Michigan, absent in Czechoslovakia at that time because of national policy.

In the decades since this plan, dramatic political and economic changes have occurred. After Communist rule terminated in 1989, the general population and leaders advocated division of the country into two, based on ethnic, traditional, and many other factors. In 1993 the Czech Republic and Slovakia were formed from the earlier Czechoslovakia. Both nations have aggressively sought tourism development, primarily by the private sector. Both have become popular travel targets based on improved air, highway, and train access; their historic architecture (not damaged during World War II); and their mountain scenery and recreational activities.

The city of Prague has blossomed into a major target for global as well as European visitors. Even though the surge of capitalism has had its ups and downs, new tourism profits have fostered massive protection and renovation of historic sites (Innaurato 2001). St. Savior Church, National Museum, and Royal Summer Palace as well as many other sites have become major attractions. Foods, nightlife, and festivals now add to this

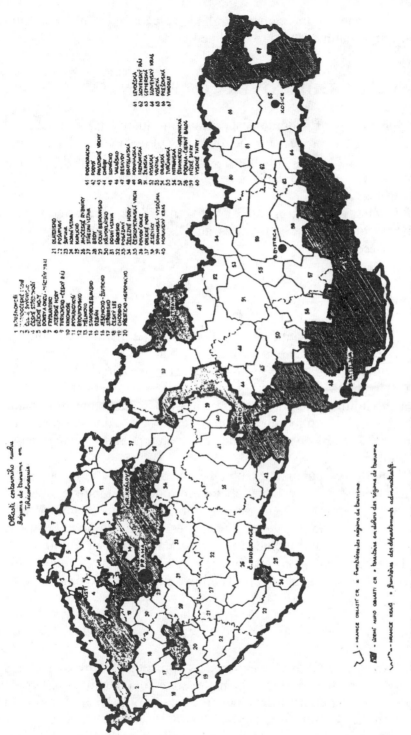

Figure 6-13. Destination Zones of Czechoslovakia, 1968. An example of zone identification based on political (Communist) policy. The aim was to develop only those areas of significance for health and welfare of local citizens. Since the split into the Czech and Slovak republics in 1993, tourism planning and development have been driven by the abundant natural and cultural resources and international as well as domestic travel markets (Prikryl 1968 and Inmaurato 2001).

unusual city's appeal to travelers through its wide diversity of music from symphonic and genuine Czech to localized jazz.

CASES: DISCOVERY OF POTENTIAL

As new generations of geographic information systems (GIS) appeared, the regional planning process for the discovery of destination zone potential improved. The basic planning steps as outlined in the Upper Peninsula project remained valid but applications demonstrated the value of newer techniques.

The following cases in regions of the United States are briefly summarized here to illustrate similarities and differences in purposes and degrees of success.

Central Texas

A 20–county region in south central Texas was studied for its tourism potential as part of a graduate student exercise at Texas A&M University. It was reported in *Tourism Development: Assessment of Potential in Texas* (Gunn and McMillen 1979). Key factors studied and mapped were: water/wildlife, topography/soils/geology, vegetative cover, climate, aesthetics, existing attractions, history/archeology, communities, and transportation. The objective was to identify potential in two categories—touring and long-stay.

Employing the Synagraphic Mapping System (SYMAP), the computer composite of overlays of weighted factors for potential long-stay tourism is shown in figure 6-14. When research information was combined with this composite map, the five generalized zones were determined as shown in figure 6-15. The report included recommendations for potential development in these zones based on market trends and resource foundation. Because of class constraints no provision was made for implementation.

Oklahoma

In 1987, Price Waterhouse was engaged as consultant to investigate and prepare a master plan for tourism in the state of Oklahoma. Clare A. Gunn served as subcontractor for regional analysis of tourism potential. It was reported in *A Proposed Master Plan for Travel Marketing and Development for the State of Oklahoma* (Price Waterhouse 1987). The study foundation included field observations, review of literature, and workshop meetings.

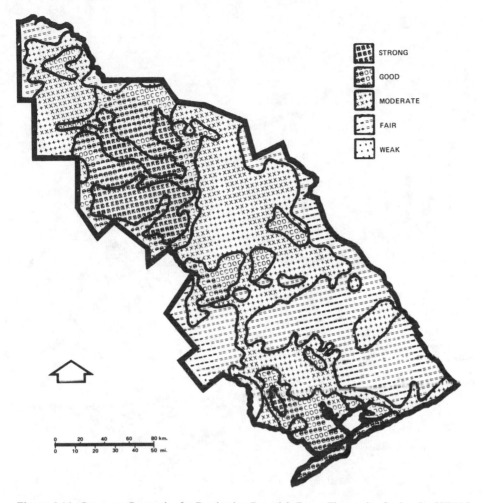

STRONG

GOOD

MODERATE

FAIR

WEAK

Figure 6-14. Computer Composite for Destination Potential, Texas. The result of using the SYMAP computer program for aggregating several resource factor maps in a 20-county region in central Texas. The darker areas represent strongest resource support for destination tourism development (Gunn and McMillen 1979, 10).

A regional analysis of potential was part of the "Product Development Plan." It utilized a process similar to that of the Central Texas project but employed a more contemporary computer mapping program called "Compass," developed by the College of Architecture, Texas A&M University. The complete report of its application to the Oklahoma project is described in *Tourism Potential—Aided by Computer Cartography* (Gunn and Larsen 1988).

Study of travel markets revealed those seeking outdoor recreation, touring, city amenities, close-to-home attractions, and visiting friends and relatives. "Pass-through" tourists were particularly important because of the state's geographic position between markets of the eastern part of the

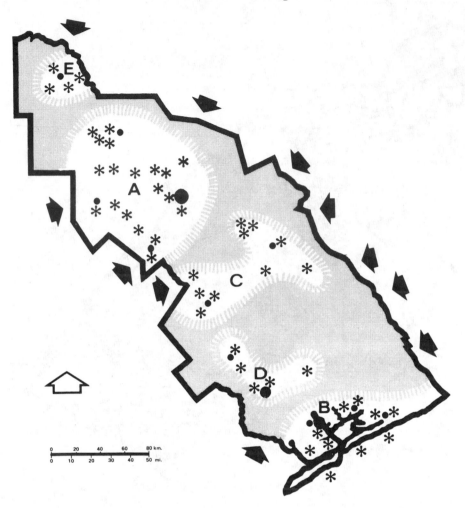

Figure 6-15. Interpreted Zones of Destination Potential. Based on resource studies and the composite map of Figure 6-14, five destination zones were identified. Each revealed distinctive potential for many development projects (Gunn and McMillen 1979, 11).

United States and attractions in the West. "Business-convention" travelers included those going to business meetings, conventions, and trade shows.

Seven resource factors were studied based on review of documents, interviews, inspection tours, and several local citizen workshops. These factors were: water, topography, vegetation/wildlife, history/ethnicity, attractions, cities, and transportation. First, generalized maps were prepared by hand to a common scale showing areas of "high," "medium," "low," and "poor" resource importance. Then, these factors were weighted, as shown in table 6-1.

Through use of the Compass overlay program, these factor maps were aggregated, producing the mosaic illustrated in figure 6-16. This shows

TABLE 6-1

WEIGHTED FACTORS FOR OKLAHOMA PLAN

Factor	Index
Water	15
Topography	8
Vegetation/Wildlife	8
HistoryEthnicity	15
Attractions	12
Cities	20
Transportation	22
Total	100

Source: Gunn and Larsen 1988, 30

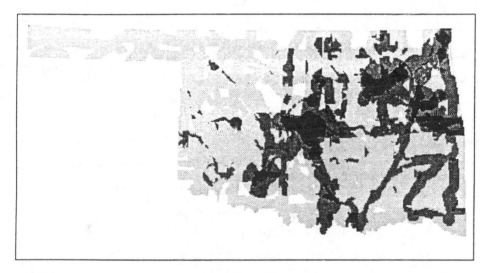

Figure 6-16. Composite of Resource Potential, Oklahoma. A computer map resulting from aggregating seven overlays of resource factors important to tourism development. The shading from dark to light represents the best to poorest foundation for potential (Price Waterhouse 1987, Ex. B).

the distribution of the combined resources and for total tourism (not separated for natural or cultural factors). By means of study and interpretation of this map and other research material, destination zones were identified, as illustrated in figure 6-17. "Primary" zones were areas where the best and most abundant resource factors are in greatest combination. "Secondary" zones are those areas where several factors show significant potential for development based on combination of several factors. "Tertiary" zones have a combination of a few factors in sufficient strength to indicate some potential. Brief summary of tourism potentials are listed below for Primary Zones A-1 and A-2;

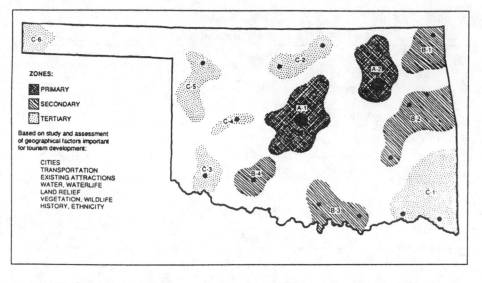

Figure 6-17. Interpreted Zones of Destination Potential, Oklahoma. By means of reviewing resource documents and interpreting the computer composite (Figure 6-16), primary, secondary, and tertiary potential development zones were delineated (Price Waterhouse 1987, Ex.8).

Primary Zones

A-1 Oklahoma City Plan

- Major improvements in conference, convention facilities.
- Better massing of existing attractions, tours.
- Greater exploitation of "pass-through" tourists.
- Major enhancement of downtown amenities.
- Major cowboy entertainment complexes.
- Agricultural/range history museum and center.

A-2 Tulsa Zone

- Major national-class "folklore institute" (Indian and other ethnic center, entertainment, events).
- Enhanced development of historic sites (Indian, Civil War).
- Improved existing attractions, tour packaging.
- Expanded meeting, conference programs.
- Greater exploitation of "pass-through" travelers.
- New Will Rogers entertainment center.
- Redevelopment of Osage, Chickasaw capitals.
- Osage ethnic museum and visitor center.

The director of the project, Eugene Dilbeck, then director of the Department of Tourism and Recreation of Oklahoma, reprinted the overall Price Waterhouse report plan, distributed it widely, and held tourism development workshops throughout the state. This action has stimulated much greater interest in Oklahoma's tourism potential than even before.

Regarding the regional analysis phase, some evaluation should be noted. This application again exposes the need for a better means of weighting the several factors. Researchers need to accept this challenge. Because tourism is dynamic, the zones delineated must not be considered permanent. They suggest to the local political and tourism officials that there are areas where the potential is better than other locations. Finally, areas indicated as having greatest potential need to be examined carefully to determine capacity limitations. Development may already have reached saturation.

The project gave priority to development recommendations as follows:

- Emphasis on enhancement and new development in primary zones and, to a lesser extent, secondary zones.

- Emphasis on enhancement and development of festivals and events, rather than attractions, in tertiary and non-zone areas.

- Development of one or several lake development areas.

- Emphasis on grouping attractions within city or area.

- Increase usage of existing facilities and attractions by extending hours of operation, extending seasons, and identifying other methods to utilize facilities and attractions more intensively during prime seasons.

- Improve quality, and number of attractions overall, and specifically those based on Western Heritage/Native Americans for which Oklahoma may become a travel destination (Price Waterhouse 1987, 23).

Although no study of implementation has been performed, there is ample evidence that the plan stimulated environmentally sound tourism development.

Delaware

For the state of Delaware a similar process of analysis for determining zones of potential was employed. The state agency, however, sought potential based only on natural resources (Gunn 1991).

Review of documents and 30 interviews with knowledgeables throughout the state revealed basic information on natural resources and travel market trends. The geographic factors used for map overlays are shown in Figure 6-18. Here is the resulting composite computer map, indicating four degrees of support for tourism development: optimal, high, moderate, and little or none. Interpretation of this resulted in identifying "primary," "secondary," and "tertiary" potential development zones (Figure 6-19). Table 6-2 summarizes the importance of resource factors for each zone. Types of suitable development were listed for these zones.

For implementation, several strategies were recommended:

1. Establish zone development programs.

2. Establish a strong private sector tourism organization.

3. Support programs to prevent water pollution.

4. Establish a tourism extension education program.

5. Increase quantity and quality of information for visitors.

6. Initiate public-private ventures.

7. Initiate planning/design competitions.

8. Increase natural resource market research.

9. Initiate new land use regulations to protect tourism assets.

10. Improve transportation and access.

11. Establish a scenic highway program.

12. Establish a statewide tourism planning/development council.

The conclusion of this study identified several types of tourism development that were compatible with the resource base and travel market segments. These included: environmental conference centers; environmental resorts; resource culture centers; ecotours; waterfowl, bird interpretive centers; scenic tours; nature hike-bike trails; hunting resorts; farm vacations; RV camping resorts; nature-oriented festivals, pageants, entertainment; and beach resorts.

It finally concluded that the next step—implementation—must be carried out by private enterprise, public sector developers, and local communities.

This plan was enthusiastically accepted by the state but, soon after its completion, the director of tourism for Delaware was replaced. It is not known to what extent this plan became policy or if implementation was carried out.

Illinois

In 1992, the state of Illinois engaged the consulting firm of A. T. Kearney, Chicago, to study and make recommendations for future

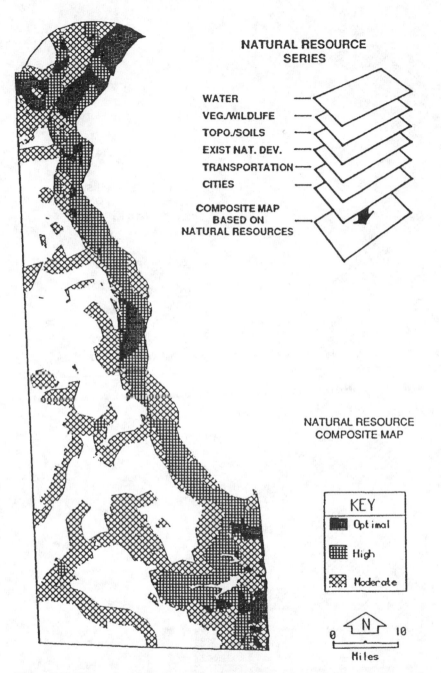

Figure 6-18. Composite of Resource Potential, Delaware. Review of natural resource documents, reconnaissance of the state, and interviews provided the foundation for this aggregation of map overlays. Shown are areas of optimal, high, and moderate support for tourism development. Also illustrated is the process of overlay maps resulting in this composite (Gunn 1991, 22).

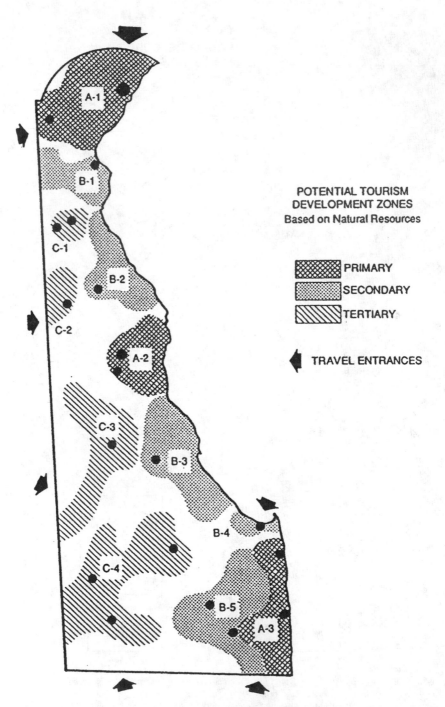

Figure 6-19. Interpreted Zones of Destination Potential, Delaware. From the analysis of natural resource foundation, the author interpreted primary, secondary, and tertiary zones of tourism development potential. (The contract for this project did not include cultural resource assessment.) (Gunn 1991, 21)

TABLE 6-2

RESOURCE BASE FOR DESTINATION ZONES, DELAWARE

RESOURCE ZONE	SERVICE CENTERS	WATER-WATERLIFE	WILDLIFE	VEGETATION	TOPOGRAPHY, SOILS, GEOLOGY	EXISTING NAT. RES. DEVELOP.	TRANSPORTATION	CITIES
A-1	Wilmington, Newark, New Castle, Arden							
A-2	Dover, Camden							
A-3	Rehoboth Beach, Dewey Beach, Bethany Beach							
B-1	Delaware City							
B-2	Smyrna, Dayton							
B-3	Milford, Milton							
B-4	Lewes							
B-5	Millsboro, Dagsboro, Frankford							
C-1	Middletown, Odessa							
C-2	Kenton							
C-3	Harrington							
C-4	Georgetown, Seaford, Laurel							

Legend:
- ■ Strong support
- ▨ Moderate support
- ▧ Some support
- □ Little or no support

tourism development in Illinois. For this project, two subcontracts were let—a marketing program and review of existing community projects by Grey Advertising, and analysis of potential destination zone development by Clare A. Gunn. Following is a discussion of *Illinois Zones of Tourism Potential* (Gunn and Larsen 1993).

This project was similar in process to those for Delaware and Oklahoma:

1. Define purpose
2. Review market demand
3. Research geographic factors
4. Synthesize foundations
5. Map and aggregate key factors
6. Identify zones of potential
7. Prepare recommendations

Figure 6-20 illustrates the two map series, natural resources and cultural resources, that were digitized for GIS computer aggregation. For example, figure 6-21 represents the composite resulting from overlays of the natural resource geographic factors. When added to the cultural resource geographic factors and interpreted, the primary and secondary zones were identified, as illustrated in figure 6-22. For each of these zones, the area included, the key service cities, and development opportunities were identified. As an example, the following information for Zone A-1 was derived from this study, analysis, and interpretation. (The prefix "P" indicates opportunities already recommended by public and private developers.)

ZONE A-1

Area: Portions of counties: Winnebago, Boone, Ogle, DeKalb

Strongest Resource Support: vegetation/wildlife, topography, prehistory, economic development, cultural attraction, access, cities.

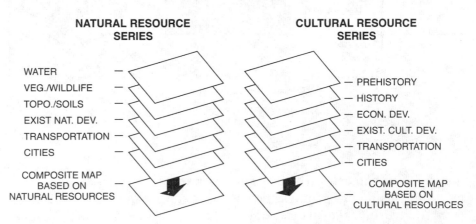

Figure 6-20. Composite Map Overlay Process. For the discovery of potential development zones for the state of Illinois, the computer map overlays included both a natural resource and a cultural resource series. Following this, an overall assessment resulted from combining the two composites (Gunn and Larsen 1993, fig.1).

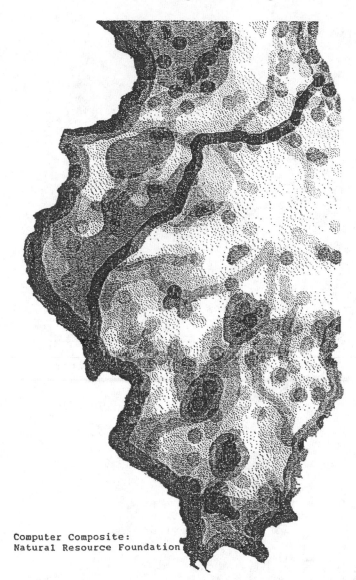

Figure 6-21. Composite Natural Resource Base, Illinois. By combining the six overlay maps for the natural resource series for Illinois, this composite was the result. The darker areas indicate strong support from several factors, influenced largely by water and topographic factors (Gunn and Larsen 1993, Fig.12).

Key Cities: Rockford
Development Opportunities:

- (P) Expand sports facilities
- (P) Establish more cross-country skiing trails
- (P) Establish a zoo

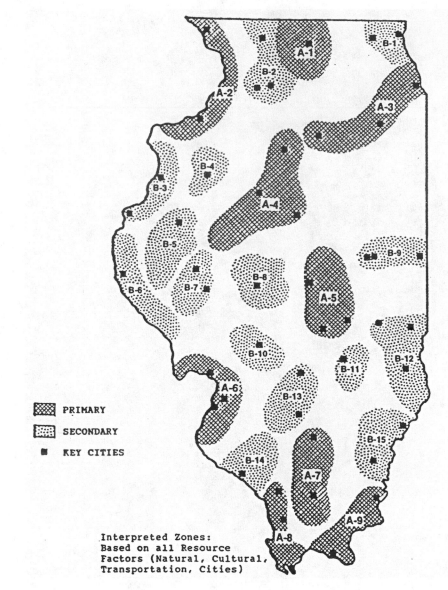

Figure 6-22. Interpreted Zones of Destination Potential, Illinois. When the composite resulting from combining natural and cultural resource composites was studied, the result is shown here—primary and secondary zones and key cities. These are the areas where local leaders can work together to take action toward appropriate tourism development (Gunn and Larsen 1993, Fig.18).

- (P) Establish a major aquarium
- (P) Create more state parks, natural areas
- (P) Develop new golf courses
- (P) Develop new botanical garden
- (P) Establish mountain bike trails

- (P) Expand recreational activities on Rock River, Pecatonia River, Leaf River canoeing, riverboat
- (P) Establish a Native American Culture Center
- (P) Establish outdoor car shows, museum
- (P) Historic restoration in Rockford; early settlements
- (P) High tech corporate meeting center
- Archeological site development and interpretation—Woodland, Archaic, Paleoindian
- Shopping tours
- Expanded Indian historic sites
- Pageant and interpretation, Blackhawk Battle Site
- Increase wildlife viewing and interpretation
- Establish new scenic drives and overlooks
- Expand nature trails

This report was presented to and accepted by the A. T. Kearney firm and the main client, the Illinois Bureau of Tourism. However, an evaluation of their project illustrates a weakness within such a contracted consulting project—implementation. The contract terms of reference did not include the subsequent action by the state tourism agency. It was assumed that it would take on the responsibility of holding workshops in each potential zone so that tourism leaders there would follow up with their own action programs to implement the recommendation. There is no evidence that this was done. In this case, the very important planning step of local participation was omitted and the only portions of the overall project acted upon were those related to promotion, not guidance for future development.

FINGER LAKES

In 1989, the Finger Lakes Association, a nonprofit tourism agency for tourism development and promotion, through its executive director, Conrad Tunney, initiated a project to plan the future for this 14-county region in west-central New York State (figure 6-23). The project was a follow-up of an overall statewide tourism plan prepared by Price Waterhouse in 1987.

This project (*Finger Lakes* 1991) differed from many others because of its approach. First, it was to be performed by a Finger Lakes Association Planning Development Committee, including representatives of the Finger Lakes Association and four regional planning and development

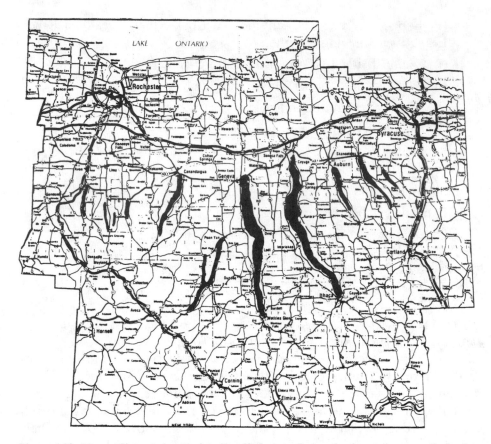

Figure 6-23. Map or Finger Lakes Region, New York. A tourism evaluation study was made for the fourteen counties of New York, identified as the Finger Lakes Region. Planning was sponsored by the Finger Lakes Association Planning Development Committee (Finger Lakes Region 1991, 20).

boards having jurisdiction over counties within the region. Second, leadership was assigned to Bill Hess, the director of the Southern Tier Central Regional Planning and Development Board. Third, public input from over 125 representatives of local tourism through survey questionnaires and area-wide workshop meetings was obtained. Fourth, tourism consultant Clare Gunn was engaged to provide technical assistance. And, finally, the costs were to be shared by the planning and development boards, the Finger Lakes Association staff, but primarily by a grant from the New York State Urban Development Corporation.

Purpose and Process

The region is approximately 9,000 square miles in size and is one of the "vacationland" regions designated by the state tourism authority. The purpose of the project was to: (1) document the economic impact of tourism, (2) identify those involved in tourism, including their roles, responsibilities, and resources, and (3) identify the constraints and opportunities for enhancing regional tourism (attractions, events, and supporting facilities). This study focused primarily on identifying opportunities and action needed but did not involve geographic analysis to determine locations and characteristics of destination zones. The process included the following five tasks:

1. Obtain consensus on goals.
2. Inventory attractions, facilities, services.
3. Identify organizations and their roles.
4. Identify constraints and opportunities.
5. Recommend implementation strategies for goals and objectives.

Task 1. The goals agreed upon were:

1. Enhance the provision of visitor satisfactions.
2. Increase the rewards to the region's economy and all sectors involved in tourism development.
3. Protect the region's natural and environmental resources.
4. Integrate tourism into community social and economic life throughout the region.

Task 2. This task included summarizing existing market information and inventorying the following:

Attractions	553
Lodging facilities	458
Food and beverage	1605
Marinas	67
Events	1601
Transportation (highways, airports, rail)	

Task 3. The several private, quasi-public, and public organizations at the state, subregional, county, and local levels were identified and their tourism roles were analyzed. These roles were identified as planning, developmental, operational, and promotional.

Task 4. Through surveys and workshops, the four major areas of constraints impacting tourism in the region were:

A. Inadequate financial assistance.
B. Poor local support of tourism.
C. Lack of cooperation among tourism sectors.
D. Insufficient promotional and marketing efforts.

Task 5. Study of the findings and deliberations among the principals of the study resulted in naming *objectives* and *strategies* to meet the challenge of the four constraints identified in Task 4. These are identified in the following description excerpted from the Executive Summary of the report, *Finger Lakes Region Tourism Development Opportunities* (1991), pp. 6–12. Table 6-3 presents a matrix of the main organizations that would be responsible for leadership for each objective.

Development Opportunities

The Study's survey of organizations impacting the Region's Tourism Industry identified many opportunities to enhance tourism in the Finger Lakes. Some of these opportunities were development projects such as a marina or a museum at a specific site. Others focused more on the processes involved in attracting and serving visitors. Frequently, these broader opportunities were related to the issues and constraints identified earlier.

The review of survey information with Study Participants in work-

TABLE 6-3

ORGANIZATIONS RESPONSIBLE FOR ACTION

ORGANIZATION	ECONOMIC A			SUPPORT B			COOPERATE C			PROMOTE D		
	OBJECTIVES											
	1	2	3	1	2	3	1	2	3	1	2	3
REGIONAL TOUR ASS'N	*	*	*	*	*	*	*	*	*	*	*	*
REGIONAL PLANNING BDS	*	*	*	*	*	*	*	*	*			
TOURIST PROMO ORG'S			*	*	*	*	*	*	*	*	*	*
VISITORS & CONV BUR			*	*	*	*		*	*	*	*	*
OTHER PLANNING ORG'S			*	*	*	*	*		*			
ECON DEVEL ORG'S	*	*	*	*	*	*		*				
CHAMBERS OF COMMERCE	*	*	*	*	*	*		*	*	*	*	*
PUBLIC OPEN SPACE AGY			*	*			*	*	*			*
NYS DEPT OF ECON DEV	*						*				*	*
TOURISM BUSINESSES	*	*	*	*						*	*	*
LEGISLATURES		*				*	*	*	*			
COOP EXT SERV						*	*	*	*		*	*
CIVIC ORGANIZATIONS					*	*		*	*			
REGULATORY AGENCIES		*					*					
FINANCIAL INSTITUTION	*		*									

shop settings and with the Project Consultant led to the formulation of the following goals, objectives, and strategies for enhancing the region's tourism.

Goal A: Improve Economic Viability of Tourism

Objective 1. Enhance Business Finance.
STRATEGIES

- Create investment incentives and tax incentives for tourism physical plant upgrading, expansion, and new development.
- Legislate dedicated tax revenues for tourism development.
- Increase public seed monies for attractions and service businesses.
- Hold educational seminars for the financial community (banks, loan agencies, financial institutions and tourism entrepreneurs) for the purpose of broadening their understanding of tourism economics.
- Modify credit and banking requirements to encourage feasible tourism development.
- Establish a venture capital fund and loan program, including the expansion of existing matching funds programs.
- Encourage existing funding agencies to give higher priority to tourism funding, at least equal to other industrial development.

Objective 2. Reduce Regulatory Constraints.
STRATEGIES:

- Review and amend laws, regulations, and administrative procedures (local to federal) that conflict, overlap, are obsolete, and unnecessarily constrain tourism development.
- Initiate local, state, and federal action to ameliorate and eliminate constrictive regulations.
- Review permitting processes to identify and propose elimination of excessive "red tape."
- Suggest new legislation that is needed for more viable protection and development of tourism in the region.

Objective 3. Expand the Supply Side of Tourism.
STRATEGIES

- Encourage existing travel-oriented attractions and service businesses to upgrade and adapt to shoulder and all-season markets. This will automatically encourage return visitations.

- Stimulate each county to review all proposed development opportunities recommended by participants in this study.

- Encourage local political jurisdictions to upgrade the infrastructure to current standards, especially those features that assure the quality of the public water supply.

- Encourage each county to study the potential for more and better attractions, services, information, and promotion.

- Improve, coordinate, and expand transportation access (including signage) to attractions and services wherever there is evidence of need.

- Encourage each county and community to determine how well present tourism supply meets the needs of market demand, especially changing market trends. An enhanced product will also tend to lengthen stay.

Following is a summary of the remaining goals and objectives:

Goal B: Advance Local Support of Tourism

Objective 1. Create an Area Tourism Plan and Obtain Local Commitment to its Implementation.

Objective 2. Integrate Tourism into all Local Plans for Development.

Objective 3. Establish a Local Tourism Awareness and Educational Program.

Goal C: Increase Cooperation Among All Sectors

Objective 1. Foster Better Inter-Public Agency Cooperation.

Objective 2. Develop More Effective Private-Public Sector Involvement.

Objective 3. Establish Better Joint Planning.

Goal D: Enhance Promotional and Marketing Efforts

Objective 1. Integrate and Coordinate all Promotional Programs.

Objective 2. Improve Tourist Information.

Objective 3. Target Marketing to Specific Segments.

In addition to these objectives and strategies, the principals of the study presented two more overall recommendations. A Policy Statement charged the Finger Lakes Region Planning and Development Committee with the responsibility for leadership on future implementation because it is the sole regional tourist association and the major supporter of this project. The Regional Planning Development Committee agreed upon a process of implementation that built upon present conditions but took action on the stated objectives and strategies to meet them. This would be fostered by the increased awareness and cooperation exhibited during the process of this project. Planned were a series of area workshops directed to adapt recommendations to local needs, establish action teams, and monitor progress.

Within one year, several steps had been implemented (Tunney 1992). A subsidiary of the Finger Lakes Association, a Revolving Loan Fund, was established to assist development of tourist businesses and attractions. In cooperation with the U.S. Travel Data Center, a Regional research and Information Center was established. A special study of the attraction potential of the Seneca/Cayuga Sections of the New York Canal System was initiated. New cooperation became a reality among the several entities important to tourism: Regional Planning Boards, all fourteen county administrations, the three Economic Councils, the Thruway Authority, and the Regional Economic Development Partnership Program, Urban Development Corporation.

This project differed from one performed by an outside consultant because it was led and performed by organizations within the region. (C. A. Gunn acted only as a catalyst and report writer.) It promises a high degree of implementation because of the commitment by these organizations. It represents a bold and innovative approach by a tourist agency that in the past was focused primarily on promotion.

SCOTLAND

In recent years, national tourism agencies such as the Scottish Tourist Board (STB) have entered into a period of transition. Although their past policies had concentrated primarily on promotion, they have begun to examine their roles and need for change.

In 1996, a Special Strategic Group made pertinent recommendations to the STB (Pollock 1996). Emphasized was the need for establishing a new integrated network of communication that modern computer technology would allow. The focus was on new techniques, databases, and networks to improve tourism marketing, sales, management, and customer services. Recommended was an integrative system that would link all entities such as businesses and the several Area Tourist Boards. Typically, this effort was directed toward marketing rather than physical land and resource development.

The STB has stated its mission as: "generation of jobs and wealth for Scotland through the promotion and development of tourism" (Corporate

Plan 2000). However, the concept of "development" had remained primarily the responsibility of the private sector and local levels, such as the Area Tourist Boards.

Realizing the need for an objective evaluation, the consulting firm of Price Waterhouse Coopers was engaged to make a thorough investigation of the policies and practices of the STB (STB Management Review 2000). This evaluative report observed several new trends in tourism that suggested a new approach was in order: greater tourist choices of destinations and greater demand for quality, greater competition, changes in technology, stronger emphasis by governments on the economy, greater need for accountability, and need for greater partnerships.

This evaluative report was approved by the STB November 3, 2000. A summary of this detailed study revealed five main categories of change needed for the STB:

- The needs and interests of tourist businesses require higher priority.
- Needed is an entirely new organizational structure.
- External parties need greater emphasis by STB.
- A strengthened framework for the tourism sector is needed.
- New funding will be needed to meet newer objectives.

The process used by the consultants was intensive and thorough rather than politically superficial. It included observations of a board meeting, meetings with five focus groups, telephone interviews, meetings with trade unions, and research of literature and Internet.

Several important conclusions were reached. Major was the need for the STB to sharpen its past ambiguous role into greater focus on the tourist business sector and its relationship to the tourist. Although marketing was still in the agenda, new functions would include a business advisory service, new support programs, and new cooperation with other support organizations. Recommended were greater ties to outside public and private entities, such as Forestry Commission, Scotland Arts Council, Scotland Museum Council, Scottish Enterprise, Scottish Executive and Parliament, Scottish Tourism Forum, Historic Scotland, Scottish Natural Heritage, National Museums for Scotland, Local Authorities, and Sportscotland. Emphasized was the need for much closer linkage with the several Area Tourist Boards. Another very important recommendation was the need for new measures of STB's achievements rather than only visitor numbers and expenditures.

This new awareness of Scotland's tourism opportunities and environmental needs was followed by the *Tourism and Environment Forum—Operational Plan 2000 to 2003*. Its mission statement:

To bring long term business and environmental benefits to Scottish tourism industry through encouraging sustainable use of our world-class natural and built heritage.

This comprehensive and realistic plan has four main aims:

1. To deliver market research and environmental capacity information to the industry.

2. To ensure the industry adopts good environmental practices and capitalizes on the advantages they bring.

3. To promote sustainable use of key national resources.

4. To ensure a national and local integrated approach to tourism and environment opportunities.

For each of these aims, specific doable objectives and their "provisional tactics and measures" are identified along with naming responsible action parties, public and private. This outstanding approach to creating and maintaining sustainability provides a blueprint for bridging tourism economics and resource protection, a lesson with universal application. Table 6-4 illustrates priority objectives and lead partners for these aims and objectives.

Of special importance to tourism in Scotland are its built heritage, including prehistoric, stone structures, Georgian houses, Victorian factories, castles, formal gardens, and World War II defenses. A major measure was enacted into law in 1991 to protect this heritage, the national agency of Historic Scotland (*Historic Scotland* 2000). Responsible to the Secretary of State, the purpose is to safeguard the national built heritage and promote its understanding and enjoyment. It owns and manages over 330 properties visited by over 2.9 million visitors a year. Although supported primarily by government funds, the agency obtains approximately one third of its revenues from retail sales, corporate events, and admission fees. Most of the work is performed jointly with local authorities, other organizations, and educational institutions. Of great concern is underwater as well as terrestrial archeology. Staff provide technical assistance to those who engage in historic reuse, particularly for tourism. Grants and awards are substantial and are not only protecting Scottish heritage but also enriching the traveler and resident experiences.

Of great aid to Scotland as well as Wales, England, and Northern Ireland is the United Kingdom Heritage Lottery Fund (HLF) established in 1993 (*Safeguarding Heritage* 2000). Its stated mission is:

To improve the quality of life by safeguarding and enhancing the heritage of buildings, objects, and the environment, whether man-made or natural,

TABLE 6-4

PRIORITY OBJECTIVES FOR TOURISM, SCOTLAND

AIM	PRIORITY OBJECTIVES	LEAD PARTNER
ONE	1 Develop approaches that indicate the carrying capacity and management requirements for visitor use of the natural and built heritage.	SNH
	2 Influence and co-ordinate existing consumer research to include assessments of visitor environmental preferences and satisfaction. Deliver new research on consumer environmental preference.	-
	3 Communicate to the tourism industry the visitor expectations of a quality environment.	STB
TWO	1 Enhance the quality of environmental information provided by the tourism industry for their visitors.	T&E
	2 Increase environmental elements to tourism training. These should be based around managing and improving the appeal of environment and heritage as part of the tourism project.	SE
THREE	1 Help ensure tourism in valued landscapes develops along sustainable principles.	Scottish Executive
	2 Keep issues of water and beach pollution and terrestrial litter to the fore as key tourism concerns.	KSB
	3 Influence local authority planners, SEPA and water authorities. Encourage the development and promotion of more sustainable 'tourist friendly' transport services.	COSLA
FOUR	1 Provide direction on more effective integration of environment and heritage through A New Strategy for Scottish Tourism, Tourism Strategies and local frameworks.	The New Strategy Implementation Group
	2 Encourage a tourism industry contribution to framing relevant policies.	Scottish Executive

which have been important in the formation of the character and identity of the United Kingdom, in a way which will encourage more sections of society to appreciate and enjoy their heritage and enable them to hand it in good heart to future generations.

In only seven years, this program has proven itself with action by providing £1.5 billion for 6,000 projects throughout the United Kingdom. The recipients have been public and nonprofit organizations. In such a significant area of historic interest, this program is a powerful force for identifying, restoring, and adapting heritage to tourism.

> Managers of this fund cite the following criteria for awarding funds:
> - heritage attractions (museums, libraries, archives, historic buildings, nature reserves, industrial, maritime, and transport heritage sites) of all sizes;
> - historic buildings and structures for non-attraction uses (either as individual grants or through our Townscape Heritage Initiative) which enhance the fabric of the U.K.'s villages, towns, and cities and serve to underpin their interest and charm for visitors;
> - historic urban and country parks;
> - the countryside (either at nature reserves or through area-based schemes which, involving a number of separate small projects, boost the appeal of rural landscapes).
>
> We can also support marketing initiatives where they are targeted at broadening access to heritage sites by groups which are currently under-represented.

Another clue to new awareness of the environmental impact of tourism is the policy of the STB regarding capacities (Dear 2001). The STB and several Area Tourist Boards are recommending geographical and seasonal dispersal of travel markets as a first step toward limiting impacts. The Scottish National Heritage, one of 14 partners of the Tourism and Environment Forum, has initiated a pilot visitor monitoring program at several key sites before entering into a demarketing policy.

Another recent program of the STB, established in 1998, is its "Green Tourism Business Scheme" (GTBS) (*Green Tourism* 2001). This program takes the new policy of greater environmental responsibility one more step by awarding tourist businesses for their environmental accomplishments. Run by the Quality Assurance Program of STB, awards are issued at three levels. The *Bronze* award is based on good environmental practices. The *Silver* award recognizes more practices as well as commitment to continuous improvement. The *Gold* award is issued only to those who demonstrate a high level of environmental compliance, including "green" purchasing and supplier screening. Elements of the inspection criteria include: waste, energy, water and effluents, transport, suppliers and sub-contractors, staff, storage, communication with the public, noise, and wildlife opportunities.

By 2000, hundreds of tourist service businesses had received awards based on their environmental accomplishments. Among the results have been reduction of energy costs, improved waste management, increased recycling, and cleaner, safer, and improved landscapes. This program is putting policy into actionable objectives. Although initiated by government, it is being carried out voluntarily by the private sector.

An example is the Ashdene House, a small hotel on the outskirts of Edinburgh. Its many environmental accomplishments have been rewarded by the receipt of a GTBS Gold Award. Among the credits are: recycling waste, low-energy lighting, thermostat heating controls, and use of local food products.

This case demonstrates a shift at the national tourism policy level from exclusive promotion to inclusive environmental awareness. These innovative programs are demonstrating more than policy by supporting actual development that is more sensitive to the environmental assets of Scotland and the region.

NEW ENGLAND TOURIST-RECREATION TRAIL

An example of the complexity of participatory planning at the regional scale involving many jurisdictions, planners, designers, and citizens is the trail plan for the six states of New England. With the intent of doubling tourist use and enhancing environmental values, an extensive trail system is planned for this region (Ryan and Lindhult 2000). Encompassing the states of Maine, New Hampshire, Vermont, Massachusetts, Rhode Island, and Connecticut (66,608 square miles, 172,514 square kilometers), the rich and abundant natural and cultural resources are not only appealing to tourists but also highly defended by citizens. Mountains, forests, valleys, rivers, lakes, and the Atlantic coast provide a great variety of appeal within a relatively small region.

Landscape architect Julius Fabos has provided leadership for this innovative greenway trail of 12,700 miles that links parks, lakes, streams, cultural resources, and historic sites throughout the region. Over 100 citizens, nonprofit organizations, and public agencies have participated. The project statement focuses on recreation trails and areas, resource protection of critical areas, and historic and cultural resources.

Already a great many trails and conservation areas exist in the region and this plan recommends an additional 19,300 miles of greenways and eight million more acres of protected land. Because of the success of Rhode Island's greenway legislation, such action is recommended for the entire region of New England (Project Overview 2000).

Of great benefit to this project is the strong New England land design ethic, a cultural norm for over a century. It was here that the American Society of Landscape Architects was founded, credited to landscape architect Frederick Law Olmsted, the noted designer of Central Park, New York. He planned the first major greenway in America, "The Emerald Necklace," for Boston and Brookline in 1867. Charles Eliot, a pupil of Olmsted, planned a 250-square-mile open space/greenway system for metropolitan Boston. His nephew created an open space plan for Massachusetts in 1928, the first statewide plan in the United States. Certainly, this precedent for a resource protection ethic fostered greatly the acceptance of such a bold and innovative plan for New England.

The planning process followed five main steps. First, *a resource evaluation* included identification, description, and mapping of existing greenways, trails, and established preserves, including a significant portion of the Appalachian Trail. Second, a *survey of trail and greenway proposals* revealed plans in most states for several million acres of trail and protected zones, including a portion of the East Coast Greenway from Florida to Maine. With input from nine workshops that encompassed representatives from government agencies, trail advocates, and landscape architects, the needed *new linkages* were identified. A fourth step created *state plans* that were then combined as a fifth step into the *overall New England plan.*

Land analysis for this project was facilitated by employing geographic information systems (GIS). State planning offices provided much data and the Environmental Research Institute (ERI) donated $200,000 worth of ArcView GIS software.

Leaders recognized that a project of this magnitude and involving such a complexity of decision makers needed an extensive publicity program. Part of this educational plan was a group of landscape architecture students at the University of Massachusetts (Amherst) who created a Web site. Planners, legislators, and the general public could get quick and informative access to the plan. This stimulated awareness of many funding sources from community, state, and federal agencies already in place.

This commendable and environmentally sound tourism-recreation project demonstrates the value of a participatory planning process. Not only did professional designers and consultants but also government agencies, nonprofit organizations, and citizens participate at several stages in the planning process. Furthermore, the publicized purposes were clearly stated from the beginning and captured the interest of residents of the entire region. However, the leaders are well aware that in spite of intensive and dedicated effort for many months, the critical step of implementation was yet to come. Implementation may take many years or even decades, a typical fundamental of regional tourism planning.

LITHUANIA

Lithuania is a nation of 25,174 square miles (65,200 sq. km.) located between Poland and Latvia, with frontage on the Baltic Sea. Following freedom from Soviet control in 1991, the nation now seeks tourism as a boost to its economy. Its main appeal is derived from its market proximity and its resources, such as coastal beaches and sand dunes, forests, over 3000 lakes, many streams, and rich cultural background. The nation is now open to tourists from Europe and beyond, a new potential for development. For many years, professional planners and landscape architects have been developing plans for environmentally sound tourism expansion (Stauskas 1995). Vladas Stauskas, planner and landscape architect of the Institute of Architectural Research, Vytautus Magnus University, Lithuania, assisted in the preparation of the national tourism plan and advocated the following seven points:

  • identification of needed landscape preservation
  • cultural and natural resource evaluation
  • identification of relationship between resource areas and service centers
  • potential of social tourism opportunities
  • for resource protection, tourist facilities should be located in urban areas
  • coastal zones should remain free from development
  • maintain only low-rise architectural development

The National Tourism Development Programme, 1999–2002, is founded in the Law on Tourism of the Republic of Lithuania (*Lithuanian Tourism Policy* 2001). Its main purposes are: to assess recent changes in the tourism sector, to elaborate a tourism policy, establish state investment, enhance the scope of domestic and foreign tourism, promote Lithuanian economic export, and enhance the Lithuanian tourism image.

This program outlines several *preconditions* for tourism development: The favorable location; ease of land, sea, and air access; and a new revival of political and social relations favor tourism expansion. The new tourism law commits the government to tourism involvement, regionally and locally. Obstacles to be overcome include a shortage of quality accommodations, poor convention facilities, lack of development of cultural and natural resources, and border crossing restrictions.

Opportunities for improvement include increasing sea access from Scandinavia, amending legislation to allow new growth of campgrounds and motels, and identifying zones of best tourism development. Also recommended is encouragement of local municipalities and parks to become involved in tourism.

The plan lists many *resources* that favor further expansion of tourism. The forests, lakes, rivers, the Baltic Sea, and aesthetic landscapes offer a foundation of natural resources. Castles, old towns, ethnic architecture, museums, art centers, and folk art provide many opportunities for upgrading and expanding cultural tourism. Rural tourism has potential. Needed are new initiatives to improve information centers, add historic interpretation, and eliminate bureaucratic formalities that hinder new private investment.

New training and educational programs are needed. This requires new coordination among Ministers of Education and Science, Social Security and Labor, Land and Forestry, and the Tourism Department.

Recommended are new business regulations that emphasize consumer service. Better marketing and cooperation with major tourism organizations, such as WTO and the European Union (EU), must take place.

Finally the law on tourism, passed by the Seimas of the Republic of Lithuania, March 1998, (*Law on Tourism* 2001) establishes priorities and principles of tourism development, requirements for tourism services, increased competence of public tourism administration, and terms of reference for use of resources for tourism.

This case is cited as representative of the significance of new changes in politics and leadership that foster tourism planning and development.

MOZAMBIQUE

The nation of Mozambique, in its quest for greater economic stability, has created a *National Policy and Strategy of Tourism* (2001). This nation of 302,330 square miles stretches along the southeastern coast of Africa for about 1,500 miles. Because of tourism stagnation during twenty years of civil war it hopes to modernize facilities, focus development on zones best able to support tourism, and expand in a sustainable manner.

Because Mozambique has cooperated with Agenda 21, it is receiving aid from the Sustainable Development Networking Program (SDNP) 1995. As the nation enters a new era of peace and reconstruction, this program will be of assistance in implementing its tourism plan. Training and distribution of sustainable development aims and principles will assist governments and communities throughout the country.

The plan involves stronger public-private cooperation. Since the peace accord of 1992, over 900 state-owned enterprises have been privatized. It cites its main assets as beaches, wildlife, culture, historic sites, and a relatively mild climate.

Objectives

Five major objectives are stated in the plan. Important is the *generation of new employment* in order to improve the quality of life for the people of Mozambique. This will require a great amount of new investment by both public and private sectors. Another economic objective is to *improve the balance of trade*. Intended is the increase of international visitors and investors. Cited is a desire to provide a *more equitable development of the nation* by a better distribution of tourist benefits, and yet focus on zones with greatest capability. The main geographic zones are: Ponta do Ouro to Cape Santa Maria (tourism zone of Inhaca), Pomene zone (and its game reserve), Bazaruto Archipelago (and its national park), the National Park of Gorongosa, and the Archipelago of Quirimbas zone.

In order to enhance its tourism growth and value to its people, Mozambique seeks *greater national unity through tourism*. This objective will require upgrading existing facilities and creation of new attractions, such as golf courses, conference centers, beach facilities, entertainment, and cultural development. At the same time that rural tourism is encouraged, main tourism facilities such as hotels, lodges, and related tourist services will be directed toward urban centers. Finally, the plan recommends stronger *protection of cultural and natural resources* to protect regional architecture, safeguard monuments, balance planning of areas, conserve resources, and provide incentives for development of handcrafts and folklore.

Anticipated is a series of important strategies. In order to work toward these objectives, needed is a better travel market-attraction relationship. High priority will be given to seashore areas for their power in attracting tourists from Europe, America, and neighboring countries. Waterfront attractions should complement the wildlife appeal of countries nearby (e.g., Kruger National Park, South Africa).

New access and the encouragement of attraction clusters will facilitate tourist use of their travel objectives. The advantage of grouping attractions is to provide a more complete "African adventure." These zones of attractions enable more efficient uses of infrastructure. In order to be more competitive with other African appeals, the minimization of costs is essential.

More careful planning of target zones is necessary to curtail sprawl and prevent depletion of resource assets. With limited sources of funds, public and private investment must be directed to areas of greatest need and value for tourism.

An example of the guidelines for each zone development is the case of the Ponta do Ouro Zone, including Ponta Malongane, Ponta Mamoli, and the Maputo Game Reserve up to Cape Santa Maria. Its geographical location is especially desirable for nearby travel markets as well as overseas tourists. Its proximity to the city of Maputo and beautiful beaches is an additional asset, particularly to attract medium to higher income visitors interested in resorts. Guidelines for other zone development are included in the plan.

Organization

For the future of tourism development in Mozambique, it is anticipated that four major units at the federal level will be required.

The *National Board of Tourism* is charged with a great many functions that are focused mostly on needed research, creating funds, licensing, educational proposals, project master planning, and technical advice. Additional functions include promotion, coordination of tourism information, and greater international cooperation. The plan does not indicate the extent of new staff, funding, nor integration with existing federal agencies.

The *National Fund of Tourism* is responsible primarily for funding infrastructure, carrying out studies, holding public meetings and workshops, and providing education on tourism's relationship with resource conservation. Cited are topics of ecology, ethics, and cultural heritage.

The *National Tourism Company* is the leading executive branch charged with implementing the national tourism policy. This includes engaging in joint development ventures with the private sector, managing national tourism finances, and becoming a national tour operator.

A *National Committee for Facilitation of Tourism* is primarily an interagency council for coordination of tourism within the general Ministries. It will recommend needed legislation, examine master plans of zones, establish subcommittees for special projects, and serve as a forum among governmental agencies and between government and the private sector.

Action Plan

The plan encompasses basic principles of tourism development important to Mozambique. Concerns are expressed over areas that may be saturated with tourism, semi-saturated, or underutilized. Overcoming seasonality is an issue. Recognized is the dominance of tourism in South Africa and that Mozambique must enhance its entire program in order to compete. Especially needed are new attractions to compete with photo-safaris, game parks, and beach resorts of South Africa, Seychelles, Mauritius,

Zimbabwe, Botswana, and the Comoro Islands. The market potential seems abundant at all levels, international, nearby countries, and domestic. Necessary is the implementation of actions outlined in the plan.

Because Mozambique has cooperated with Agenda 21, it is receiving aid from the Sustainable Development Networking Program (SDNP 1995). As the nation enters a new era of peace and reconstruction, this program will be of assistance in implementing its tourism plan. Training and distribution of sustainable development aims and principles will assist governments and communities throughout the country.

Markets

The intercontinental travel markets are a challenge for the planners. The hope is to tap the existing visitor flows to nearby regions such as South Africa, Botswana, Zimbabwe, Zambia, and Malawi. To do this will require major improvements in attractions, access, and pricing. Today's largest market is from nearby nations but facilities for upscale tourists are not yet plentiful. Most visitors are at the lower socioeconomic rank and use tents, trailers, and boarding houses. It is believed that the domestic market has far *greater* potential than is available today. Although upper and middle class travelers are not plentiful in this comparatively poor country, there is potential within business leaders, politicians, and resident foreigners. Although the urban centers now attract visitors for business contacts and meetings, opportunities abound for travel packages to tour the country, visit the beaches, and engage in other attraction activities

As part of an overall African initiative for new "peace parks," planned is a Gaza-Kruger-Gonarezhou joint park between Zimbabwe, South Africa, and Mozambique (Godwin 2001). Although the main motivation is conservation of wildlife, this enlarged park will be a boost for tourism. The future of this and other beach park proposals is dependent upon new political stability, increased multinational cooperation, and new transfusion of financial input, not only for investment but for management. Especially critical will be the planning and funding for new infrastructure.

This example is cited here as representative of the trend around the world for previously impoverished nations to begin to plan and establish policies of tourism development. Rather than only promotion, this plan provides for its positive impacts and yet protects its basic resources. The blueprint for Mozambique has been put forth and only future action within and surrounding the country will demonstrate its effective implementation.

NEPAL

Nepal, a nation of 54,362 square miles, is a mountainous region of southern Asia. It now has updated its plans for tourism. Nine-tenths of the nation is dominated by the Himalaya mountain range. A mountain climbing challenge continues to be Mount Everest, a peak of over 29,000 feet. Many rivers arise in the range and flow through Nepal into India.

Planning Background

The first national tourism plan was prepared in 1959, with primary emphasis on promotion. Included also was support for new infrastructure such as roads, water, electrical power, and airport construction (Shrestha 1999).

For a third five-year plan (1965–70) infrastructure again was given emphasis. In addition, new plans for conservation of temples, museums, and historic sites were put forth.

A fifth five-year plan focused on promotion and increasing the balance of payments. Again, conservation of historical and cultural attributes and extension of tourism development beyond the Kathmandu area gained attention. Other planning purposes were to extend the season of visitors and expand religious tourism.

The tourism policy of 1995 responded to tourism's increasing recognition as a major aspect of the economy and societal value. The limited governmental support stimulated greater involvement in investment and operations by the private sector. In spite of some progress, leaders of tourism have been dissatisfied with the lack of a clear vision, poor coordination among government agencies, and low commitment of resources by the government. Overcrowding, lack of airline capacity, and increasing pollution are cited as issues that need resolution.

Modern Tourism

A critique of modern tourism in Nepal has been presented by D.P. Dhakal (1999). He cites the wide range of cultural and natural tourism assets but is critical of overinvolvement by government. He emphasizes government's role as catalyst, facilitator, motivator, and monitor rather than a prime developer. Progress on private sector involvement has been slow but by the Visit Nepal Year of 1998, new air access and tourist accommodations had been developed. The past monopoly of these services was finally broken, opening up new opportunities for private investment. The governmental Department of Tourism was replaced by the Nepal Tourism Board that included private sector representatives. Many services have expanded for trekking, rafting, and mountaineering. Yet to be created are more cultural developments.

This example of a relatively poor nation's planning for tourism demonstrates the difficulties and progress in utilizing its abundant and spectacular natural and cultural resources.

CONCLUSIONS

These nine case studies of planning tourism at the regional (national, state, provincial) scale are offered here for several reasons. Although they are not necessarily representative of all world-wide tourism planning at this level, they offer a diversity of process and content. But even with their differences, several common threads run through all.

Regional tourism planning roles are shifting.

New regional plans for tourism show major changes in the goals and objectives of planning as compared to the past. The singular focus of promotion in the past is being amended by concerns over environmental, social, and political issues. Although the governmental role in tourism planning continues, the relevance of private sector roles is increasing. Throughout, there is a common goal of seeking a match of demand and supply.

Tourism foundations are consistent.

There is a consistent trend toward awareness of basic tourism planning foundations. The essential roles of natural resources, cultural resources, transportation, infrastructure, and cities are now being cited in regional tourism plans. Environmental impacts are of increasing concern. How these resources function and the need for their nurture and protection are just beginning to be acted upon.

Tourism planning politics are powerful.

Regional policies (national, state, provincial) take precedence over all other aspects of tourism planning and its processes. Political ideologies, traditions, and controls override research and professional planning practices. Tourism plans and planning are subject to what and how a nation places its political priorities. Some believe that the primary roles of government in tourism should be as a catalyst, facilitator, motivator, and monitor rather than developer.

Regional tourism has a strong role at the destination level.

A major regional function is the identification of zones of greatest potential. Because these zones contain the best and most abundant factors in support of tourism, they are the most efficient units for planning and development. Important is the integration of these zones into all regional plans. Usually, this action requires new public and private cooperation because of the many political and jurisdictional barriers when the overall breadth of tourism planning is considered.

New awareness of the need for greater intergovernmental and public-private cooperation is evident.

The complexity of tourism is now being understood as involving a great many governmental and private entities. In the past, most of these bodies had not identified their roles in tourism planning even though their policies and actions influenced tourism. Slowly, this new realization is suggesting that there may be need for new legal mandates, policies, and practices to allow greater cooperation on tourism matters.

The greatest obstacle in the tourism planning process is implementation.

It is at this final step that most tourism plans fail. A new approach requires factors of implementation to be part of the very first steps in planning. Engaging key development actors, public and private, at the very beginning is essential. Decision makers in tourism development must be involved at the start in creating the planning vision, purpose, and those responsible for implementing action. Such participatory planning can foster not only better plans but also actual implementation of final recommendations.

DISCUSSION

1. Discuss the main planning differences between Czechoslovakia's plan of 1968 and that of modern Scotland.

2. Discuss ways in which integration of regional and destination tourism plans can be accomplished.

3. What are the main limitations to the search of a region for its potential destination zones?

4. Discuss the problems likely to occur if government dominates physical and operational tourism development.

5. From a contemporary and global perspective, what topics should be included in a national tourism plan?

6. Consider the event of overplanning tourism development at the regional scale and identify potential outcomes.

7. Discuss the role of professional designers/planners in the process of tourism planning and development

REFERENCES

Blank, Uel, Clare A. Gunn and Johnson, Johnson & Roy, Inc. (1966). *Guidelines for Tourism-Recreation in Michigan's Upper Peninsula.* Cooperative Extension Service. East Lansing: Michigan State University.

Corporate Plan 2000/01 to 2002/03. (2000). Edinburgh: Scottish Tourist Board.

Dhakal, D. P. (1999). "Tourism Development in Nepal—a Critical Analysis." From www.nepalinfo.com/imagenepal/juneplans.htm as of July 20, 2001.

Dear, Sandy (2001). Correspondence, Sandy Dear, director, Tourism and Environment Forum, Inverness, Scotland, April 18.

Finger Lakes Region—Tourism Development Opportunities (1991). "Final Report, Executive Summary." Penn Yan, NY: Finger Lakes Association.

Godwin, Peter (2001). "Wildlife Without Borders." *National Geographic* 200 (3), pp. 2–31.

The Green Tourism Business Scheme (2001). From www.greentourism.org as of April 26.

Gunn, Clare A. (1952). *Planning Better Vacation Accommodations.* Cir: R–304, Tourist and Resort Series. East Lansing: Michigan State University.

Gunn, Clare A. (1964). "U. P. Tourism Expansion." *The Michigan Economic Record,* 6 (7) July–August.

Gunn, Clare A. (1965). *A Concept for the Design of a Tourism-Recreation Region.* Mason, MI: B J Press.

Gunn, Clare A. (1973). "Vacationscape: A Case Study—Government, University, Landscape Architects." presentation. Annual Meeting, American Society of Landscape Architects, Mackinac Island, Michigan, July 11.

Gunn, Clare A. and Terry R. Larsen (1988). *Tourism Potential—Aided by Computer Cartography.* Aix-en-Provence, France: Centre des Hautes Etudes Touristiques.

Gunn, Clare A. and Terry R. Larsen (1993). *Illinois Zones of Tourism Potential.* For A.T. Kearney, Inc. and Illinois Bureau of Tourism. College Station, TX: Self-published.

Gunn, Clare A. and J. Ben McMillen (1979). *Tourism Development: Assessment of Potential in Texas.* MP–1416. Texas Agricultural Experiment Station. College Station: Texas A&M University.

Gunn, Clare A. (1991) *Delaware's Natural Resource Potential for Tourism.* For Price Waterhouse and Delaware Tourism Office. College Station, TX: Self-published.

Historic Scotland (2000). Edinburgh: Historic Scotland.

Innaurato, Albert (2001). "City of Alchemy." *Conde Nast Traveler,* October, pp. 126ff.

Law on Tourism (2001). Republic of Lithuania Law on Tourism. Vilnius, Parliamentary Record 1999, No. 1. From www.lrs.lt/cgi-bin/preps2?conditionl as of July 23.

Lithuanian Tourism Policy Overview (2001). From www.tourism.lt/services/htdp as of July, 23.

National Policy and Strategy of Tourism (2001). From www.mozambique.mz/turismo/political/general.htm as of July 18.

Pollock, Anna (1996). *Planning for the Digital Economy—An Implementation Plan for Scotland's Tourism Industry.* Edinburgh: The Strategy Group.

Positive Impact (2001). Tourism & Environment Forum, Inverness, Scotland.

Price Waterhouse (1987). *Proposed Master Plan for Travel Marketing and Development for the State of Oklahoma.* Contributors: Grey Advertising and Clare A. Gunn. Washington, DC and Oklahoma City: State of Oklahoma Tourism and Recreation Department

Prikryl, Frantisek (1968). *Tourist Regions in Czechoslovakia.* Ser. No. 196. Papers in Tourism. Aix-en-Provence: Centre d'Etudes du Tourisme.

Project Overview. From www.umass.edu/greenway/Project/1PO-Prj, April 19, 2000.

Ryan, Robert L. and Mark Lindhult (2000). "Knitting New England Together." *Landscape Architecture* 90 (2), pp. 50–55.

Safeguarding Heritage, Helping Tourism Grow (2000). London: Heritage Lottery Fund.

Scottish Tourist Board Management Review Report and Recommendations. (2000). Price Waterhouse Coopers: From www.scotexchange.net.

Shrestha, Hari Prasad (1999). "Tourism in National Development Plans—An Evaluation." From www.nepalinfo.com/imagenepal/juneplans.htm as of July 20, 2001.

Stauskas, Vladas (1995). "Tourism Development and Landscape Protection," 49. *Proceedings, IFLA World Congress.* Bangkok, October 21–24. Thai Association of Landscape Architects.

Sustainable Development Networking Program (1995). From www.sdnp.undp.org/countries/af/mz/mzpdoc as of July 22.

Tourism and Environment Forum (2000). Operational Plan 2000 to 2003. Inverness, Scotland: Scotland & Environment Forum.

Tunney, Conrad (1992). Correspondence of November 4, 1992 from Executive Director, Finger Lakes Association.

Chapter 7

Destination Planning Concepts

INTRODUCTION

Tourism planning, as conceived throughout this book, is facilitated greatly by applying planning concepts and processes at three geographical scales—region (nation, state, province), destination (community and surrounding area), and site (land for development). Although the same tourism development goals apply to all three scales, this spatial approach has much merit. Generally, the policy makers and action agencies are different among the three levels. However, it must be emphasized that all plans and recommendations at all levels require integration, usually demanding new cooperation and even collaboration. For example, the value of identifying potential destination zones as described in chapter 5 will not bear fruit unless leaders and developers at the destination and site scales follow through with their own implementation. The vision, the policies, and the integration at the regional scale provide the blueprint for a systems planning approach, essential to the elimination of many obstacles and the determination of desirable directions. But, eventually, commitment and land use decisions at the destination and site scales are necessary in order to translate these guidelines into reality. It is at these levels that economies may be strengthened and visitors may be enriched.

DESTINATION DEFINED

In this book, the term *destination* is arbitrarily defined as illustrated in figure 7-1. It is equivalent to other terms such as "community tourism" and "destination zone." The basic functional and spatial elements include access, gateway, attraction complexes, one or more communities, and

DESTINATION ZONE

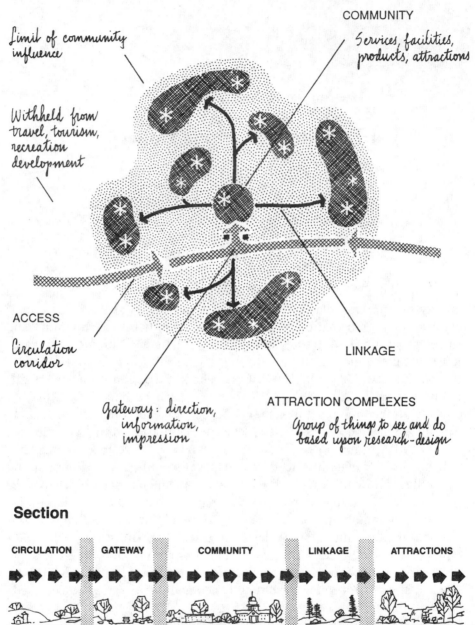

COMMUNITY

Limit of community influence

Services, facilities, products, attractions

Withheld from travel, tourism, recreation development

ACCESS

Circulation corridor

LINKAGE

Gateway: direction, information, impression

ATTRACTION COMPLEXES

Group of things to see and do based upon research-design

Section

| CIRCULATION | GATEWAY | COMMUNITY | LINKAGE | ATTRACTIONS |

Figure 7-1. Concept of Destination Zone. A destination zone, as defined here, includes: major access and gateway, community (with its infrastructure services and attractions), attraction complexes, and linkage corridors (between attraction clusters and community). When these elements are integrated, tourism is most successful. The section illustrates the diversity of supply development typically encountered by the traveler (Gunn 1972, 47).

linkage between attractions and the community. The diagram "Section" illustrates the tourist's flow to and through these elements, suggesting the important need for planning them together.

The engine that powers the destination for travelers is composed of the *attraction complexes*. These are geographic places, rooted in resources that have been developed to provide for visitor activities. These attractions serve two functions—drawing people to the places and fulfilling their expectations from a visit. The term "complex" is used to imply that there is value in clustering compatible attractions together, either physically or by tour. These attraction complexes may be within the focal city, nearby, or reasonably remote, such as a national park. (A national park is usually a complex unto itself because of the great number of compatible attractions it contains.)

Several other components of a destination zone function as facilitators. The *linkage* corridors between the key city and attraction complexes are important planning elements. These corridors require careful design consideration in order to provide a visual prelude to the attraction objective. For rural and remote attractions, self-guided and guided tours should provide the visitor with interesting explorations of the background and characteristics of the landscape being traversed. Key to planning these corridors are elements such as signage, maps, and other wayfinding information. Long linkages may require travel stops for restrooms and food services and interpretation of the travel corridor. Often, these are designed scenic highways. Again, the principle of adapting to the land resources and development, as well as visitor desires and needs, is paramount.

For all destination zones, one or more *cities* (communities) are essential. They provide several critical functions. All travel modes lead to terminals at cities. These terminals—train depots, airports, seaports, bus terminals, highway exits—perform an entering function important to travelers. The quality of physical planning, development, and management can set the psychological setting for further visitor activities. Cities offer the preferred location for most travel services such as hotels, restaurants, post offices, drug stores, shops, health services, and communications. These are preferred settings because they offer greatest financial feasibility, catering to both resident and traveler markets. Cities contain basic infrastructure that would be costly to develop at remote locations—water supply, waste disposal, police protection, fire protection, power. Cities have an organized management structure providing public services and amenities important to tourism. Cities often contain existing and potential attraction complexes—entertainment, parks, exhibits, festivals, historic sites, sports arenas, convention centers, trade centers, industries, institutions (medical, religious, organizational), and homes of friends and relatives.

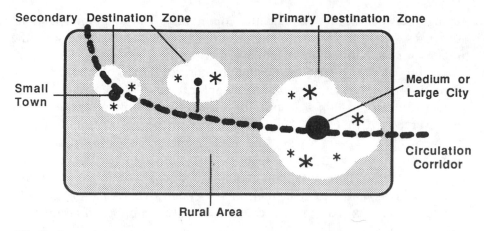

Figure 7-2. Rural-Urban Destination Zones. Cooperation between a major city and surrounding rural area and towns benefits all. The tourist cuts across many political jurisdictions in the search for attractions and services (Gunn 1988).

Many destination zones will be served by a major city surrounded by several small towns in rural areas as illustrated in figure 7-2. Advantages to both small and larger cities can be derived through cooperation, such as tour efficiency, increased mass of attractions, and greater promotional impact.

Another important component of a destination zone is *access* from markets. Too often, communities internalize their tourism planning to the extent that cooperation and assistance are not provided to developers and managers of transportation systems. Many transportation agencies focus policies primarily on resident or nearby markets, thereby developing routes and signage not easily mastered by outside visitors. For example, the main places of transport route penetration of a destination zone and city deserve special planning attention for visitors. It is at these sites that directional information is critical. A major information center at the *gateway* that provides maps, brochures, and personal guidance is essential.

Although the actual configuration of destination zones will vary around the world, these same components and relationships will require planning and integration for best service to the traveler. All components must function in concert, each depending on the successful rendition by the other.

It is no coincidence that this destination zone concept agrees with contemporary views of urban development. At one time, apparently in response to urban congestion, the principle of dispersion was put forward. Many conservation proponents will adhere to this principle. But, dispersion can reduce the indigenous value of both community and hinterland whereas combining a central city with its periphery into a single unit allows each to function in a symbiotic relationship. Gradus (1980, 418) calls this a "regiopolis." The regiopolis combines the advantages of social, human, and economic development within and around communities. Likewise, the tourism destination concept allows retention of a community's identity but also recognizes the important interdependency of the surrounding area.

It is the intent of this chapter to put forward the fundamental of place for planning at the destination scale and offer modes, concepts, and processes for planning tourism at this level.

THE IMPERATIVE OF PLACE

For all of tourism, the greatest geographical imperative is *place.* Throughout history, place has played a dominant role in society. Sovereignty of place continues to foster strong land protection and even stimulate wars. For many peoples of the earth, the qualities of place mean survival or death. In spite of man's gargantuan efforts to reshape the earth, the land and its geographical differences remain. Every place on earth posses its own peculiar characteristics, both as the result of natural physical forces and acts of mankind.

And so, as the topic of planning tourism shifts from the macro (the region) to the micro (destinations and sites), the importance of place dominates every concern. As tourism developers and promoters tend toward homogenizing the world with sameness, the opportunities for extracting the unique qualities of place, so important to the traveler, are being missed.

In many countries and for many decades, residents tended to remain anchored to a community. In spite of changes in local policy and economy, the family homestead remained a sacred icon. The street pattern was so familiar that street signs were really redundant. Landmarks, such as churches, schools, and shops, were stable images in the minds of local residents. The community's landscape of flat lands, hills, mountains, waters, and vegetation were not only fixed as mental images but also the substance of poetry, prose, and music.

Then came global change. Perhaps due to the outburst of electronic communication, air travel, and diversified economics, devotion to local values began to erode. For many, especially younger generations, the appeal of a different locale became so strong that they would transplant themselves. Their new place had none of the ties exhibited by the earlier residence but in spite of this, new spatial and societal patterns were learned.

Today there appears to be a groundswell of reversing this trend. Especially small villages, that were destined to become ghost towns, are being restored not only for nostalgic reasons but also for tourism potential. And, in spite of outmigration, many remain to take on these new responsibilities. An example is the renascence of small villages throughout the province of Newfoundland-Labrador, Canada (Gunn 1994). The study of tourism potential revealed that many third- and fourth-generation residents believed so intensely in local place qualities that they adapted to tourism enterprises following the demise of the cod fishing economy. One individual had converted his great-grandfather's home into a bed-and-breakfast facility, established a restaurant for residents and tourists, and managed a tour boat concession for Gros Morne National Park. It is this attachment to place and its contribution to sustainability that authors Timothy Beatley and Kristy Manning (1997) state are critical to a community-building process.

Characteristics of Place

Place, as defined by Motloch (2001, 242), "is the mental construct of the temporal-spatial experience as the individual ascribes meaning to settings, through environmental perception and cognition." Explicit in this definition are important meanings for tourism. All travelers must have a place to go. Simple as this may be, it is not well understood. Place qualities are absolute for all developers of services, facilities, transportation, information, and promotion—the components of supply.

Temporal aspects of place are critical to tourism planning. Visitor impressions and experiences vary greatly with the time a place is visited. When visitors report they have already visited a location and see no point in visiting again, they may miss an entirely different experience at a different time. Weather conditions—brilliant midday sun, cloudy sunset time, snowclad winter, cold, heat, stillness, wind—can give places quite different meanings. Places seen at night may be in absolute contrast when viewed in daylight. Not only will the landscape differ but also the patterns and colors of activities and users be quite different. For example, a dramatically different experience results from visiting the French Quarter of New Orleans in the day, with its rich history and famous food, as compared to night with its raucous entertainment. Places of a singular year-round climate have different appeals to travelers as compared to places with seasonal change—spring flowers, summer green, autumn foliage colors, and a white landscape of winter. The new market surge of nature tourism selects places as much on the time of viewing wildlife as on their habitat location. Animal migrations are critical to viewing wildlife.

An equally important dimension of place is its *age*—ancient to new. Generally the older the place, the greater its amenities. An author of the southern region of the United States has stated, "Older places have accumulated more meaning, but they also very often are more tied to a specific setting or are simply more humane in scale or character—and thereby more engaging—than new places" (Morris 1990, 7). Viewing the ruins of Delphi on the wooded slopes of Mount Parnassus cannot help but evoke dramatically different images for the visitor than visiting the more recently planned cities of Canberra or Brasilia. Even within a lifetime, "The special places of childhood are not sacred but the memory of them is necessary for attaching sacredness to place" (Shepard 1967, 37). An enlarging travel segment seeks the experiences of visiting historic sites and places of prehistory. Planning for visitor use of these sites creates challenges never before faced by designers and managers—maintaining the integrity and patina of age at the same time providing for mass visitor use. For example, the removal of ancient skeletons from grave sites for display in museums is opposed by many native populations.

Spatial distribution of places around the earth has deep meaning for tourism and its planning. Each place has its own unique relationship to all other places, far or near. Analysis of geographic position and geographic content described in the last chapter for regional planning are attempts to identify spatial characteristics of importance. In modern context with changing air fares and access, spatial distribution of destinations takes on new meaning. What once was *far* may now be *near* in terms of ease of access, changing dramatically the competitive position of places.

Another important characteristic of a place is its *name*. "Landscapes without place names are disorienting; without categorical forms, awful" (Shepard 1967, 43). Place names call up fantasy and imagery. The names of places from northern Michigan to the Arkansas River in the United States—Wisconsin, Peoria, Des Moines, Missouri, Osage, Omaha, Kansas, Iowa, Wabash, and Arkansas—stimulate recollection of French Jesuit explorers Louis Joliet and Jacques Marquette's trip through this land in 1673 (Stewart 1967,91). For settlers of Texas, religious themes dominated place names: San Antonio, San Jacinto, San Saba, Corpus Christi, and the Brazos River. Deeper analysis of market characteristics would probably prove that few travelers seek out places with unfamiliar or no names.

In spite of today's great mobility and migration of peoples, certain landscapes retain *people-place qualities*. A travel writer stated, "I believe you could exterminate the French at a blow and resettle the country with Tartars, and within two generations discover, to your astonishment, that the national characteristics were back at norm" (Durrell 1969, 157). Thus is the dominance of the landscape as place. Landscapes have special

values, even when some cultural labels are the same. Durrell points out
that the spirit of place has modified Catholicism to the extent that it is
different in Ireland, Italy, Spain, and Argentina. The theology and practices result from modification due to place. Although these subtleties are
not always readily apparent to travelers, with some time and introspec-
tion they become real. Visits to the Scottish landscape evoke the adven-
turous landscapes of Stevenson.

Of course, places can be described by *technical and scientific facts*.
For example, Mount Everest is five and one half miles (8.9 kilometers)
above sea level, is located in the Himalaya range, and was named for a
British surveyor-general of India, Sir George Everest. Madrid is located
at latitude 40° 24' 30" North and longitude 0° 14' 45" West and occupies
about 550 square kilometers. In resource analysis the quantities of natural
and cultural resource facts are listed: acres of forest land, sizes of lakes,
lengths of streams, numbers and species of wildlife, and so on.

These and other characteristics of place are of great importance to
planners of future tourism development.

Human Attribution of Place

Although such inventorying is necessary in describing place, it is of little
value unless interpreted and given meaning. An understanding of the
human attribution of resource facts is essential for planning. "Neither the
environment as such nor parts or features of the environment *per se* are
resources; they become resources only if, when, and insofar as they are,
or are considered to be capable of serving man's needs. In other words,
the word 'resource' is an expression of appraisal and, hence a purely sub-
jective concept" (Zimmerman 1933, 3). Resources are not; they become.
For tourism and travelers, "The environment is encountered in a way in
which self and place are related" (Shepard 1967, 34). Statistics of
resources are of value to the planner only as anticipated visitor contact
can be assessed. This intimate amalgam of place and human experience is
again expressed by Shepard (1967, 44):

> The desert is the environment of revelation, genetically and physiologically
> alien, sensorially austere, esthetically abstract, historically inimical. It is
> always described as boundless and empty, but the human experience there is
> never merely existential. Its solitude is not an empty void, a not-quite
> silence. Its forms are bold and suggestive. The mind is beset by light and
> space, the kinesthetic novelty of aridity, high temperature and wind. The
> desert sky is encircling, majestic, terrible ... The moon, sun and stars are per-
> ceptually exaggerated lower in the sky. Apparent motion in the horizontal
> plane is always greater.

An excellent example of place-attribution is the concept of scenery. No such thing existed until after the Middle Ages. In early history, forests were filled with demons and were of no value until felled for settlement and agriculture. Not until the painters and writers of the nineteenth century romanticized nature did vistas of landscapes become scenery. "The terrible awe of God was made into an esthetic—or, if you prefer, the forests and mountains of the earth came to be revered with religious intensity. The enjoyment of primeval wilderness had not been possible before" (Shepard 1967, 188).

Another aspect of human relationship to place is *social*. Cheek and Burch's (1976) early work emphasized the many social dimensions of leisure and place relations. There may be activities, such as picnicking, where the social group has higher priority than the location of place, as long as the needed amenities are present. Wilderness buffs often subscribe to Thoreau's tenet of seeking such places to find one's self. Certainly, the social context—solitude, friend, family, lover—can evoke different visitor meanings from the same place.

Perhaps the greatest lesson from this brief examination of place is understanding of its powerful and almost inexplicable necessity for tourism. And if its power and necessity are accepted, planning and design action must follow: *placemaking*. Areas that seek to become destinations for tourism have no choice but to become involved in placemaking. But, this process is not to be interpreted as artificially contrived. Quite the opposite. The challenge before the designer is to have sufficient understanding of both physical and human dimensions of place in order to design. Landscape architectural scholar John Motloch states (2001, 252): "The landscape designer facilitates positive placemaking by designing preferred relationships between anticipated place, intended behavior, and setting characteristics." Throughout the following discussion of planning destinations for tourism, the importance of place is a continuing theme.

DESTINATION PLANNING ISSUES

Opportunities

Although much marketing of tourism is done at the regional (national, state, provincial) scale, for most travelers, this is too generalized. That size of geography is often too large to comprehend as being able to satisfy travel objectives. It may be open to question whether travelers really seek their rewards in the United States, Canada, and Australia or in destinations such as in and around New York, Montreal, or Sydney.

Communities and their environs—destinations—more frequently carry with them images of appeal for both business and pleasure travel.

At this period in the history of tourism, it would appear that so many destinations have already been developed that there are few opportunities left. This is a half-truth because of the dynamics of both the market and supply sides of tourism. Changes in markets—demographics, economics, life styles, fads, interests—are constantly opening up new areas for development. Changes in transportation, attractions, information, services, and promotion are introducing new areas as having potential.

The destination discovery process described in chapter 5 demonstrates how one can identify potential destination zones. In most instances, when current market trends are examined, communities will discover much greater opportunity than popularly considered.

The geographic model, illustrated in figure 7-1, dramatizes a basic fundamental of all destination zones. They function for tourism only because of a *symbiotic relationship between focal cities and the surrounding area.* A destination function does not stop at the jurisdictional city limits. The surrounding rural area and small towns are an integral part of a destination zone. This symbiosis derives from tourist use, which involves visiting attractions both inside cities and in surrounding areas and using tourist services, such as for lodging, food service, car service, primarily in the cities.

A great many traveler activities take place in the surrounding rural area, such as:

Picnicking	Canoeing
Camping	Cross-country skiing
Hiking	Swimming
Horseback riding	ORV use
Bicycling	Resorting
Hunting	Retirement residence
Fishing	Historic touring
Boating	Scenic touring
Waterskiing	Festivals, events

(Gunn 1988, 238)

On the other hand, when travelers have participated in these activities, they seek lodging, food service, and other amenities and services of a focal city. Furthermore, feasibility of these service businesses favors city locations because they serve both resident and traveler markets.

Issues and Constraints

If opportunities for destination development are so abundant, why haven't more destinations occurred? One might glibly answer that this is due to lack of planning, but that is too simple, because destination tourism planning is complicated. Because the community (or several) is the focal point for tourism destinations, it appears that there are many constraints for community action toward tourism development.

Blank (1989) outlines several salient limitations:

- Lack of comparative advantage (location, quality of potential).
- Carrying capacity limitations.
- Lack of community's acceptance of change—preference for status quo.
- Power structure's preference for other development.
- Myopic view of tourism.
- Fear of tourism—erosive characteristics.
- Environmentalist resistance to any development.
- Narrow and inflexible policies on public lands.

Canada has experimented with much planning for tourism, most of which has stimulated valuable development. Yet, in working with destination planning, a few researchers have observed some important constraints at the local level. Frank Go et al. (1992) have analyzed implementation of tourism strategies there and noted the following resistances:

- Tourism involves such a diversity of actors that clear mandates for development are lacking.
- Lack of local guidance and will.
- Mandates are unclear or in conflict.
- Local jurisdictions seek a quick fix and fail to provide financial and human resources to do the job.
- Lack of monitoring system to measure success.

A worthwhile initiative for ecotourism community tourism development in Uganda has made progress but planners cite some important lessons learned (Victurine 2000). A major need was training and building the skills needed for entrepreneurs to accomplish their objectives. Businesses gained by outside advice and information. Although some funding was essential, especially for start-up, too much money overwhelmed some enterprises. Local culture, customs, products, foods, and traditional architecture were of great interest to visitors. A long-term view demands capacity considerations. And involvement by the entire local community and for a *long* period is essential.

In another case, the effort to plan tourism in an area of Canada's far north met with many obstacles (Addison 1996). Inuit populations began to group with settlements in the 1960s for economic opportunities and greater governmental assistance. In 1981 a blueprint for tourism development in the Baffin region was commissioned by the Territorial Government, *Baffin Regional Tourism Planning Project* (BRTPP). Among the purposes of this project were community involvement, resource potential, economic feasibility of projects, and market potential. However, a review of progress ten years later revealed many obstacles that hampered the success of the BRTPP. Key issues included:

- Local people resented dominance by government, that they lost control of their own future.

- No one seemed to understand who was responsible for implementation.

- It was apparent that the project had been hampered by lack of local education on tourism, lack of tourist business needs, and political turf protection by business and public sector.

- Adequate human resource development did not exist.

- The project had presented an inadequate and distorted view of native culture.

This example demonstrates the need in all community tourism development for overcoming these pitfalls.

The process of using community-wide workshops can be productive, such as a local citizen workshop directed toward tourism planning and development in greater Victoria, British Columbia. Here, it was successful in creating a task force organization for future guidance (Murphy 1985). It provided new coordination and direction for several purposes: local fundraising, local tourism theme, coordinated calendar of events, staff training, market research, and information centers. Task force committees were named: product development, funding, research and education, and community awareness. This event is cited here because it illustrates a recognition of several years ago that destination tourism planning reaches beyond the city boundaries and that a broad representation of action groups can reach consensus on future planning for tourism.

A consequence of the geographic reality of a destination zone is that it *transcends political jurisdictional boundaries*. This fact complicates planning and development. Tourism leadership and political decision making, such as for promotion, taxation, and development of amenities, also usually stops at such boundaries. Within the several small towns, rural areas, and focal cities there may be different population characteristics and different traditions, political allegiances, attitudes, customs, and even conflict or animosity.

If tourism is to be developed, these barriers against cooperation must be removed or at least ameliorated. Tourism leaders and constituencies of all jurisdictions within a zone must be equally committed to tourism development or it will not take place. Conversely, *tourism may be the catalyst to bring peoples of these separate parts of the zone together for a common cause.*

Misunderstandings

The concept of destination, even as defined here, is plagued by several misunderstandings. They are related here only so that those involved in planning may avoid these problems (Gunn, 1982).

1. Fallacy: *Destination zones are singularly defined.* Some writings refer to nations or continents as destinations. Certainly, this is a half-truth because images of these areas are often so defined in the minds of travelers. For example, Africa may have a stronger image than Kruger National Park. Destination zones are not uniformly defined. Some governments have divided regions into destination zones on the basis of administration. Such zones are suited to governance but have no relation to marketable or potential development areas. Sometimes marketing zones have been delineated but these lack consideration of resources yet to be developed. Zones based only on existing development and travel trends have been identified by Ferrario (1979). Ruest (1979) and others prefer to base destination zones on geographical resource factors. Such emphasis is given in this book.

2. Fallacy: *Destination zone boundaries are fixed.* This fallacy needs special emphasis. As markets change and as development grows or decays, destination zones can take on new size and shape or even disappear. It is an error to publish maps of zones with the implication that the edges are well defined. Zones are generalized areas that have broad and soft edges. Even though some resource characteristics appear to be fixed, new interpretations will cause change in the future, such as today's new emphasis on ecotourism.

3. Fallacy: *Destination zones are of one type.* When one kind of development becomes popular and successful, there is a tendency to copy this development at other locations. There is room for repetition but within a

different market range; witness the Disney attraction in Paris. However, there is usually a stronger competitive edge when each destination builds upon its unique characteristics of place creating a tourism theme of its own. Although the elements and principles of tourism development may be the same, they can be expressed differently, depending on the special resources of each place.

4. Fallacy: *The best zones are developed by the private sector.* Those who support a "tourism industry" philosophy are inclined to believe that private investment is the only solution to destination development. Although no one can deny the very important role of private investment and development, it represents only part of the formula for successful tourism destination zones. Even in capitalistic and industrialized nations, governments and nonprofit organizations continue to play important roles by providing basic infrastructure and many historic and natural resource attractions. Most isolated resorts, for example, fail not necessarily because of poor management but because they do not benefit from nearby attractions and government input in the form of roads, water supply, waste disposal, police, fire protection, and a governed community nearby. The best destination plan is created jointly by nonprofit organizations, government and the private sector.

5. Fallacy: *Zones succeed the best where tourism is the only economic provider.* Experience has clearly demonstrated the fallacy of this statement. Areas dependent only on tourism are plentiful and many continue to survive, but they are very vulnerable. Fads, fashions, politics, wars, competition, and economic changes can be devastating for tourism. Industrial developers for many years have promoted the principle of economic diversity. Such diversity provides a buffer against exigencies of change. A tourism destination can remain much more stable if the area includes a diversity of industry and services.

6. Fallacy: *Zone identification assures success.* Because basic factors are geographically clustered in a destination zone, all three developer sectors do have better chances of success. However, it should not be construed that individual project feasibility is assured. If, for example, it is indicated that an historic site might be developed into a major attraction, the question of developer feasibility remains. If it fits the policies and abilities of governmental park and historic agencies, one may see that it is feasible. Or, it may be more feasible for a nonprofit organization or commercial enterprise. Mere zone identification does not assure success. Many investment and management factors remain to be resolved.

Planners of tourism should be aware of these misunderstandings of destination zones.

DESTINATION PLANNING GUIDES

In recent years, governments, consultants, and other organizations have prepared guideline processes for the planning of destinations. These include "community" planning manuals but in the context of definitions used in this book, they generally include an entire "destination zone." The guides offer step-by-step processes that can be of great value to leaders and local population constituencies in their search for better tourism development.

A Planning Process in Ontario

Observation of the several difficulties of planning tourism destination development has given rise to proposals for better processes. For example, an experience in planning tourism for the Port Hope/Cobourg area, Ontario, resulted in putting forth a desirable process of four phases (Joppe 1996).

Preparatory Phase. Already, the government in its plan for the Lake Ontario Greenway had recommended service nodes rather than dispersing business all along the greenway. A beginning step was creating a local committee for planning participation. It included local politicians; community official staff; representatives of cultural, natural, and environmental organizations, and performing arts; local businesses; business associations; and tourism coordinators. Their charge was to review existing plans, obtain a consensus on objectives, improve access, protect resources, provide for public use, and relate to existing settlement. They were to consider market implications. They concluded that a long-term strategy was needed, but to stimulate new development a short-term initiative was needed to demonstrate proper direction. It was agreed that all involved must be informed frequently on planning progress. This step took approximately six months.

Phase I. Framework for Long-Term Strategy. It was in this phase of about four months that several local workshops were held. Because information on specific topics was needed, several subcommittees were formed and meeting schedule was prepared. Discussions resulted in name iden tion and a plan for bicycle trails was developed as the initial the work progressed, regular reports were made to the four city councils involved in the area.

•

Phase II. Presentations. For the next six months, the several short-term initiatives were planned and development was begun. There was agreement on planning principles for these projects: must provide healthier, cleaner, open, green, accessible, and attractive opportunities; all planning must be cooperative with all governmental authorities; and regular reporting should be made that includes all aspects of implementation, monitoring, and evaluation.

Phase III. Implementation. This section was not reported because implementation had just begun at the time of the writing. However in spite of this the planners reported that several lessons had been learned at this point:

- A community seeking tourism must assume leadership, not give it to others.
- A careful selection of representatives in the planning process must be multidisciplinary and multisectoral.
- In the beginning, the purposes of planning must be clearly stated and approved by all.
- Political support throughout the planning process is essential.
- A key factor for planning and implementation is adequate funding.
- Successful planning takes time.

This practical application contains elements of destination tourism planning   e to others.

iters and agencies prepared destination (commu-
. An excellent example is the Community
l (Alberta 1988) developed by the provincial
anada. It was prepared over several years of
nts, and representatives of the private sec-
:

ction begins by defining tourism to
astructure, hospitality, and services. It
f these components can limit tour-

ism. The content includes answers to questions such as: where do we fit in? who benefits? who doesn't benefit? and, is it for us?

Book 2. Organization. This chapter opens with emphasizing the need for organization but only after full commitment to tourism. There must be first a strong desire by residents, businesses, and the municipal council. The first step following expression of commitment is to create a Tourism Policy and Tourism Action Committee. A sample policy would be stated as follows:

> Tourism will be encouraged within _____ and its surrounding area in ways that will attract more tourists, increase their length of stay, increase the amounts of money they spend here, and ensure that any adverse social, economic, and/or environmental effects are minimized as a result of activities to improve tourism.

According to the manual, The Tourism Action Committee should be mandated by local government with support from organizations such as the Chamber of Commerce and local tourism zone association (if one is in place). Members should include representatives of the following:

Chamber of Commerce Economic Development Board
Hotel/motel operators   Service station operators
Restaurant operators   Historical society
Service clubs   Youth groups
Tourist zone representatives Municipal administration
Recreation board   Tourist attraction operators
Tourist event organizers

Members should meet the following criteria: knowledge of the community, commitment to tourism, ability to work in a group, ability to invest sufficient time in the Committee, and reliability.

Book 3. Process. Figure 7-3 illustrates the suggested flow of 24 action steps to develop a tourism plan at the local level. Step 1 is market analysis. Step 2 is gaining input from organizations and agencies. Steps 3–7 involve study of area to determine assets and concerns. Steps 8–10 include identification of goals, objectives, and strategies for reaching objectives. Steps 11–18 involve completion of plan, obtaining input from several publics, and gaining official approval. The final steps, 19–24, consist of action implementation and reporting to the city council.

Book 4. Appendices. This book contains sources of assistance. Included are private and governmental sources of grants as well as individuals and agencies that can offer technical and educational assistance.

Book 5. Workbook. The manual includes outlines for use in detailing the results of the several steps in the planning process.

As of September 1, 1990, approximately 250 of 429 eligible Alberta

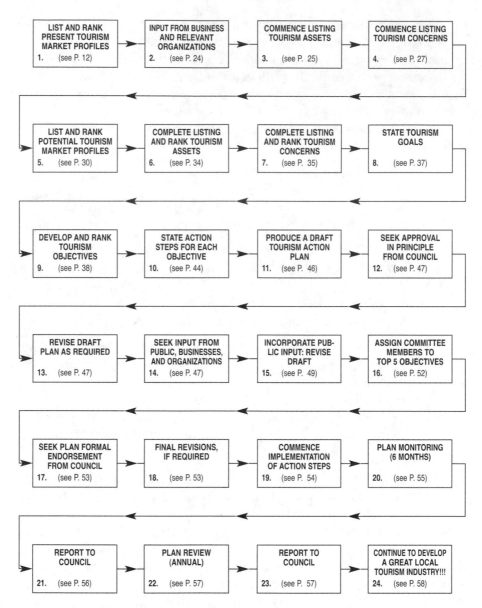

Figure 7-3. Community Tourism Planning Steps. The tourism agency of the Canadian province of Alberta developed these 24 steps for community tourism development. Note public involvement in the process (Alberta Tourism 1988, 3–1).

communities had developed tourism action plans. Go et al. (1992), based on the Alberta experience, have identified five conditions for destination tourism development:

• a clustering of communities, each supporting the others

- avoiding duplication
- key players who are ready, willing, and able to cooperate
- resident cooperation
- finance for start-up is available

Wight (1992), as a member of the provincial tourism agency, further emphasized the need for local stakeholders to integrate their ideas with governmental agency initiatives. When communities waver in their involvement, progress bogs down. In recent years, developers see the new regulations of the Alberta Environmental Protection and Enhancement Act (AEPEA) and the Natural Resources Conservation Board (NRCB) as "red-tape" hurdles that are costly in both time and money.

A plan for the Calgary-Canmore destination zone (Pannell 1985) identified many opportunities for tourism. Table 7-1 lists these together with potential impacts and funding support. A 1992 update (Wight) expands on the analysis and information required for public and private tourism development decisions. They could be summarized as: resource analysis, market analysis, development opportunities, development constraints and issues, and public-private collaboration.

The SDC Tourism Process

Developed by community tourism researchers from Brazil and Canada is the excellent guide, *Tourism and Community Development: An Approach* (Palermo 2001). The process is labeled the "SDC Approach" (supply, demand, consequences) and includes what the community provides, why visitors come, and what difference tourism makes.

Recognizing that communities vary greatly, this guide begins with an analysis of three key dimensions: community type, tourism development indicators, and a strategy for intervention. Examination of type indicates impressions of building characteristics and uses and the community's size, history, and economic status.

Although difficult to analyze, some description of traveler characteristics (demand)—interests and motivations—is an essential step. Finally, a community can assess positive and negative impacts from increased tourism development. How this concept was applied to twenty communities in Nova Scotia and Brazil are described in a companion publication, *Tourism and Community Development: Case Studies* (Palermo 2000). These are valuable resources for leaders and citizens of communities worldwide that seek greater tourism development.

TABLE 7-1

TOURISM DEVELOPMENT OPPORTUNITIES IN ALBERTA, CANADA

Opportunities	Capital or Operations Emphasis	Potential Impacts			Cost	Primary Action or Funding Emphasis
		Social	Economic	Environmental		
GENERAL						
Attract world class sports events before and after the Olympics	O	2	2	−1	C	Pr
Develop an annual winter festival	O	2	2	1	C	Pr
Develop a western heritage theme	O	2	2	0	C	Pr
Promote the scenery by making highways "visitor friendly"	C	2	1	0	D	P
Capitalize on linkages to attractions outside the zone	O	1	2	0	A	Pr
Prepare circle and exploration tour publications	C/O	1	1	0	B	Pr/P
Develop and promote major attractions in locations outside of Calgary	C/O	2	2	−1	X	Pr
Provide tax incentives and concessions for tourism developments	C/O	1	2	0	X	P/M
Develop and promote tours of Olympic facilities	O	1	1	1	B	Pr
Expand the variety of fixed roof accommodation available	C	1	2	−1	X	Pr
Improve and develop recreational vehicle (RV) parks	C	1	1	−2	B	Pr
Develop transportation services to destinations in Rural Zone 10	O	1	2	0	C	Pr
Adjust zone boundaries to include Cochrane, Highway 1A, and Stoney Indian Reserve	O	0	1	0	A	Pr

Rural Tourism Guide

Based upon general decline in U.S. rural economy, the U.S. Congress in 1988 directed the Travel and Tourism Administration to sponsor a study of tourism potential in rural areas (Edgell 1990, 32). As a consequence the report, *The National Policy Study on Rural Tourism and Small Business Development* (ERA), was issued in 1989. Another product was the preparation of a planning and development guide, *Rural Tourism Development Training Guide* (Koth et al. 1991), in cooperation with the Tourism Center of the University of Minnesota.

This is a destination planning and development guide for rural areas and small towns and contains 13 parts, including description of five cases of self-help. Although the manual covers other aspects, a planning process is outlined:

1. Getting organized.
2. Identifying community values.
3. Attraction inventory.
4. Attraction assessment.
5. Attraction packaging.
6. Organizational funding strategies.
7. Business inventory.
8. Marketing situational analysis.
9. Identifying a community tourism product.
10. Identifying target markets.
11. Setting market objectives.
12. Selecting promotional strategies.
13. (Optional) Tourism business retention and expansion.
14. (Optional) Community appearance.
15. Evaluation.

The educational materials within the manual are directed toward a five-step model, paraphrased as follows.

1. *Values.* Community groups must first explore the importance of their values, such as their geography, history, culture, and life style. This is necessary to make sure tourism is not approached as an overlay or merely a cosmetic treatment. Tourism should play an important role in enhancing and preserving basic community values.

2. *Attractions.* It is essential to identify the activities, cultural, arts, and historic resources, and developments that have the power to attract visitors. Both surrounding as well as internal community attractions are

important. Attraction quality, drawing power, grouping, and management must be assessed.

3. *Services.* Both public and private services need to be evaluated. Water supply, sewage disposal, drainage, snow removal, fire protection, health control, police, and other public services by government need to be checked. Service businesses—lodging, food, transportation, entertainment, and retail trade—need to be offered in high quality. Entrepreneurship and business retention factors are important. New business should reflect the character of the community as well as fulfill tourism needs.

4. *Marketing.* Marketing is driven by traveler needs. Claims must be honest, accurate, and consistent with product delivery. Analysis of current programs should reveal efficiencies and need for change. Identification of market segments and use of most effective marketing strategies should be made.

5. *Organization.* Because tourism involves so many facets and political jurisdictions, it requires special organization and leadership. The tourism organizations need to:

- Create a vision.
- Have a clear understanding of goals.
- Develop consistent leadership.
- Be adequately funded.
- Conduct periodic evaluation.

Cooperation with all other public and private groups is essential.

Tourism USA

Sensing a need for guidelines for communities and areas to develop their tourism, the U.S. Department of Commerce, U.S. Travel and Tourism Administration, and the Economic Development Administration requested the University of Missouri to gather information and produce a manual. The first edition was published in 1978 and a second edition in 1986. A revised and updated third edition was published in 1991 (Weaver 1991). The following discussion summarizes the contents of this guide.

Appraising Tourism Potential

This section of the guide provides information on how tourism can benefit communities with incomes, jobs, and tax revenues, especially as an aid to a diversified economic base. It also identifies costs of development: transportation, roads, parking, signs, water, sewage, restrooms, safety, health, and welfare. Included is the relationship of the community to attractions, services, and markets.

Planning for Tourism

Included here is recommendation on leadership and organization. The need to coordinate the many components of tourism is addressed. The following planning steps are suggested:

1. Inventory and describe the social, political, physical, and economic development.
2. Forecast or project trends for future development.
3. Set goals and objectives.
4. Study alternative plans of action to reach goals and objectives.
5. Select preferred alternative(s) to serve as a guide for recommending action strategies.
6. Develop an implementation strategy.
7. Implement the plan.
8. Evaluate the plan.

It is recommended that all segments of the community participate in all steps.

Assessing Product and Market

This chapter offers methods of market analysis—characteristics of visitors, expenditures, and activity preferences. Ways of inventorying and evaluating the match between market preference and attractions are presented. Forms for inventorying other elements of the supply side are included.

Marketing Tourism

Development of a promotional plan, target market advertising, local advertising and promotion, public and community relations, cooperative promotion, and souvenirs and promotional mementos are the main topics of discussion in this chapter.

Visitor Services

Visitor services are defined as all the normal city services together with those needed for hospitality. Recommendations on anticipating and planning service needs, coordination of visitor services, training for visitor services, hospitality training, public awareness, establishing tourist information centers, and evaluating visitor services are offered.

Sources of Assistance

Suggested help sources from federal, state and local agencies as well as private consultant aid are recommended in this section. Examples of tourist organizational structures and tax legislation are included in the Appendix.

Community Tourism Guide

Another helpful guide for planning tourism development at the destination scale is *The Community Tourism Industry Imperative: The Necessity, the Opportunities, the Potential,* authored by Uel Blank (1989). This book is organized in twelve chapters and is directed to local individuals and organizations interested in development of tourism. Although all chapters provide basic guidance, of special value in planning are the following final three chapters.

Chapter 10, "Getting it Together—The Matrix of Decision Making," includes important topics such as the process of decision making, how to assess opportunities, the importance of long-term planning, and how to move from knowledge to action.

Chapter 11, "Getting it Together—Tourism Development Policy," emphasizes leadership and relationship to state and federal tourism policy. The section on planning cites the need for several agencies, organizations and individuals to exercise planning roles. Planning guidance must include:

• attraction development
• hospitality services
• activities
• promotion/advertising
• transportation systems
• community ambience and aesthetics

- local information, direction, and interpretive system
- community infrastructures
- financing of development
- resource quality and management
- agency responsibility and coordination
- special market thrusts
- seasonality
- residents' living quality

Chapter 12, "Getting it Together—From Policies to Plans to Actions," provides guidance on preconditioning for action, bringing out the genius of the community, and the following action steps:

1. Initiative—recognition of need.
2. Set goals and objectives.
3. Collect needed information.
4. Analyze information.
5. Develop concepts for future development.
6. Develop specific strategies.
7. Carry out the plan.
8. Monitor and evaluate on regular basis.

DESTINATION ZONE PLANNING MODEL

As was recommended for regions, the authors present two approaches to destination planning appropriate for today. Periodically, it is important for a destination to prepare its own tourism plan. Such a plan includes content and recommendations based on the time of the project. Equally important is the constant process of regular planning because tourism is dynamic. Put forward first is the authors' concept of a desirable plan, based on their own experience and that of others.

Destination Planning Project

A project should include the following basic steps:

1. *Identify sponsorship and leadership.* Because the focus of the destination zone will be on the principal community, it may provide the best

organization and leadership. Although a chamber of commerce, convention and visitors' bureau, or industrial development agency or organization may initiate destination tourism planning, a new ad hoc or permanent council or commission may be needed. The organization and leader should be drawn from a wide cross-section of the community and surrounding region. Again, it is important to have representation from the greatest diversity of constituencies possible, not only the primary tourism businesses. Commitment to tourism and the desire to collaborate on planning are more important than expertise in tourism.

2. *Set goals.* The same goals as were stated for regions apply to destination planning—enhanced visitor satisfactions, protected natural and cultural resources, improved economy, and integration into the life and economy of the entire destination area.

3. *Investigate strengths and weaknesses.* Local people, with perhaps input from a tourism specialist or consultant, should gain a good understanding of the area's strengths and weaknesses. Each destination will pose different problems but an objective study of the following in the entire zone would be useful:

- natural resources: location, kinds, quantities, qualities, problems, issues, viability for attractions;
- cultural resources: location, kinds, quantities, qualities, problems, issues, viability for attractions;
- potential environmental impact, need for capacity control;
- transportation and access: capacities, access, quality, deficiencies;
- service business: quality, suitability to all markets, problems, issues;
- information about area for tourists: quality of maps, guidebooks, descriptions, hospitality;
- promotion: effectiveness of advertising, publicity, public relations, incentives, use of Web sites;
- organizations: sectors, organizations, agencies best suited to take leadership and implement development;
- present commitment by public and private sectors—resident attitude toward tourism growth.

4. *Develop recommendations.* From the above investigation, those performing it will be able to conceive of how the positive factors can be enhanced and the negative issues be ameliorated or corrected. Specific recommendations should be expressed on the same list of topics included in the investigation:

- natural and cultural resource potential

- transportation improvement
- service business improvement
- information improvement
- promotion improvement
- key organizations to take action
- how to improve commitment.

5. *Identify objectives and strategies.* This step is a refinement and expansion of the last step. It should identify specific objectives and how to reach them for each of the recommendations above.

6. *Assign priorities and responsibilities.* The entire list of objectives and strategies should be reviewed for assignment of priorities. Short-range objectives are critical and deserve highest priority. They should be of small enough size and cost to demonstrate immediate improvement. But long-range objectives need to be kept in mind so that each increment of shorter range accomplishment will build toward a well-planned overall destination zone. At this stage, it is important to assign responsibilities for action—who and what organizations are most logically the ones to get the job done?

7. *Stimulate and guide development.* With the identification of specific project development needed, derived from steps 1 through 6, these opportunities should be publicized for action by business, nonprofit organizations and governments. It is their responsibility to develop feasibilities, plan and design, build and manage the needed development within the destination zone.

8. *Monitor feed back.* Regularly, all implementation of action should be monitored. Enthusiasm and commitment may wane if it appears that no one is concerned about whether the objectives and strategies are working. Each increment of development, whether it be a newly built project or a new program, will change overall relationships and demonstrate new market-supply experience. Part of this feedback is to check on the relationship between this and other competitive destinations. Especially for touring circuit markets who visit the destination en route, it is important to know about planning and action in the zones that come ahead and after in the touring sequence. Also, the relationship between zone and regional promotion should be understood. A major issue today is to monitor capacity and the threat of oversaturation of tourism. Related to this feedback step may be the need for new research and education. Seminars, workshops, conferences, and hospitality training may be needed.

Finally, for all planning models it must be emphasized that the respect for the foundation of land characteristics is paramount. The need for environmental sustainability is not mere altruism but good tourist business.

For example, the Caribbean Coalition for Tourism has endorsed clean air and water, tropical vegetation, and beautiful beaches as prime lures for traveling there (Holder 1996). Any degradation of these factors damages not only the area's ecology but also tourism success in the Caribbean region. This fundamental of resource protection is true for every other destination in the world. It is for self-preservation that the business sector of tourism should support all public and private policies and practices for sustainability.

Destination Continuous Planning

Although a destination planning project for tourism development can stimulate specific action, accomplishment is likely to lag without a continuing planning function in place. The destination zone plan can provide guidelines for development as of that time period but regular updating is essential. For example, the destination plan may have called for creating new parks and historic sites. But, because sponsorship for an interesting outdoor recreation park became available first, it was established first. In its first year of operation, it demonstrated that there was a strong travel market interested in natural resource-based activities. This suggested the potential for expanding existing parks, developing adventure travel, nature trails, camping, and natural resource interpretation.

However, unless there is a permanent body charged with tourism leadership and continuous planning, new opportunities may be missed. Such a body would represent a diversity of interests within the destination zone and also be capable of carrying out specific functions on a regular basis.

As was done in Alberta, a Tourism Action Committee may be the best organization for continuous planning at the destination level. This committee should have representation from all governing bodies within the destination area: major city, satellite towns, counties, and regional government. This committee should also represent important constituencies, such as developers, businesses, nonprofit organizations, and residents.

Among the duties of such a committee would be the following:

- Foster the renewal of a Destination Zone Planning Project every five years.
- Monitor each increment of supply development to determine how well it is succeeding, resolve issues that impede its success and examine its integration with other tourism development.
- Promote the development of new and expanded attraction development based on the special natural and cultural resources of the area.

- Promote the establishment of needed service businesses to meet need of travelers.
- Encourage resource protection, not only by tourism supply development but other threats to the environment.
- Integrate tourism planning with city, county, and regional plans.
- Integrate all plans for information services and promotion.
- Monitor changes in market trends in order to determine new supply needs.
- Gain cooperation and integration of tourism development by all three sectors—governments, nonprofit organizations, commercial enterprise.
- Cooperate with regional tourism planners, managers, and policymakers to assure a continuing viable role for the destination.
- Monitor the societal, environmental, and economic impacts to determine oversaturation and the need for demarketing.

Planning actions at the destination level need regular evaluation. Communities and their surrounding regions are involved daily in a great many public and private actions other than tourism—services, education, sanitation, land use, policing. Whatever is done to plan tourism will need to be evaluated frequently in context with these actions. If the destination area has a tourism plan in place, it will need monitoring and frequent assessment. Frank Go (et al. 1992, 33) have developed a taxonomy (Table 7-2) for implementation of a community tourism action plan.

Organic/Rational Planning Process

The process outlined above for a "Destination Planning Project," can be classified as a "rationalist" approach. It focuses on problem solving, using a process that promises to implement specific objectives. However, this approach has its critics who claim that at a scale such as a tourism destination, the problems are too complex to resolve in such a direct manner. Within such an area there are too many factors—decision makers, influences, resource conditions, trends in public opinion—to deal with in a strictly rational manner. Called for is a more "organic" approach.

Steiner (1991, 520) has proposed a process that "reflects a middle ground approach to physical planning, somewhere between a purely organic and a truly rational one." Its basic premise is a flexible, iterative method that has the merits of a project as well as a continuing process that allows for contingencies. Again, it represents an integrated and

TABLE 7-2

TAXONOMY FOR IMPLEMENTATION OF COMMUNITY TOURISM ACTION PLAN

Level of Analysis	*Activity and Questions*			
	Interacting	*Allocating*	*Monitoring*	*Organizing*
Actions	How can community leaders and residents be encouraged to see their destinations through the eyes of a tourist?	How can communities best achieve their tourism action plan objectives?	How are community resource persons best evaluated by community tourism action members?	How can the local community become involved in the tourism action plan and get a feeling of ownership of the process?
Programs	How can the level of service and hospitality be developed to sustain and enhance the image of the community; specifically, the existing attractions, facilities, and infrastructure?	How should communities be selected for the funding of their tourism action plan?	How is research into present and potential tourism target markets best conducted by the tourism action committee?	How can sound local tourism strategies be built into a regional strategy?
Systems	How should the community promote attractions and facilities to travelers and tour operators?	How should a provincial government ensure that the local council is directly involved in plan development proceedings?	How is quality of tourism action plan evaluated?	How should the community organize its own unique tourism strategy?
Policies	How should the tourism action plan be managed on the community level to coincide with existing master plans and economic development strategy on the provincial level?	How should dollar and manpower resources be allocated to ensure that a community covers all important steps toward plan preparation?	How does a community regularly check the effects of tourism on its social structure?	How should the industry be restructured so that it will have a broader base, making it more representative of an responsible to the community?

Source: Bonoma, Thomas V. 1984: Go et al. 1992, 33.

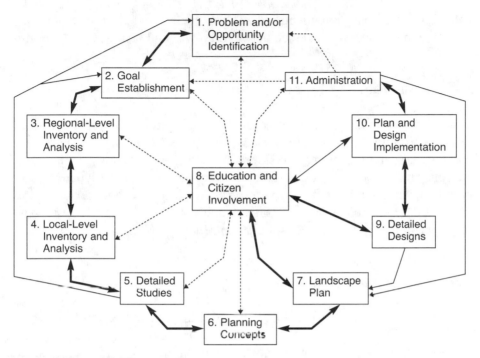

Figure 7-4. An Organic/Rational Planning Process. Landscape architect Frederick Steiner has put forth this 11-step process that incorparates involvement by professional designers, developers, and local citizens (Steiner 1991, 520).

interactive planning approach. The model, illustrating the sequence of steps is illustrated in Figure 7-4. The following summary paraphrases this process for a tourism destination zone.

Step 1. Problem and Opportunity Identification.

By means of workshops and other public participation, concerns, opportunities, and issues within the destination area can be identified. For example, issues of growth impact on the environment, job and tax potential accompanying growth, quality of new development, and ability of governments to respond with services may be major concerns in a destination zone.

Step 2. Goal Establishment.

Local people and governments as well as outside investors should develop a consensus on the kind of area they seek in the future. Based on the issues and concerns, goals are identified. These goals are long-range and for tourism would involve a balance between growth and resource protection.

Step 3. Area Analysis.

For a destination zone, analysis of factors is needed for the focal communities, surrounding area, and sites within. Key biophysical factors are examined, such as geology, physiography, climate, hydrology, soils, vegetation, and wildlife. Cultural factors, including historic settlement, economic development, land use patterns, demographics, and services are identified. Documentation in narratives and maps represents this analysis step, performed jointly by consultants and local people.

Step 4. Local Land Analysis.

For this step, more detailed documentation is obtained. Changes taking place and their environmental impacts are assessed. Population trends, relationship to physiographic features, and wildlife management issues are parts of this more specific descriptive analysis.

Step 5. Detailed Studies.

These studies focus primarily on suitability analysis—how suitable portions of the destination are for projected development. Policies of public and private landholders are an important part of such studies. Although a focus may be on tourism, this step allows equal consideration of overall future growth such as housing, recreation, business and industrial expansion, and transportation.

Step 6. Planning Concepts and Options.

Based on the foregoing steps, consultants can begin to conceive of planning concepts and options that these studies suggest could lead toward the desired goals. This step is based on logical and imaginative processes and suggests allocation of potential land uses. These concepts would include ideas for future tourism development projects, natural resource and cultural interpretation, transportation, and settlement appropriate to the interests of the public and resource base. Included could be appropriate themes for sub-areas.

Step 7. Guidelines.

This step produces what may be described best as guidelines rather than a plan because it contains much flexibility. Local officials, property owners, investors, and consultants jointly identify options for growth and development. In some areas, such guidelines must be formalized into a "master plan" or "comprehensive plan" in order to meet legal mandates for planning.

Step 8. Education and Citizen Involvement.

Although public input has characterized this entire process, this step assures that the plan concepts and guidelines are disseminated widely. Interaction between consultant planners and the many constituencies affected is essential. Use of the press and public meetings can stimulate open discussion on how well the concepts are directed toward all perceived goals.

Step 9. Detailed Designs.

In some instances, the timing is appropriate for initiating project designs. Such design plans will require owner-developer initiative. These projects carry the destination planning guidelines to actual case development. Often, these are not possible until some later date when there are sufficient owner-developer interests and support for actually building these projects.

Step 10. Implementation.

This step includes all policy and management implementation measures including specific projects. By means of close collaboration among local representatives, tourism businesses, and governmental agencies, initiation of land easements, zoning, and initiatives for purchase of private and public lands for development can be made.

Step 11. Administration and Evaluation.

Public and private organizations and agencies that initiated the planning process now monitor progress of implementation. As projects are selected and begun, feedback may indicate changes needed in the overall plan. In cases where the plan was legalized by the city and county, official administration and monitoring can then take place.

Steiner (1991, 528) concludes that such a design and planning process at this scale must be dynamic and reflect newly gained incremental experience. And any land planning process must begin with understanding landscapes and then making changes only in ways that protect and conserve these important foundations.

Integration with Community Planning

Because communities play such a critical tourism role in destinations, all plans and planning processes at this level need to be integrated.

Official community plans traditionally focus on physical public needs, especially for updating and enlarging public structures and systems. These needs are often for resident transportation, water supply (potable and industrial), sewage disposal (solid and liquid waste), power (electrical and gas), fire protection, and for police and public safety. Regulations for land use and structures, such as zoning ordinances and building codes, are included in most city plans. Included also are concerns over housing, education, trade, amenities (zoos, parks, recreation areas), and industry.

Unfortunately, in most communities, these traditional plans do not include issues related to visitors even though their decisions do affect tourism and vice versa. Too often, planning for the five components of the supply side of tourism—attractions, services, transportation, information, promotion—are not seen as responsibilities of city planning officials.

Dredge and Moore (1992, 8) have examined this issue as found in Queensland, Australia. They cite several inhibitors to the integration of tourism planning into traditional community plans. Much of tourism involves private sector facilities and services, often outside the perceived role of local planning. Local understanding of the complicated multi-owner supply side of tourism is not helped much by their perception of industry involving only a few physical plants. The overlap between the needs of visitors and residents, as well as their differences, are not well understood. The dynamics and interdependencies of the components of the tourism functional system are foreign to their day-to-day decisions relating to residents. Finally, the training and education of planners and designers have not usually encompassed tourism as a curriculum topic.

The conclusion of Dredge and Moore (1992, 20), equally applicable elsewhere in the world, is that town planners have not only great opportunities but responsibilities to incorporate vision, guidelines, and specific plans for tourism into their traditional local roles.

Even though the suggested process steps for destination tourism planning as outlined here are effective in a general sense, each case will have to be planned in greater depth and in its own context. The process must be modified to reflect local factors that cannot be generalized here. Every community has its own attitudes, policies, and experience regarding tourism that will influence actions.

Tyler and Guerrier (1998), after careful study of many community tourism development experiences, have concluded that more research is needed on the many factors that favor or deny destination planning and implementation of tourism action. They cite three major factors.

The *politics of decision making* are influenced greatly by local motivation for tourism development. In some cases it may be simply economic growth whereas in others it may be to raise local prestige or enhance local quality of life. At issue in some cases, and a difficult one to resolve, is inclusiveness. Frequently, certain local sectors feel left out of the loop of tourism planning, policy, and action. In other instances, local aims and policies may be at odds with those of higher levels of government. The proponents of tourism can meet with local opposition on the simple issue of resistance to change, a preference for the status quo.

A second factor revolves around the *process of change,* an important local distinction. A locality may resist the conversion of itself into a new form, that of a tourist commodity. By following only market pressures, it may be metamorphosed into a city entirely different from what once attracted settlement. And this touristic conversion may eventually create a new image no longer appealing to certain travel market segments.

The third factor cited is the *use of space.* From a landscape and land use perspective, tourism often converts portions of the city into exclusive tourism enclaves. Residents may question whether this spatial isolation is really their image of why they supported tourism development. The market may be satisfied but whether such development benefits the entire community is questionable. Finally, the use of space exclusively for tourists may displace locals who frequented certain community features such as bars, theaters, and parks. Cultural heritage may well be threatened by tourism.

These issues endorse the authors' desire for better research of the many elements of destination planning. Certainly each community must take an introspective view of tourism's implications before accepting the simplistic conclusion that tourism is always good for a community.

SCENIC HIGHWAYS

Driving for pleasure, implying scenic roadways, continues as one of the top travel market activities. Eight state departments of transportation have approved policies of "context-sensitive design." These policies are directed toward appropriate landscape adaptation of roads in cities and towns. Over 80 National Scenic Byways and All-American Roads have been established (Maguire 2002). Although some scenic highways are the result of specific site design, most lie within the context of a destination zone. Many governments have designated scenic highway segments of travel routes, but in doing so have discovered that what appeared to be a simple task became very complicated. The dual function—transportation and scenic appreciation—poses a great challenge to policy makers and designers.

Scenic Highway Issues

A report of the U.S. Department of Transportation (Federal 1991) pertaining to "scenic byways" identifies some of the key issues of design and designation for these special places:

Ambiguity of Definition. Although there may be common public opinion of what a scenic road is, the issue of defining it in order to plan, build, and protect its assets becomes very difficult. Generally accepted is the principle that it embraces both the visual countryside as well as the roadway. But there is no accepted definition of design standards—visual depth beyond the roadway, roadside maintenance, vertical and horizontal curvature, and vehicle speed and capacity. Each case will need its own definition and standards.

Corridor Protection. Because the entire highway corridor is involved, the protection of the natural and cultural resources that caused it to be so identified becomes an issue. Owners of adjacent lands have property rights that sometimes conflict with the scenic highway concept. Special land use planning, acceptable to adjacent owners, must be employed.

Traffic Issues. When high-speed commuters and truckers are mixed with slower-speed, roadside-viewing tourists there may be increased conflict and accidents. Roadway design standards for these two types of users may need to be different, even requiring entirely separate roadways.

Signing and Classifying. At present, there is considerable proliferation and diversity of policies on how scenic highways are classified and signed. Travelers are never sure of the policy rules and identification of attractions and driving hazards. These need to be integrated.

Community Acceptance. Because many scenic highway designations have been put forward by others, local people may not agree that the rewards (tourist revenues) are worth the added traffic and land use restrictions.

Bicycle Conflict. Because both scenic bicyclists and automobile tourists often like the same qualities of scenic highways, their use is often in conflict. More automobile traffic creates hazards for safe and slower use by bicyclists.

Funding and Management. Because scenic highways traverse many political jurisdictions and a multiplicity of adjacent property owners, questions of who will fund and manage these special routes are not easily answered.

But, even though these issues are widespread, the movement toward establishing more scenic road designation continues. For example, by 1990, there were 51,518 miles of designated and potential scenic byways in the United States, Virgin Islands, and Puerto Rico (Federal 1991, 8). All but seven states have some kind of scenic highway program. Germany has more than 70 scenic highways, including the "Castle Route," "Fairy Tale Route," and "Route of Emperors and Kings" (Federal 1991, 36).

Although the criteria for designating scenic routes vary, most are similar to those adopted by the North Carolina Board of Transportation in 1990 (Federal 1991, 13):

- They should be at least a mile long.
- The development along the byway should not detract from the scenic character and visual quality.
- There should be significant visible natural or cultural features along its borders. These include agricultural lands, historic sites, vistas of marshes, shorelines, forests with mature trees or other areas of significant vegetation, or notable geologic or other natural features.
- There should be preference for roads that are protected by land use controls.
- There should be provisions for de-designation should the character of the road change over time.

Smardon (1987) describes a survey process used in analysis of scenic highway potential for the 450–mile Seaway Trail, located along the Lake Ontario and St. Lawrence River waterfront in northern New York State. Investigative panels of 50 college students and 45 area residents, guided by professional planners, documented and evaluated the entire corridor. Positive and negative attributes, listed by priority, included the following:

Negative Attributes	*Positive Attributes*
1. Utilities	1. Views/presence of water
2. Trailer parks	2. Vegetation

3. Screening development

4. Signage

5. Excessive vegetation

6. Flat topography

7. General clutter

8. Boats, docks

9. Poor field maintenance

10. Fences

3. Natural landscape

4. Rural image

5. Water features

6. Views to opposite edge

7. Unique landscapes

8. Edge variety

9. Superior view

10. Nearness of water

11. Fences

12. Dirt roads

Arizona Example

A state program, enacted in Arizona in 1982, provides for the establishment of parkways, historic and scenic roads (ADOT 1992). As of 1992, 17 roads had been so designated. The purpose was to provide a procedure that would insure that future travelers are able to enjoy important historical, cultural, and scenic resources along highways. Key to the process is evaluation and approval by Parkways, Historic and Scenic Roads Advisory Committee (PHSRAC) for final approval and support by the Arizona Department of Transportation (ADOT). This advisory committee is made up of eleven members, including six citizen appointees named by the governor, and one each from ADOT, Arizona State Parks Board, Arizona Historical Society, Arizona Office of Tourism, and Tourism Advisory Council.

The PHSRAC reviews, prioritizes, and evaluates the requests from individuals and organizations for designation. Figure 7-5 illustrates the flow chart for designation.

Documentation of an inventory process is included in each request and application for approval. Historical road applications emphasize cultural resources, and scenic road applications emphasize natural and visual resources. Following is a list of the features for which information is to be determined and described.

Natural Resources

Geology

Hydrology

Climate

Biota

Topographic

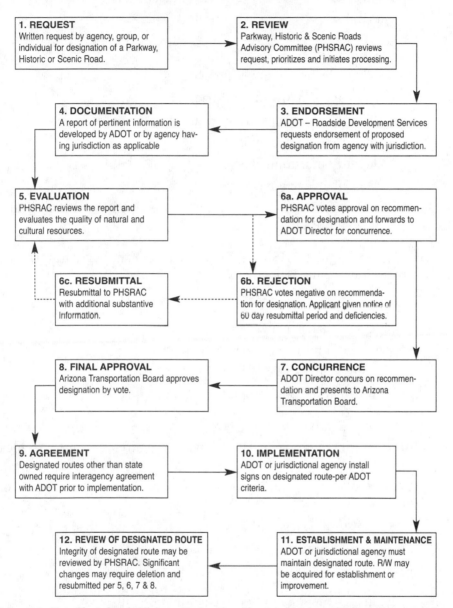

THE ARIZONA PARKWAY, HISTORIC AND SCENIC ROADS DESIGNATION PROCESS

1. REQUEST
Written request by agency, group, or individual for designation of a Parkway, Historic or Scenic Road.

2. REVIEW
Parkway, Historic & Scenic Roads Advisory Committee (PHSRAC) reviews request, prioritizes and initiates processing.

4. DOCUMENTATION
A report of pertinent information is developed by ADOT or by agency having jurisdiction as applicable

3. ENDORSEMENT
ADOT – Roadside Development Services requests endorsement of proposed designation from agency with jurisdiction.

5. EVALUATION
PHSRAC reviews the report and evaluates the quality of natural and cultural resources.

6a. APPROVAL
PHSRAC votes approval on recommendation for designation and forwards to ADOT Director for concurrence.

6c. RESUBMITTAL
Resubmittal to PHSRAC with additional substantive Information.

6b. REJECTION
PHSRAC votes negative on recommendation for designation. Applicant given notice of 60 day resubmittal period and deficiencies.

8. FINAL APPROVAL
Arizona Transportation Board approves designation by vote.

7. CONCURRENCE
ADOT Director concurs on recommendation and presents to Arizona Transportation Board.

9. AGREEMENT
Designated routes other than state owned require interagency agreement with ADOT prior to implementation.

10. IMPLEMENTATION
ADOT or jurisdictional agency install signs on designated route-per ADOT criteria.

12. REVIEW OF DESIGNATED ROUTE
Integrity of designated route may be reviewed by PHSRAC. Significant changes may require deletion and resubmitted per 5, 6, 7 & 8.

11. ESTABLISHMENT & MAINTENANCE
ADOT or jurisdictional agency must maintain designated route. R/W may be acquired for establishment or improvement.

Figure 7-5. Flow Chart, Scenic Highway Designation. The state of Arizona Highway Department has created this sequence for establishing parkway, historic, and scenic roads. Note involvement by jurisdictions involved (ADOT 1992, 3).

Cultural Resources

Architectural	Archaeological
Historical	Cultural

Visual Resources

Visual Quality Assessment Procedures

- Landscape Classification Process

    Identify on map proposed highway segments.

    Identify biotic communities.

    Identify transition zones.

    Describe vegetative cover.

- Landscape Inventory (a Visual Assessment Inventory and a Viewpoint Rating).

Criteria used by the PHSRAC for evaluating parkways, historic and scenic roads include the following:

Parkways:

One mile distance between access roads.

Meet criteria for historic or scenic roads.

Interpretive area space is available.

Controlled access and property rights obtained.

Historic Roads:

Impact of route—importance within national, state, or local framework.

Impact of area—contribution to exploration, settlement, or development.

Proximity—physical and/or visual access to historical place.

Uniqueness—relative scarcity or abundance.

Scenic Roads:

Vividness—memorability of visual impression.

Intactness—integrity, freedom from encroachment.

Unity—harmonious composite.

Important criteria and standards of operation include protection of vegetation, freedom from negative impacts, and compatible development (if any) along all designated routes. Recommended are local and county protective zoning and design review overlay along all designated highways in order to protect, maintain, and enhance the quality of the highway corridor environment. Special emphasis is placed on the establishment of pullouts for interpretation. Normal highway standards of design and construction may be amended in order to protect and enhance special features or unique resources important to such designated highways. Final approval lies with the Arizona Transportation Board.

This examination of scenic highways, their importance and planning issues, demonstrates the need for close collaboration among decision makers, local people, and designers/planners. It seems that the answer to better planning lies within each situation rather than a set of rigid standards and policies.

THE DESTINATION ZONE PLANNERS

Who plans tourism destinations? Increasingly, this is seen as a public-private cooperative venture. There is little question that this process must involve at least the following groups: tourism developers, public officials, resident groups, existing tourism businesses, organizations, and planners. This great number is needed if decisions on a planned future are to be meaningful and have a good chance of being implemented. Such a process requires a great amount of leadership and coordination, especially because several political jurisdictions are included in a tourist destination. Of the many planning groups, the following are cited for the promise they hold for effective destination zone planning.

Consultants

Increasingly, the design and planning professionals are taking on tourism planning assignments. Because of tourism's complexities, a team approach is preferred. Overall planning and integration with official planning is often assumed by professional planners. These are individuals with education and experience in urban and regional planning. Land use planning and site design, as well as detailed construction plans, are usually performed by professional landscape architects. They are individuals with training and experience in site planning. Specific structural design is usually handled by professional engineers. For the design and specific planning of buildings, architects are engaged because they have the training and experience for this specialty.

Some firms have all or most of these specialties represented by a team of staff members. Such firms are best capable of the larger scale planning projects for destination tourism. For tourism, often additional specialists are needed—golf course architect, museum design specialist, zoo designer, historical restoration specialist, archeologist, interpretive visitor center designer, market researcher. The advantage of these firms is that they can perform a wide range of services, from general studies of tourism potential to complete working drawings and from acting as a catalyst to resolve conflict to detailed specifications for project construction.

Increasingly, other types of consultant firms are doing tourism planning work. Firms traditionally focused on accounting, finance, and feasibilities have often performed tourism planning projects. These usually are directed more toward policy, economics, and marketing than to physical planning. Sometimes marketing firms become involved in tourism planning. Professors at universities are sometimes called upon to engage in tourism planning projects.

Task Forces

Task forces for tourism planning vary greatly in their makeup but provide an effective mechanism for a variety of needs. Some are made up entirely of planning professionals whereas others are composed of public and private representatives of groups interested in tourism.

For several years, the Pacific Asia Travel Association (PATA) has been co-sponsor of many tourism task force studies including planning. Generally, the request for outside overview assistance comes from the nation or city within the Pacific basin. Their needs are detailed and submitted to PATA. In response, PATA contacts four or five tourism specialists who have the competence and experience to address the issues of concern, and requests their services. PATA provides a leader. A rapporteur maintains close contact with the team and is responsible for the final manuscript of a written report. Travel expenses (but no professional fees) are provided by the host area. This is an intensive on-site evaluation with a tightly scheduled agenda. The process includes site inspection, rapid review of documents, and interviews with individuals and representatives of public and private organizations. By the end of the week, consensus on recommendations is reached. Reporting to the sponsor includes an on-site oral and a written report after the site visit. Compared to a major project by consultants this technique is much less costly and provides immediate input. It is particularly effective in resolving a smaller number of issues and those requiring no long term in-depth research.

The task force concept is also used in initial stages of planning to crystallize interest and organization. As an example, after several studies by the National Park Service and Pennsylvania State agencies, the state created a task force to develop coordination and guidelines for the industrial heritage corridor in the southwestern part of the state (*Plan for Allegheny Ridge* 1992). It was called the State Heritage Interagency Task Force and was made up of 70 representatives of culture and heritage groups, businesses, and public agencies having jurisdiction in the area. These agencies included state departments of Community Affairs, Commerce, Environmental Resources, Education, Transportation, the Pennsylvania Historical and Museum Commission, the Pennsylvania Heritage Affairs Commission, and the Pennsylvania Council on the Arts. This task force studied reports, evaluated the resource base, reviewed several alternatives, and selected a preferred approach. Allied with the task force was a consulting tcam of planners. Together, they held meetings, open houses, and workshops as planning concepts were developed. After two years, a general conceptual plan emerged. The use of a task force demonstrates an effective approach to integrate the ideas and policies of the many constituencies that often are involved in destination and regional tourism development.

Extension Tourism Planning

An effective method of public education about tourism and guidance for planning has been adult educational programs sponsored by governments and universities. In the United States, the Cooperative Extension Service (CES) was created by Congress in 1914 (Smith-Lever Act) to carry educational information directly to farmers of the nation. An adaptation of this first took place in Michigan in 1945 with the Tourist and Resort Service Program, based at Michigan State University. The program was designed to assist tourist and resort businesses with their technical problems. The program had three themes: business management, building and grounds planning, and food service. The methods used were preparation of technical bulletins, workshops, conferences, news releases, and personal on-site consultation. Through this program, carried out by three Extension Specialists (Robert W. McIntosh, Clare A. Gunn, and Gladys Knight), thousands of motels, hotels, restaurants, campgrounds, marinas, and parks benefited from better location, site and building design, operational management, accounting, financing, and food production and management for over two decades.

Many states patterned new programs after this pioneer effort. Several, such as Minnesota, have established University research and extension centers to develop new information and disseminate it to those in need. Approximately 15 of these centers are located across the United States (Gitelson 1989,9). Their main functions are: to foster multidisciplinary efforts, to provide on and off-campus educational programs, and conduct applied research. In addition, adult education specialists in many states are providing tourism technical assistance, supported primarily by governments (federal, state, local) with additional input from private organizations and businesses. Some 38 presentations have been published as *Using Tourism & Travel as a Community and Rural Revitalization Strategy* proceedings of the national workshop of May 10–12, 1989. The USDA-CES now publishes *Update Tourism & Commercial Recreation,* as a newsletter of activity by CES in tourism.

Reviewing files of accomplishments by CES specialists also reveals much activity in tourism planning. The following illustrate some of these projects.

Boundary Waters Canoe Area

In the United States, "wilderness" areas are designated and managed by the U.S. Forest Service by law of 1964. In such areas, several regulations of use apply in order to fulfill the criteria demanded by wilderness designation. Because of controversy over this change in one area, the Agricultural Extension Service, the University of Minnesota was requested to investigate the impact of these new administrative controls in the Boundary Waters Canoe Area, an area in northeastern Minnesota along the Canadian border (Blank et al. 1988). The economy, supporting the 13,000 residents, was based on timber production, iron mining, and tourism. In the Extension tradition, action to resolve issues is as important as identifying them. Following the federal legislation that designated the area as wilderness, there was much negative local reaction on two issues. Private holdings were to be eliminated and motorized vehicle access was to be prohibited. Over a period between September 1, 1979, to September 30, 1982, the Extension Service fulfilled a leadership and catalytic role to resolve issues and stimulate planning in this area. This effort included the following subprojects:

A Needs Assessment—determining needs of firms impacted by the area.

Educational and Technical Assistance to Tourism Related Firms—152 occurrences: physical plant, business management.

Educational and Technical Assistance in the Management of Community Grants—catalytic function.

Educational and Technical Assistance in Marketing Programs—improved materials, strategies, and delivery.

Education and Technical Assistance in Special Project Management—Canadian custom station closings, licensing outfitters, etc.

Applied Research Efforts—fostering data collection, management.

Building Communication Flows—integrating many agency and firm plans and actions.

The Extension role has resulted in new understanding of public and private roles, new cooperation, new investment in facilities and services, much greater visitor use, greatly increased tourist business, and a significant increase in resource protection. One example was the marketing assistance for the Lake Vermillion area, generating $360,000 in new business. Another is the case of catalytic action between U.S. Forest Service and the National Park Service for establishing The Cook Visitor Center (Simonson 1992). The Extension program in Minnesota has greatly enhanced tourism there for several decades.

Southern Kentucky

An example of constructive extension education is offered by Allan Worms, Recreation and Tourism Extension Specialist, University of Kentucky (1992). In 1986, a steering committee of representatives from a 27-county area was formed, including members of a newly formed Southern Kentucky Tourism Development Association. Midwest Research Institute prepared a Market Study that included information on unfulfilled potential, relationship to the I-75 highway corridor, natural and cultural resources, and suggested project development. These results were then disseminated by means of several extension techniques: meetings of key leaders, presentations to civic clubs, press releases, workshops. Study tours were held to expose local participants to development opportunities.

In a region of economic decline, these activities changed public attitudes, encouraged entrepreneurism, initiated new programs, and stimulated new interest in developing local resources into viable attractions. New cooperation between communities was begun, resulting in awareness of opportunities overlooked before. In addition to socio-cultural improvements, following are some of the tangible results totaling over $72 million from local and outside effort, coordinated by the Extension Specialist as a leader-catalyst.

A major first project: $14.5 million Jamestown Resort & Marina on Lake Cumberland.

Renfro Valley Music Park & Museum ($7.5 million).

Five new motels in London, one each in Corbin, Somerset, James-
town, and Russell Springs.

Sycamore Island boat sales business and several dry storage businesses.

Several new resort and/or retirement complexes.

In addition, an estimated $46.7 million in marinas and a great many
other projects are in the planning stages: houseboat construction, new
restaurants, and new festivals and events.

As a result of Extension guidance and technical information, the volume of visitors has increased greatly (one festival drew 200,000 in 1991),
stimulating the local economy with jobs, incomes, and taxes paid.

Delaware Bays

Similar in function to CES but focused on coastal areas is the Sea Grant
College Marine Advisory Service that includes tourism and recreation
among its objectives. An example was a project in cooperation with the
Delaware Department of Natural Resources and Environmental Control
to determine carrying capacity issues and recommendations for management of the state's inland bays (Falk 1992).

The study included on-site interviews of recreational users, mail ques-
tionnaire survey of shoreline residents, and reconnaissance of boating
activity on Little Assawoman, Rehoboth, and Indian River Bays.

Results indicated that:

- High boating density was creating safety hazards.
- Marine debris and litter were becoming a problem.
- Environmental degradation was taking place.

From the findings of the study several planning and management
options were put forward and reviewed by the constituency groups. Main
conclusions reached were:

- Discharges of pollutants into the bays should be stopped.
- Boat speed and jet ski use should be controlled for safety.
- Number of marinas should be limited.
- Fishing management should be improved.
- Water use should be zoned.
- Boat users should accept funding support for bay improvement.
- Monitoring and enforcement need improvement.

• Boat use and safety education programs are needed.

In many coastal locations throughout the United States, the Sea Grant Program has provided educational and technical guidance for better planning of tourism.

A current Extension tourism development program in North Carolina is described in the next chapter.

CONCLUSIONS

Several very important conclusions regarding destination tourism planning can be drawn from past approaches and examples of development. As communities look toward the field of tourism for greater economic support, they need to be aware of the issues that accompany the opportunities. Following are some of the key conclusions that may be of assistance to communities and their surrounding areas as they plan for destination tourism development.

Destination plans must be integrated with regional plans.

Communities and their surrounding areas must plan their tourism expansion within the context of regional plans. If regional tourism planning is not in place, lobbying for such planning may be the first item on the community's agenda. Destinations are dependent upon regional (federal, state, provincial) policies and action on such matters as: transportation network, national parks and protected resource areas, incentives for community tourism development, tourism promotion, and cooperative marketing and promotion.

Placeness is a fundamental of destination planning.

In today's competitive travel market, excessive replication of the same theme of tourism development dampens rather than fosters economic success. Travelers seek destinations because of the special qualities of place; otherwise why travel? Every destination has a different set of geographical factors, cultural and natural resources, traditions, relationship to markets, and different host characteristics. Analysis of these factors can lay the foundation for building upon the uniqueness of place.

Destination planning requires cooperation between community and surrounding area.

Because a destination includes cities and their surrounding areas, the planning for future tourism must include such a geographic area—not just the city. Tourism's attraction potential lies within the nearby rural area as well as the cities within a destination zone. It is likely that the abundance of cultural resources will be found within communities whereas most natural resource assets are located in the surrounding area. The logical location for most travel service businesses is within communities where they can benefit from public services and both the residential as well as the travel market. However, the reason travelers will come to the destination will depend on the quantity and quality of attractions in the surrounding area as well as within the city.

Opportunity: public-private cooperation.

At issue in many potential destinations is the lack of cooperation between the public and private sectors. Public participation is essential throughout destination planning. City and county governments are preoccupied with public services, such as water supply, waste disposal, police, education, and related functions. Essential as these are for the residents, they are of equal concern in tourism planning and development. Too often city councils believe tourism is the prerogative of business and promotion only. Even though the business sector does play an important role in tourism development, a successful destination is one in which policies and actions of both public and private sectors are complementary rather than competitive or divisive. In addition to the need for government-business cooperation on tourism planning is the equal need for them to cooperate with nonprofit organizations as they plan and make decisions on historic restoration, parks and preserves, and festivals and events.

New leadership required at destination level.

becoming clear that for effective tourism planning at the destination
ditional functions of a Chamber of Commerce or tourism
too narrow. These organizations usually focus only on
they lack representation from needed contingency
usually restricted to within the city limits.
tourism council with competent leadership is usually
uncil will need the official support from the jurisdic-

tions encompassed within the zone—cities, counties. Its membership must include representatives of government, tourist businesses, civic groups, nonprofit organizations, planners, and environmentalists. Planning guidelines have a much greater chance of implementation if these influencers and decision makers are represented.

Engage in destination planning.

Destination areas will be less successful, less fulfilling to visitors and less sensitive to environmental stress if not planned in an orderly step-by-step sequence. A combination of both approaches—planning project and continuous planning—is likely to produce best results.

Although planning processes vary, all have similar basic elements: setting goals, objectives; analyzing present situation; identifying issues, constraints, opportunities; developing alternative concepts for development; and identifying action strategies. This is the only way that communities can foster sustainable tourism development in a manner satisfactory to everyone.

Prevent environmental degradation.

Because the majority of pollution of air and water and erosion of land resources occurs within and around cities, the challenge for tourism planners is to foster environmental improvement and continued protection.

This responsibility at the destination scale must address all sources of environmental degradation, not only from tourism. Municipal sewage and industrial waste must be brought under control if tourism is to thrive. Land use regulations are necessary to avoid overdevelopment, excessive congestion, and incompatible development. They are necessary to protect environmental resources, so critical to tourism's success.

Utilize special cultural and natural resources.

A search of natural and cultural resources within a destination can reveal abundant opportunities for tourism development. Because they are anchored to a specific location they have greatest competitive potential. When these opportunity ideas have been named, local leaders can evaluate their impact on the community and create feasibilities. Then it is essential to gain public participation. The results of a logical planning process can be beneficial to all.

DISCUSSION

1. Because a tourism destination encompasses both a community and surrounding area, it covers several political jurisdictions. Discuss how barriers to cooperation can be overcome.

2. Why is the concept of place so important in destination planning?

3. Discuss why some communities may not opt for any tourism development. If too many destinations develop tourism what are the consequences?

4. If public-private cooperation is so essential, what are the barriers and how can they be overcome?

5. Discuss the several cited destination tourism guides to planning and a step essential today that generally was not emphasized or was even omitted in earlier guides. Justify your conclusions.

6. Although leadership can come from a Chamber of Commerce or a special Tourism Council, discuss which may have the greatest scope for destination tourism planning and management.

7. Discuss why even the best destination plan needs updating from time to time.

8. Are destination tourism plans generally incorporated into the scope of city planning departments and what are the consequences?

9. Discuss planning solutions to the many issues inherent in creating scenic highways.

10. Discuss the limitations and advantages of the several kinds of planners, especially professional designers/planners.

REFERENCES

Addison, Liz (1996). "An Approach to Community-Based Tourism Planning in the Baffin Region, Canada's Far North." In *Practicing Responsible Tourism*. L. C. Harrison and W. Husbands, eds. New York: John Wiley & Sons, pp. 296–312.

Alberta Tourism (1988). *Community Tourism Action Plan Manual*. Edmonton: Alberta Tourism.

Arizona Department of Transportation (ADOT) (1992). *Application Procedures for Department of Parkways, Historic and Scenic Roads in Arizona*. T.D. Walker, ed. Phoenix: ADOT.

Beatley, Timothy and Kristy Manning (1997). *The Ecology of Place*. Washington, DC: Island Press.

Blank, Uel et al. (1988). "Contributing to Tourism Industry Vitality of a Natural Resource Based Region Through Educational/Technical Assistance." *Staff Papers Series*, P83–20. St. Paul, MN: University of Minnesota.

Blank, Uel (1989). *The Community Tourism Industry Imperative: The Necessity, the Opportunities, Its Potential.* State College, PA: Venture.

Bonoma, Thomas V. (1984). *Managing Marketing Text, Cases, and Readings* (Teacher's Manual). New York: Free Press, p. 14.

Cheek, Neil H. Jr., and William R. Burch, Jr. (1976). *The Social Organization of Leisure in Human Society.* New York: Harper & Row.

Dredge, Dianne and Stewart Moore (1992). "A Methodology for the Integration of Tourism in Town Planning." *Journal of Tourism Studies,* 3 (1), pp. 8–21.

Durrell, Lawrence (1969). "Landscape and Character," In *Spirit of Place,* A.G. Thomas, ed. New York: E.P. Dutton.

Edgell, David (1990). *Charting a Course for International Tourism in the Nineties.* Washington, DC: U.S. Travel and Tourism Administration.

Falk, James et al. (1992). "Recreational Boating on Delaware's Inland Bays: Implications for Social and Environmental Carrying Capacity." Prepared for Inland Bays Estuary Program, Delaware Department of Natural Resources and Environmental Control, Division of Water Resources. Lewes, DE: University of Delaware Sea Grant College Program.

Federal Highway Administration (1991). *National Scenic Byways Study.* (Pub. PD-91-010). Washington, DC: U.S. Department of Transportation.

Ferrario, Franco F. (1979). "The Evaluation of Tourist Resources." *Annals of Tourism Research,* 17 (3): 18–22/24–30.

Gitelson, Richard (1989). "What's Happening with Centers in the U.S.A.?" In *Using Tourism & Travel as a Community and Rural Revitalization Strategy.* Proceedings, National Extension Workshop, May 10–12, Minneapolis: University of Minnesota.

Go, Frank et al. (1992). "Communities as Destinations: A Marketing Taxonomy for the Effective Implementation of the Tourism Action Plan." *Journal of Travel Research,* 30 (4), pp. 31–37.

Gradus, Yehuda and Eliahu Stern (1980). "Changing Strategies of Development: Toward a Regopolis in the Negev Desert." *APA Journal,* 46 (4), pp. 410–423.

Green, Peter (1986). "New Hampshire Adapts Interstate." *Engineering-News Record,* Nov. 20.

Gunn, Clare A. (1972). *Vacationscape—Designing Tourist Regions.* Austin, TX: Bureau of Business Research, University of Texas.

Gunn, Clare A. (1982). "Destination Zone Fallacies and Half-Truths." Presentation, "International Conference on Trends in Tourism Planning and Development." Surrey, U.K.: University of Surrey.

Gunn, Clare A. (1988). *Vacationscape: Designing Tourist Regions.* 2nd ed. New York: Van Nostrand Reinhold.

Gunn, Clare A. (1994). *An Assessment of Tourism Potential in Newfoundland and Labrador.* Prepared for Hospitality Newfoundland and Labrador and Canadian Heritage. Conference Proceedings, September 9–19.

Holder, Jean S. (1996). "Maintaining Competitiveness in a New World Order." In *Practicing Responsible Tourism.* L. C. Harrison and W. Husbands, eds. New York: John Wiley & Sons, pp. 145–173.

Joppe, Marion (1996). "Everything Must be Connected to Everything Else." In *Practicing Responsible Tourism*. L. C. Harrison and W. Husbands, eds. New York: John Wiley & Sons, pp. 313–329.

Koth, Barbara et al., eds (1991). *Rural Tourism Development Training Guide*. St. Paul, MN: Tourism Center, University of Minnesota.

Maguire, Meg (2002). Communication from Meg Maguire, President, Scenic America, Washington, DC.

Morris, Philip (1990). *Southern Places*. Birmingham, AL: Oxmoor House.

Motloch, John L. (2001). *Introduction to Landscape Design*. 2nd ed. New York: John Wiley & Sons.

Murphy, Peter (1985). *Tourism: A Community Approach*. New York: Methuen.

Pannell Kerr Forster (1985). *Calgary-Canmore Tourism Destination Area Study,* Vols.1 and 2. Edmonton: Travel Alberta.

Plan for the Allegheny Ridge (1992). Prepared by the Allegheny Ridge Industrial Heritage Corridor Task Force, Hollidaysburg, Pennsylvania.

Palermo, Frank (2000). *Tourism and Community Development: Case Studies*. Halifax, Nova Scotia: Cities and Environmental Unit, Dalhousie University.

Palermo, Frank (2001). *Tourism and Community Development: An Approach*. Halifax, Nova Scotia: Cities and Environmental Unit, Dalhousie University.

Ruest, Gilles (1979). *The Tourism Destination Concept*. Ottawa: Canadian Government Office of Tourism.

Shepard, Paul (1967). *Man in the Landscape*. New York: Alfred A. Knopf.

Simonson, Lawrence (1992). Personal correspondence, August 10, former Extension Specialist of Tourism, Grand Rapids, Minnesota.

Smardon, Richard C. (1987). "Visual Access to 1,000 Lakes." *Landscape Architecture*, 77 (3), pp. 86–91.

Steiner, Frederick (1991). "Landscape Planning: A Method Applied to a Growth Management Example." *Environmental Management,* 15 (4), pp.519–529.

Stewart, George R. (1967). *Names on the Land*. Boston: Houghton Mifflin.

Tyler, Duncan and Yvonne Guerrier (1998). "Conclusion: Urban Tourism—The Politics and Processes of Change." In *Managing Tourism in Cities*. D. Tyler, Y. Guerrier, M. Robertson, eds. New York: John Wiley & Sons, pp. 229–237.

Victurine, Raymond (2000). "Building Tourism Excellence at the Community Level: Capacity Building for Community-Based Entrepreneurs in Uganda." In *Journal of Travel Research* 38 (3): pp. 221–229.

Weaver, Glenn D. (1991). *Tourism USA*. Washington DC: U.S. Travel and Tourism Administration.

Wight, Pamela (1992). "Tourism-Recreation EIAS in Alberta: A Need for an Integrated Approach in Legislation, Environmental Assessment, and Development Planning." Presentation, 12th International Seminar on Environmental Assessment and Management, July, University of Aberdeen, Scotland.

Worms, Allan J. (1992). "The Southern Kentucky Tourism Development Project." presentation, "Partners in Rural Development" conference, Washington, DC.

Zimmerman, Erich (1933). *World Resources and Industries*. New York: Harper.

Chapter 8

Destination Planning Cases

INTRODUCTION

Although the term destination is often used with several meanings, from a nation to a relatively small area, it is interpreted here to mean a community (or cluster of several) and surrounding attractions. Examples of "discovery" of potential destination zones were presented in chapter 6.

The attractions within a destination zone are located in the surrounding area as well as within the community. Although destination-type markets sometimes engage in their total activities *in situ,* such as at beach resorts or casinos, they often take trips to surrounding attractions as well. Conversely, remote attractions, such as national parks, depend on the nearest community for basic lodging, diversity of food services, travel services, communications, health services, and shopping.

As the more stringent politics of nations around the world began to lessen, communities, large and small, began to strive for control of their own destinies. Generally, their first motivation for tourism development was economic improvement with emphasis on marketing. Then followed a variety of planning approaches from the use of planning models to the creation of their own. No matter the approach, a great many community areas have truly become tourism destinations.

In the search for cases to be presented here, the criteria for selection included a variety of planning processes, a wide geographic distribution, new sensitivity to the environment, participatory planning, and community integration. Wherever available, an evaluation of planning progress is noted. Examination of even this small sampling of cases reveals many lessons of planning tourism destinations.

273

HOPI, ARIZONA

The tourism development of the Hopi Reservation in Arizona reveals successes and difficulties as the people balance cultural protection with tourism. Several sources offer insight into a case that is similar to many others around the world where native peoples are striving toward the goal of a better economy through tourism (Hopi Tribe 1997, Kelly 2001, Schroeder 2001, Shifflet 2000). Over a period of many years, the people of the 1.5 million-acre reservation have maintained their cultural integrity at the same time they have adapted to gradual tourism development. Although visitor levels are small compared to other destinations in Arizona, such as the Grand Canyon and Phoenix, approximately 150,000 tourists came to the Hopi Reservation in 1991. About 9,000 Hopi and 1,000 others live mostly within the twelve villages.

Rich in history and culture, this area's settlement predates all other surviving early cultures of the United States. Natives recognize that in addition to many physical assets, their dominant attraction is traditional culture, also their most vulnerable resource. Although the top visitor activity is seeing the spectacular scenery of arid landscapes and scenic canyons, visiting national and state parks, visiting historic sites, and shopping for crafts are also important. About 25 galleries now offer high quality indigenous creations of Katsina Dolls, pottery, baskets, and weaving.

Today's visitors represent a high scale segment of sophisticated travelers seeking educational as well as enjoyable experiences. Most are over 50 years of age, have incomes of over $50,000, and have college degrees. About 73 percent are from outside Arizona and many are retired. A significant number have visited before and personally know individual Hopis. The majority come on arranged tours. These tourists are typical of a modern trend toward cultural tourism.

Recent meetings and studies locally are providing insight into their accomplishments and problems. They take delight in hosting visitors in their homes and yet there are many instances of invasion of sacred moments, rudeness, and unacceptable behavior. At the same time they seek economic improvement, interviews show that cultural traditions and religious privacy must not be sacrificed for tourism. An additional concern is the erroneous information given tourists by tour companies before they visit. Some areas, such as First Mesa Consolidated Villages, have become active in providing their own guiding and interpreting. But other villages prefer not to have visitors and to maintain their own peace and quiet. Educational seminars reached the conclusion that each village should control its own level of tourist acceptance—how much, when, and where to drive, park, and walk.

Members of the tribe are divided on the issue of closing villages to visitors during special native events, such as Katsina ceremonies. Some fear such action would depress tourism too much whereas others believe these events give tourists a better understanding of their culture. As they plan ahead for tourism development, a major recommendation was revitalizing the Hopi Cultural Center. Included would be a learning center, demonstrations by artisans, visitor information on ethics and behavior, and a directory of guides. Also recommended was adding more accommodations and other tourist services within the reservation.

Other elements within their tourism development plan include:

- greater cooperation with Hualapai, Apache, and Navajo tourism interests
- create a new Hopi tourism vision
- agree on the role of the tribal government with reference to tourism
- better signage policies
- designate highway 264 a scenic/historic route
- create new scenic and archeological parks
- invest in off-reservation services for tourists.

They have created their own statement of tourist ethics that is a model for other destinations of native cultural tourism:

- Recording of any type is strictly prohibited, especially during ceremonies, while in and around Hopi villages. This includes, but is not limited to: picture-taking, video recording, audio recording, sketching, and note-taking. Visiting Hopi is a wonderful time to use your mind and heart to record what you are privileged to see.

- Publication of your observations and/or any recordings is both exploitative and prohibited without prior consent.

- Wear appropriate clothing, just as you would when going to a formal or religious ceremony, you should consider what you wear when you go to a Hopi ceremony. Shorts, short skirts, and tightly fitted clothes are not recommended. Also, it is preferable that you do not wear a hat at the *Katsina* Dances.

- Please note that not all ceremonies are open to the public. Often, posted signs indicate who is welcome. If no signs are posted, seek information from local shops or the village administration.

- At ceremonies open to the public, be aware that there are behavioral guidelines to follow. Do not interrupt ceremonies. Well-meaning people have nonetheless interrupted, distracted, or simply got in the way at Hopi ceremonies.

- Unless you are specifically invited, the simplest rule is to stay out of *kivas* (ceremonial rooms) and stay on the periphery of dances or processions.

- Do not touch, especially if you are not sure. Some shrines are more easily recognized than others. Hopi spirituality is very intertwined with daily life, and objects that seem ordinary to you might have deeper significance to the person who placed them. Shrines are placed by sincere individuals and should not be disturbed. If you come upon a collection of objects at Hopi, respect the wishes of the person who left the offerings and do not disturb them.

- Do not remove any articles such as potsherds or other artifacts. Possession of such items may subject you to federal prosecution. All archeological sites on the Hopi Reservation are protected by federal laws and Hopi tribal ordinances.

- Law on the Hopi Reservation strictly prohibits alcoholic beverages and drugs.

The Hopi are demonstrating that they are able to plan and guide their efforts toward sustainable tourism that protects traditional cultural elements as economic and social rewards are sought. They are striving toward integrated solutions to their tourism development issues.

BOUCTOUCHE BAY, NEW BRUNSWICK

The community of Bouctouche Bay, New Brunswick, Canada, has demonstrated successful participatory planning of tourism integrated with overall community growth (Bouctouche Bay Ecotourism Project 1996). The Vision Statement:

> To lead as the model coastal community of Atlantic Canada in developing a dynamic rural economy, based upon local capacity building, coastal management, and sound principles of sustainable development.

Characteristics of the project include:

> Economically viable tourism planned in harmony with the natural and cultural environments.

> Understanding and responding to potential benefits and adverse impacts on social (community and cultural) and bio-physical environments.

> Respect for community attitudes and aspirations, celebration of culture and heritage.

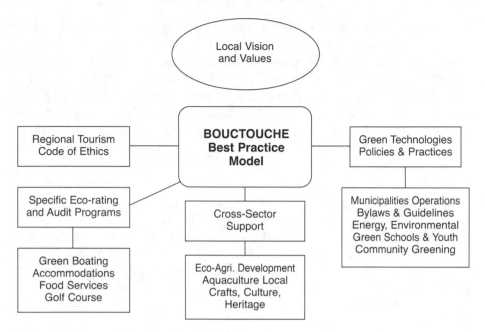

Figure 8-1. Bouctouche Bay Planning Model. This model is especially important because of its inclusiveness. Tourism is seen as an integral part of the entire community and its surrounding area (Bouctouche Bay 1996).

Respect for tourism and other stakeholders, and the community in the planning process.

Environmentally sensitive design and development guidelines.

The benefits of tourism are widely spread throughout the community.

Figure 8-1 illustrates the planning model for this innovative project.

Begun in 1996, the Bouctouche Bay Ecotourism Project has obtained worldwide acclaim because of its public-private partnerships and sensitivity to resource protection and interpretation for tourism. The Bouctouche Watershed includes several environmental zones that have special physical and cultural attributes that lend themselves to visitor enrichment. Outstanding accomplishments from this project in only three years include over $8 million investment in the Irving Eco-Centre, a 17-kilometer natural trail network, Sawmill Point Boat Basin & Park, rehabilitation of the former Marine Site at the Rotary Nature Park, and the Bouctouche Farmer's Market-Inner Bay Market Centre.

Early on, local educational workshops directed toward the vision goal and involving local "entrepreneurial/community groups" from Atlantic Canada were held throughout the area. These sessions included leadership by professionals, field trips, and a sharing of ideas for the future. These resulted in benchmarks for the best practices as applied to "green

boating, green hoteliery, eco-agri-development, and sustainable tourism."
Strong local leadership and community involvement have included key
groups such as the Chamber of Commerce, Rotary Club, governing bodies, and significant associations. The planning process was open to local
stakeholders, community groups, government, industry, aboriginals, and
others in order to tap their vision, expertise, and energy in an open forum
and democratic manner. The purpose was to enhance the quality of life in
the entire watershed. Included is a monitoring system to evaluate impacts
of developmental change on the environment.

Evidence of positive change from 1996 to 1999 is demonstrated with
increased visitors, public investment, private investment, new employment, and facility growth. Visitors to the Information Centre have increased 185 percent, to Pays de la Sagouine 36 percent, and Irving
Eco-Centre 85 percent. Public investment is over $9 million and private
investment has increased from $440,000 to $1.5 million. Employment
has grown from 321 to 437. The accommodation sector units have
increased 59 percent, campsites now total 373, and five bed-and-breakfast inns have been added.

The key features of this project include input from professional planners, local participation, a clear vision statement, and monitoring environmental indicators as progress in implementation takes place.

SANIBEL ISLAND

Islands of the world are among the top tourist destinations and yet are the
most vulnerable to overdevelopment. An outstanding case of an island
that several years ago recognized the potential for depletion of its essential resources from tourism is Sanibel Island (Bosselman 1999, Clark
1976, Plan 2001, Rogers 2001, Vision Statement 2001). It is a crescent-
shaped barrier island of about 20 square miles off the west coast of
Florida. The island is about 12 miles long with a coastline of approximately 31 miles. Of this coast, over 14 miles are beautiful sandy beaches.
The natural and cultural resources are its major appeal and yet have
already experienced considerable encroachment. Originally occupied by
Calusa Indians, after the late 1800s agriculture became the dominant land
use. Development was slow following complete inundation of the island
by salt water during the hurricane of 1926.

Following the construction of a causeway to the mainland in 1963, a
boom in development of homes and tourism began. By 1967 the environmental devastation from uncontrolled development was clear, causing
many protests against further development. But proposals for huge
expansion of apartments, resorts, and motels kept coming. Even the Lee

County Planning Commission's zoning in 1972 would have allowed a population growth of 90,000.

In spite of opposition by the Chamber of Commerce, a referendum vote in 1974 favored incorporating Sanibel as a city in order to exercise stronger planning and environmental control of development. The intent was not to stop development but to guide it in ways that would retain the local quality of life, visitor appeal, and protect the environment as well.

The Sanibel Plan

A first step by the city was to engage the land planning firm of Wallace, McHarg, Roberts, and Todd (WMRT) to create a vision statement and a plan that would guide development and yet maintain the island's quality. A principal in this firm of landscape architects and planners, Ian McHarg, had become well known for his environmental planning advocacy expressed in his book of 1969, *Design with Nature*. Along with many other organizations that participated in the development of a plan was the Conservation Foundation. Their studies of resource foundations included analysis of the island's ecosystems, identification of principal ecological zones, diagnosis of the condition of these zones and their functions, and recommendations for management to conserve the island's resources.

This case illustrates a true participatory planning process involving citizens, government officials, and consultants. In addition to the input from the design/planning firm of WMRT, engineers, traffic specialists, scientists, attorneys, and economic specialists provided valuable assistance. A diagram of the planning process is illustrated in figure 8-2.

After many months of meetings, discussions, and detailed resource and management studies, The Sanibel Plan was prepared and finally adopted in July 1976. The major segments of the Plan were:

Article 1—Preamble

Article 2—Elements of the Plan: safety, human support systems, resource protection, intergovernmental cooperation, land use

Article 3—Development Regulations: definitions, maps, permitted uses, subdivisions, mobile homes and RVs, floor and storm proofing, site preparation, environmental performance standards, paved surfaces

Article 4—Administrative Regulations: authority/purpose, definitions, standards, permits, amendments to plan, notice of hearing, and Council action

The Vision Statement of this Plan included four major categories (Vision Statement 2001). Within the topic of *Sanctuary* is insistence upon retaining

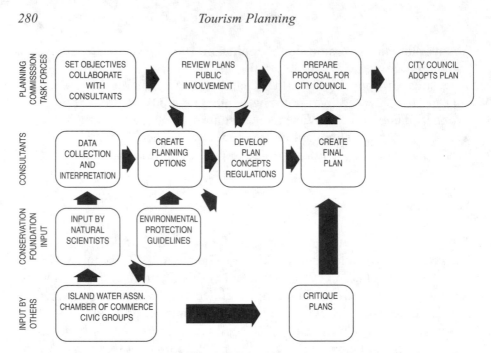

Figure 8-2. Planning Process, Sanibel Island, Florida. This area is an excellent example of comprehensive community planning that includes tourism and the environment. The process involved cooperative input from professional designers/planners, government officials, environmental organizations, civic groups, chamber of commerce, and general public involvement (Adapted from Clark 1976, 88, 89).

the integrity of the island's assets and vigilance in resisting excessive population growth. Opposed are human activities in areas that threaten plant and animal habitats and the surrounding aquatic ecosystems.

Included in the topic of *Community* are stressed five important elements: maintain its social, cultural, ecological, and economic diversity; foster quality of the island's amenities including its history, beauty, culture, and natural systems; preserve its small town uniqueness of place; retain the special rural character of the island; and continue a strong public-private land ethic that provides stewardship and accepts responsibilities of well-planned husbandry of its natural and human resources.

The Sanibel community recognizes its power of *Attraction* but will exercise all means of controlling development for visitors and residents in order to perpetuate this attractiveness.

The overarching factor is retention of the *Hierarchy of Values* needed to maintain the dominant principle of Sanibel's sanctuary quality.

Especially important is Part 3.2 of the Plan, "Protection of Natural, Environmental, Economic, and Scenic Resources." This very comprehensive section of the Plan emphasizes policies and regulations adapted to the island and within coastal management controls as stated in the Florida Administrative Code.

A major element of this section is recognition of how its intrinsic characteristics provide important functions, and at virtually no cost: buffers storm winds and flood tides; stabilizes shorelines; purifies water; and maintains a fresh water aquifer that supports a rich wildlife population and lush vegetation. These functions in turn support the health, safety, and welfare of residents and visitors.

Because environmental sustainability, for tourism as well as other development, was a major goal, six ecological zones were identified on Sanibel: Gulf Beach, Gulf Beach Ridge, Freshwater Wetlands, Mid-Island Ridge, Mangrove Forest, and Bay Beach. The Plan described in detail the main functions of each, such as flood protection, water quality control, wildlife habitat, and shoreline stabilization. Figure 8-3 illustrates the location of these and their subdivisions.

This information provided the planning foundations for all future development—residences, shopping centers, resorts, roads, recreation, conservation, offices, schools, marinas, and other land uses.

A very important issue of planning here was recognition of evacuation during a natural disaster. The only egress from the island is over the two-lane highway and causeway to the mainland. For aesthetic reasons, local policy prevents road widening, limiting the capacity of off-island flow. The City's Emergency Management Plan includes traffic control and also post-disaster redevelopment plans. All these measures are intended to protect the many assets of the island with emergency plans.

Recognized in the Plan are the many plant and animal resources. In spite of its small size it supports a surprising number of vegetative species, contributing greatly to its aesthetic appeal. Although some damage has been done by land clearing and introduction of exotics in the past, a great amount of restoration has taken place. The abundance of mangrove areas along the shoreline provide marine habitat, wave buffer, and scenic interest for much of the island. Cordgrass and sawgrass provide food and refuge for ducks, songbirds, and mammals. Sea oats, sea purslane, railroad vine, and elder are abundant behind the sand dunes.

Although most barrier islands are susceptible to saltwater intrusion into freshwater aquifers, this has not yet been as issue on Sanibel. Its river, ditches, ponds, real estate lakes, and borrow pits continue to replenish the aquifer. Vegetative cover assists in protecting this source. The capacity of this aquifer is a critical limitation to development growth.

Another attraction is the abundance and diversity of wildlife. Seashell collecting is very popular but live shell collecting is prohibited within the city limits. The area supports over 280 species of birds, 70 species of marine fish, and 48 reptiles and amphibians. The Plan's restriction on overdevelopment tends to maintain a wildlife and vegetative balance.

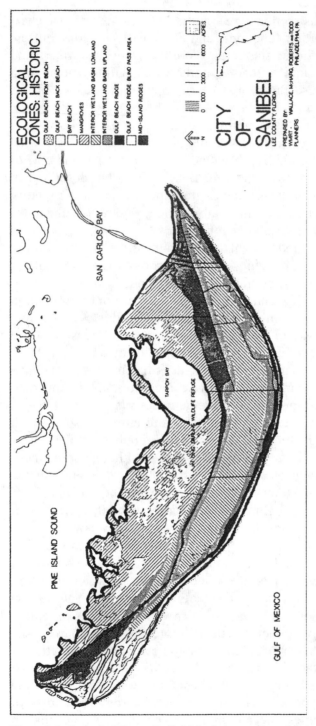

Figure 8-3. Ecological Zones, Sanibel Island. As an aid to planning for land use as well as environmental protection, research by the Conservation Foundation revealed these ecological zones (Wallace, Roberts & Todd, by permission, 2002).

A major attraction is the 6,300-acre J. N. "Ding" Darling National Wildlife Refuge. A former chief of the U.S. Biological Survey, Jay Norwood "Ding" Darling saw the need to preserve this unusual natural habitat and scenic resource. A five-mile drive allows the visitor to view alligators, roseate spoonbills, snowy egrets, great blue herons, and other wildlife. The coastal edge is ringed with mangrove trees that spread by trapping sediment.

The Sanibel-Captiva Conservation Foundation owns more than a square mile of land that is managed for conservation purposes. The 1976 Comprehensive Land Use Plan initiated regulations and performance standards that control development. This policy is reinforced by the goal statement: "Protect and appropriately manage Sanibel's natural resources to ensure conservation of ecosystems by maintaining air quality, water quality, native vegetation, and native habitats and species diversity."

Of interest to both visitors and residents is the island's history. Due to the efforts of the Historical Preservation Committee and the Sanibel Historical Society, many remnants of the past have been retained. Examples are the Lighthouse Keeper's Quarters and Old Town Sanibel Lighthouse and Boardwalk (1884), the Bailey House (1896), and the Cooper Homestead (1891).

Many regulations have provided backup for both development and environmental protection. Setback regulation protects the scenic appeal and storm protection of the shoreline as well as fosters marine life. Scenic protection is further enhanced by a building height limitation of forty-five feet. Commercial development is fostered at nodes, leaving scenic open space in between. Access roads are designed for their landscape beauty and parking areas are of shell rather than impervious paving. Signage and outdoor lighting are minimal and integrated with the environment. Much of the development is screened by native vegetation.

In the city's first two decades, a reasonable balance has been maintained between development and preservation as well as the rights of individuals and public controls.

This case is cited in detail because of its many planning lessons. Environmental sustainability would not have been accomplished without early concerted effort and leadership by the many constituencies that influence tourism and development—residents, resorts, resource protectionists, and civic leaders. Too often the planning of tourism destinations exposes conflicts and issues that become polarized, environmentalists against developers. This example demonstrates strong catalytic action by a private landscape architectural firm and the city planning department that resolved major controversy and adhered to common goals. Finally, the Plan has not met the fate of so many—lack of implementation. In these first two decades of the city's life, a reasonable balance has been maintained between development and resource protection. The planning process continues to involve political and participatory action as changes take place in residential and travel market interests.

GASCOYNE, AUSTRALIA

An important and complex issue in all planning, especially for tourism, is the implementation step. The case of the Gascoyne area in northwestern Australia illustrates how excellent planning concepts can encounter difficulties in the final action step. This is an area of approximately 600 by 300 kilometers along the Indian Ocean coast. It was identified as a political jurisdiction by the Regional Development Commission Act of 1993 to foster economic development. It has a population of about 9000, 15 percent Aboriginal (Wood 2001).

A major planning effort toward environmentally-based tourism development for the area was developed by Ross Dowling (1992). The World Commission on Environment and Development (WCED 1987) advocated integration of development and environmental protection in 1987 but tourism implications were not mentioned. Based on the need, Dowling initiated the Environmentally Compatible Tourism Planning Framework (ECT), as illustrated in figure 8-4.

This ECT planning process places emphasis on two main themes. A fundamental is the identification of environmental characteristics, especially those that are critical. Another essential is community participation that includes environmental education and land use zoning. Note the importance of feedback in this model.

This process was applied to the Gascoyne area, already an important tourism destination attracting about 201,600 tourists in 1991. The main appeals, most popular in the cooler periods from June to September, were its climate, fishing, and many natural resource attractions. The bottlenose dolphins of Shark Bay attract more than 100,000 visitors a year.

Important environmentally sensitive nodes and land use conflicts were identified. These included areas essential to ecosystem stability, rare or fragile features, and those needing protection but amenable to environmentally compatible outdoor recreation. Surveys of local residents revealed that they were as much concerned over resource protection and environmental gains as were tourism proponents. Many argued for better management of existing parks.

Also identified were potential land use areas with access, capable of supporting services and compatible attractions. Recommended were not only touring routes but also areas where tourism development would cause environmental stress.

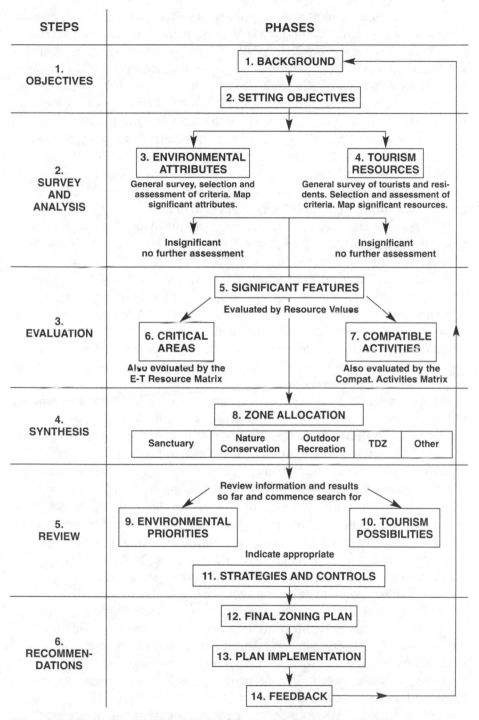

Figure 8-4. Environmentally Compatible Tourism Planning Framework. This destination planning process identifies land use zones based on setting objectives and making a thorough analysis of the environmental resources. Note feedback at all steps. This process was applied to the Gascoyne region in Western Australia (Dowling 1992, 5).

Even though this plan was originally approved by the Western Australian Tourism Commission (WATC), implementation has been difficult.

A later plan was proposed and the participatory process became the subject of a study by Wood (2001). In 1989 the Industries Assistance Commission Inquiry (IAC) identified the importance of the natural environment for tourism growth. In 1992 the Council of Australian Governments highlighted the need for better tourism planning and creation of strategies. Equally significant in this period was new awareness of cultural tourism, especially Aboriginal. Apathy was challenged by market studies that proved traveler desire to learn about this important ethnic group. A similar political native-settler issue has occurred in many nations, including the United States and Canada.

In conflict here was an interpretation of the Australian Constitution that denied all rights to Aboriginals. Until 1967 land use in the Gascoyne was dominated by pastoral leases in which Aboriginals became virtual slave labor and were denied a vote. Amendments to the Constitution in 1967 gave them voting privileges and the right to equal pay, putting the economy and politics into chaos. Gradually the Gascoyne moved into fishing, salt production, horticulture, and tourism.

Based on the rich natural and cultural assets, the Gascoyne Development Commission (GDC) finally obtained a grant for a pilot project for a regional ecotourism strategy. The intent was not only to build a stronger tourism economy but also to foster collaboration among the many governments, tourism developers, community groups, and the Yamatji (Aboriginals). The Gascoyne Regional Ecotourism Strategy was intended to allow participation by all those impacted by tourism. The planning process included many meetings by diverse representatives, often the very first time they had met. The goals were: environmental preservation, community participation including Yamatji, education, training, development of projects of best practices and demonstrations, and political collaboration among the many public and private institutions. Essential was encouragement of local areas to be involved in the process. However, work toward these desirable goals exposed the difficulty of gaining agreement among these many groups.

A detailed analysis of the participatory process by Wood (2001) involved interactions with 60 public servants, 60 tourism operators, 20 pastoralists, 30 Aboriginals, and 30 others. The study also encompassed general observation and review of pertinent documents.

Although some progress on tourism planning was made, his study revealed five important dilemmas:

1. Conflict arose between centralized and local knowledge, especially on travel market data. Local experience was denied by the Western Australian Tourism Commission (WATC). The State favored its general pol-

icy of promoting primarily group travelers whereas the locals preferred specific market segmentation.

2. Local resentment of too much governmental intervention in tourism became an issue. It created such divisiveness that one member of the Steering Committee refused to appear at a final meeting and was accused of undermining the project.

3. Even though Aboriginals had been given land rights, these rights were resisted by the government, hampering progress on the Strategy. Anger against the State often surfaced in meetings. The Strategy process showed stronger cooperation between the region and the Commonwealth than with the State.

4. Conflict arose between pastoralists and Aboriginals, to the point of pastoralists seeking removal of these natives from the process of the Strategy.

5. Specific proposals for tourism became issues and conflict arose between local concerns and the government over environmental degradation. The Strategy revealed the inherent problems associated with political and agency power overriding local needs and interests in participatory planning processes.

The Strategy was initially adopted by the executive of the Gascoyne Tourism Association but was later rejected by the Association's membership following what was purported to be intense lobbying by the two pastoralists who had expressed their objections earlier. However, it was concluded that in spite of these obstacles, over half of the recommendations had been implemented within the next four years. This case is cited here not to suggest culpability of any individual or organization but rather to demonstrate the extreme complexity and breadth of tourism planning and development. The example reveals that in spite of some difficulties, participatory planning remains a viable process for obtaining input from many levels of interest, especially local experience. Without a means of communication, issues can become polarized and collaboration on plans for the future may be difficult for many years.

THE MINERAL WELLS AREA

In 1984, the Mineral Wells (Texas) Chamber of Commerce Tourism Committee wished to discover if it had opportunities for tourism development. The city has an approximate population of 20,000 and is located in Palo Pinto county, with an area of 948 square miles. It is located in north Texas, about 50 miles west of Fort Worth. The Texas Agricultural Extension Service, Recreation and Parks Program, responded by sponsoring a project to study the potential (Gunn, Watt, Frost 1985). Specialists from Texas A&M University and graduate students performed the study in collaboration with a local Liaison Committee representing a diversity of local citizenry.

The project process was as follows:

1. Setting goals.

2. Identification of existing markets and resources.

3. Identification of key obstacles to tourism development.

4. Concepts for problem-solving.

5. Recommendations:

    A. Development: downtown, environs, countryside.

    B. Program: activities, information, promotion, research.

    C. Organization and development strategy.

1. Goals

The goals agreed upon by the planning team and Liaison Committee were:

1. Increasing economic impact (jobs, incomes, taxes paid).

2. Enhancing visitor satisfactions (more things to see and do and better orientation to visitors).

3. Protecting the basic cultural and natural resources of the city and surrounding area.

2. Identification of Markets and Resources

Travel *markets* were identified as primarily those coming for business, personal reasons, recreation in area, and passing through. (The study scope did not provide for new market surveys but interviews with motel operators and other business people provided some insight into travel markets.) The area is within 100 miles of 4 million people and over 15 million live within a day's drive.

Natural resources, such as the wooded hills in and around the city of Mineral Wells, offer an attractive scenic contrast to the dominantly flat ranch land in the surrounding area. The setting provides good wildlife habitat—whitetail deer, turkey, dove, quail, rabbit, and squirrel. The climate is subtropical and subhumid, supporting outdoor recreational activities nearly year round. The area receives about 70 percent of available sunshine. The Brazos River winds through the entire area, providing recreational opportunity as well as landscape beauty.

The *cultural resources* are dominated by historic assets. Several Indian tribes, including Ioni and Comanche, inhabited this area at one time. The settlement era left many scars of Indian conflict and frontier hardship. Hardy and vigilant ranchers and trail drivers, such as Goodnight and Loving, molded here a significant chapter of Texas history. Mineral Wells experienced two growth booms. In the early 1900s, the city became one of the most prominent health spas in the Southwest. Many artifacts of this era remain. The second boom came with Fort Wolters, providing the majority of U.S. trained helicopter pilots for three decades. The human resources show initiative, business ability, and an increasing motivation to enhance the economy and quality of life.

3. Identification of Obstacles

Through several workshop meetings, the Liaison Committee and planning team identified the following as the key restraints to tourism development:

Insufficient attractions.

Low level of attractiveness/image.

Low level of public sector involvement.

Insufficient information.

Insufficient promotion.

Lack of research data.

Insufficient organization. (Gunn et al. 1985, 15)

4. Concepts for Problem-Solving

Illustrated in figure 8-5 is a conceptual diagram of six types of development changes needed in and around the city of Mineral Wells. Improvements needed are listed for downtown, parks, retirement village, recreation, airport, and highways. Emphasis is placed on new attractions as well as a new parking plaza, new Crystal Canal Park, new miniparks downtown, and new East Mountain Park. Concepts for Palo Pinto County are illustrated in figure 8-6. These encompass recommendations for historic, scenic, and recreational attractions.

Together, the planning team and the Liaison Committee concluded that all seven issues can be solved, allowing considerable expansion of tourism in the destination. The first step recommended was review of all project ideas for their feasibility. This would require study by all three

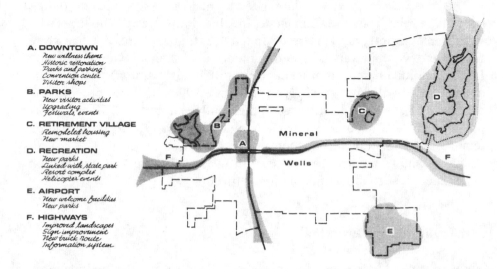

A. DOWNTOWN
New wellness theme
Historic restoration
Parks and parking
Convention center
Visitor shops

B. PARKS
New visitor activities
Upgrading
Festivals/events

C. RETIREMENT VILLAGE
Remodeled housing
New market

D. RECREATION
New parks
Linked with state park
Resort complex
Helicopter events

E. AIRPORT
New welcome facilities
New parks

F. HIGHWAYS
Improved landscapes
Sign improvement
New truck route
Information system

Figure 8-5. Development Recommendations, Mineral Wells. Based on study of existing development and resource foundation, these six areas were given major recommendations for development important to tourism (Gunn et al. 1985, 25).

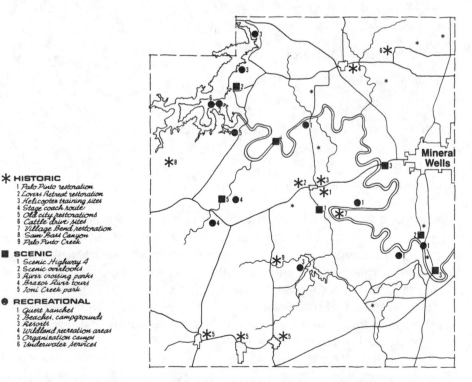

✳ HISTORIC
1 Palo Pinto restoration
2 Lovers Retreat restoration
3 Helicopter training sites
4 Stage coach route
5 Old city restorations
6 Cattle drive sites
7 Village Bend restoration
8 Sam Bass Canyon
9 Palo Pinto Creek

■ SCENIC
1 Scenic Highway 4
2 Scenic overlooks
3 River crossing parks
4 Brazos River tours
5 Ioni Creek park

● RECREATIONAL
1 Guest ranches
2 Beaches, campgrounds
3 Resorts
4 Wildland recreation areas
5 Organization camps
6 Underwater services

Figure 8-6. Concepts for Tourism Development, Palo Pinto County. Although Mineral Wells is the principal city, study revealed that the rest of the county had a great amount of tourist development opportunity. The main themes were: historic, scenic, and recreational (Gunn et al. 1985, 28).

action sectors: commercial enterprise, nonprofit organizations, and government agencies involved in development. The next step would require actual building and managing of specific projects, primarily new and expanded attractions. After these projects are in place, increased visitor demand will stimulate the need for new services—lodging, food service, travel services. This step, in turn, would increase expenditures in the local area, providing greater tax input, more employment, and greater social as well as economic impact. This increase in amenities will also enhance the local quality of life for residents.

A major concept derived from the study is adoption of the *theme,* "Wellness Capital of Texas." The great increase in market demand for health, fitness, outdoor recreation, and personal enrichment could be matched with supply development to support this theme. There are ample foundations for this locally because of its one-time prominence as a health spa. The natural resources of the surrounding area as well, such as the Brazos River, hills, wooded areas and wildlife, provide abundant support for developing this theme.

Downtown redevelopment is needed to overcome the stark, bland, and uninteresting appearance. Restoration of old bathhouses, re-orienting downtown businesses to travel trade, and addition of sidewalk cafes and arts and crafts shops would enhance the downtown for tourism. Equally important is the creation of miniparks and addition of landscape materials in all open spaces downtown. An improved and landscaped parking area to the rear of downtown shops could multiply parking availability and increase attractiveness. Proposed redevelopment is shown in Figure 8-7.

The remainder of the *city environs* requires removal of the sign clutter along main streets, new tree planting, conversion of Fort Wolters into a retirement village, and renovation and improved maintenance of parks. Plant tours, festivals, and outdoor theater productions should be considered.

Surrounding area improvements are needed. Outdoor recreation services radiating from Mineral Wells as well as land and water based activities should be added: camping, fishing, hunting, boating, hiking, canoeing, and passive outdoor recreation. Restoration of rural small towns, such as Palo Pinto, could stimulate historic tours. Highway scenic easements are needed. Ranch resorts and expanded development of Possum Kingdom Lake and Lake Mineral Wells could aid tourism.

Several concepts for *program development* resulted from the study. New entertainment, reunions of former helicopter pilots, horse shows, and historic tours could be added. Lack of local awareness of tourism needs to be countered with informational clinics. The quality and quantity of informational literature for visitor use should be improved. Increased promotion should be emphasized only after all other improvements

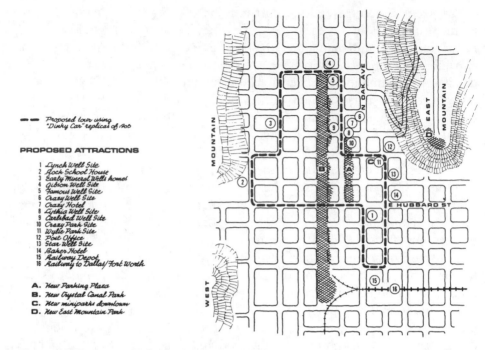

- - Proposed tour using
     "Dinky Car" replicas of 1905

PROPOSED ATTRACTIONS

1 Lynch Well Site
2 Rock School House
3 Early Mineral Wells homes
4 Gibson Well Site
5 Famous Well Site
6 Crazy Well Site
7 Crazy Hotel
8 Lythia Well Site
9 Carlsbad Well Site
10 Crazy Park Site
11 Wylie Park Site
12 Post Office
13 Star Well Site
14 Baker Hotel
15 Railway Depot
16 Railway to Dallas/Fort Worth

A. New Parking Plaza
B. New Crystal Canal Park
C. New miniparks downtown
D. New East Mountain Park

Figure 8-7. Downtown Redevelopment, Mineral Wells. Study and analysis of the downtown area revealed concepts for 16 tourism development opportunities. For circulation, revival of a 1905 minibus tour would provide linkage to all historic attractions in the downtown area (Gunn et al. 1985, 26).

have been made. A research program to regularly monitor economic impact and provide statistics in tourism should be established.

5. Recommendations

Three major categories of recommendations were derived from the study: development, program, and organization. Emphasis was placed on the responsibility of the local community through its politicians, sector developers, and tourism leaders to make the changes needed for increased tourism.

For *development,* thirty-six recommendations were made:

1. Establish theme of Wellness Capital of Texas.
2. Provide greater landscaped parking space downtown.
3. Restore original historic Lynch Well site.
4. Restore historic buildings: Baker Hotel, Crazy Water Hotel, depot, post office, bathhouses, Fannin School, Rock School.
5. Establish new truck routes around downtown.
6. Add miniparks downtown.

7. Add visitor-oriented shops downtown.

8. Add new wellness, craft, and art shops.

9. Convert Fort Wolters to retirement village.

10. Add recreational complex in Fort Wolters.

11. Increase quality of park maintenance.

12. Revitalize the clean-up, paint-up programs.

13. Establish sign control.

14. Landscape entrance highways.

15. Establish tree and shrub planting program.

16. Promote film production in this locale.

17. Add new convention center.

18. Improve hospitality services at airport.

19. Add outdoor theaters and dramas.

20. Establish wellness theme throughout area.

21. Establish new guest ranches.

22. Develop and control Highway 4 as scenic route.

23. Establish new beaches, campgrounds at Possum Kingdom Lake.

24. Provide public scenic lookouts on highways.

25. Protect and develop Brazos River for recreation.

26. Restore the community of Palo Pinto.

27. Establish year-round resorts.

28. Redevelop "Lover's Retreat" into public park.

29. Establish aesthetic easements along highways.

30. Redevelop Indian, settlement, stagecoach sites for tours.

31. Develop natural resource parks for visitors.

32. Establish new youth organization camps.

33. Add new marinas and tourism facilities at Possum Kingdom Lake.

34. Establish scuba diving and dive shops.

35. Redevelop old towns: Mingus, Strawn, Gordon, Thurber, Brazos for historic tours.

36. Develop creek valleys for nature tours.

For *program activities,* the following thirteen items were recommended:

1. Establish new festivals based on the wellness theme.

2. Expand and improve the Crazy Water Festival.

3. Use the Texas Sesquicentennial as focus for new activities.

4. Add new rodeos, horse shows, racing events.

5. Package and promote historic tours.

6. Add evening entertainment—wellness theme.

7. Add new water tours, cruises on the Brazos River.

8. Hold reunions of former helicopter pilots.

9. Establish new health spa pageants and plays.

10. Open industries to plant tours.

11. Develop loop scenic tours from Mineral Wells.

12. Package historic rural town tours.

13. Package nature tours.

For *program information,* these recommendations were offered:

1. New attractions and tour maps be prepared.

2. New brochures, videos, guidebooks be prepared.

3. Hold public educational seminars on tourism.

4. Introduce tourism talks to school system.

5. Prepare literature on destination zone basis.

6. Hold hospitality training program.

7. Place new electronic information equipment.

8. Offer novels and writings about area for sale.

For *program promotion,* the following recommendations were made:

1. Establish a Mineral Wells Area Tourism Council.

2. Create a new promotional plan to include advertising, publicity, public relations, and incentives.

3. Increase the budget for promotion.

4. Identify specific target markets for promotion.

5. Create special promotion for each segment.

6. Emphasize unique qualities of destination in all promotion.

7. Promote new attractions as they are added.

For *program research,* the following recommendations were offered:

1. Annual inventories of supply side be made.

2. Study market sources for interests.

3. Study results of hospitality program.

4. Investigate need for revision of laws and regulations pertaining to tourism

5. Study and report economic impact of tourism

6. Make conversion studies of promotion.

In addition, the following elements for a *strategy* should be prepared:

> Visit comparable destinations elsewhere.
>
> Review of pertinent literature.
>
> Participate in state professional tourism conferences.
>
> Engage professional assistance when needed
>
> Assist in removing obstacles to tourism development.
>
> Contact potential developers.
>
> Develop feasibility studies with developers.
>
> Foster implementation of recommendations.
>
> Maintain close cooperation among all stakeholders.

Implementation

In spite of a comprehensive process that involved a local tourism committee, this project had difficulty at the implementation stage. Several political and organizational changes interfered with implementation as planned. Just as the report was completed, the key local leader, director of the Chamber of Commerce, resigned his post and left. In the interim, several changes in city administration have occurred. The breach between city and county officials has widened, further limiting cooperation and joint development between the city and surrounding area. The city was accepted as a Main Street City within the state program of Texas but faltered after four years, suggesting that the economy and political support had weakened. For the first time in many years, a major civic improvement bond issue was passed, eventually helping tourism but diverting public attention for a time.

In spite of these obstacles, several improvements have been started or completed. Seven years after issuing the report, a review of the extent of implementation to date, by means of interview (Cunningham 1992 and Midkiff 1992), resulted in the following accomplishments:

- The Palo Pinto Historical Foundation has been formed and has taken an active role in restoration of the Famous Water Company, the Depot, public use of the Crazy Water Hotel, some progress on historic highways, and renovation and tours of Lover's Retreat.
- New Visitor Center has been built.
- Keep Mineral Wells Beautiful organization has become active, fostered a new park downtown, new plantings, new sign ordinance,

enhanced the entrance highway right-of-way and median, landscaped the downtown canal, developed new recreational complex near the high school, and fostered a county science tour.

- A marker has been placed at the Lynch Well site.
- Fort Wolters Recreation Building now houses many conferences.
- Clean-up, paint-up program has expanded—now semi-annual.
- Mineral Wells State Park has expanded its recreational activities and programs.
- An ostrich farm has been established—open to public.
- Scuba diving in Possum Kingdom Lake has increased.
- A new rodeo arena has opened.
- Reunions of members of the Vietnam Association are held regularly.
- New maps and informative literature have been prepared and distributed.

This destination project demonstrates the value of planning collaboration between university tourism specialists and local people. Within a few years' time, and at comparatively low cost, many worthwhile recommendations evolved from joint study and analysis. This project also shows how political changes and economic slowdown can delay implementation of needed strategies and projects. Even so, the project provided local enlightenment on tourism development and stimulated new interests in planning for tourism.

CAPE TOWN, SOUTH AFRICA

For Cape Town, tourism's promise of economic input is seen as the catalyst for its Reconstruction and Development Program (RDP). This program has, as it promised, improved infrastructure, new housing, and social amenities and in a planned development manner (Bennett, Smoot, and Popendorp 1995). New planning and development have not been accomplished by a sole developer but rather through cooperation and collaboration among several entities—City Council, Victoria and Alfred Waterfront, and planners and designers, such as Gallagher Prinsloo & Associates, MLH Architects and Planners, and Waterfront Landscape Architects. The key beneficiaries have been enhancements of Table Mountain, Central Business District, and the Victoria and Alfred Waterfront, major tourist attractions as well as local amenities.

The spectacular background for the city, Table Mountain, is part of a surrounding mountain chain. But, like many natural resource attractions,

Figure 8-8. Table Mountain-Cape Town linkage. Essential for all planning in the future is integration between Cape Town central area and the surrounding landscape, especially Table Mountain, seen in the background.

increased volumes of visitors have taken their toll on this special environment. The six floral kingdoms are being threatened by trampling, fires have gotten out of control, waters are being polluted, soils are eroded, and littering has detracted from landscape beauty. These impacts have been a challenge to the 14 public authorities and 160 private landowners. To cope with these impacts, new plans have resulted in better paving to reduce erosion, new viewing platforms, and barriers to keep visitors from destruction of the fragile vegetative cover. Trails have been reconstituted with more durable materials. The environmental assets have been restored at the same time visitor enrichment has increased.

The Central Business District has undergone many changes since the Dutch East India Company dropped anchor at this site in 1652. Aboard was Jan van Riebeeck who planted a garden corridor inland from the waterfront. Three hundred years later, this has become the setting for major civic buildings such as the National Art Gallery, Natural History Museum, South African Reference Library, Houses of Parliament, a cathedral and synagogue, and the University of Cape Town School of Fine Art. As the city grew along the foreshore, the older buildings were actually preserved by benign neglect and recently have been restored as an historic district. Adding to the visitor interest is the beginning of a Green Pedestrian Way linking the dominant landscape features from Table Bay to Table Mountain. (Figure 8-8) A spinoff has been stimulation

of investment along the corridor, becoming a social magnet for townspeople as well as visitors. Since the fall of apartheid, the informal business element has increased. Nearer the waterfront is the historic Grand
Parade, a popular marketplace. Bounded by the Old Town Hall and the
Castle, it has become the venue for important gatherings, such as the
inaugural address of President Mandela. The great popularity of this special place is becoming a challenge to manage in order to prevent excessive littering and occupation by homeless people.

The historic and functional waterfront has become the focus of many
discussions regarding its future. The Ministers of Transport Affairs and
Environmental Affairs appointed the Burggraaf Committee to consider
plans. A result was the establishment of the Victoria and Alfred
Waterfront Company (V&AW). Their study resulted in recommendations
for greater tourism use, residential and retail development, continuing
harbor activity, and measures to maintain year-round vitality and safety.

The region's population is about three million and nearly one-half million live within 15 minutes of the waterfront. In September 1987, V&AW
submitted its Development Framework Report. The focus was on 83
hectares of waterfront land. The project was to be guided by several principles. The intent was to integrate plans and actions into the rest of the
city and its townspeople. Heritage qualities are to be protected for both
visitors and residents. Leaders wished to avoid an "ersatz Disneyland"
atmosphere.

The plan divided the area into seven precincts, as illustrated in Figure 8-9.

- Granger Bay—residential, small craft harbor.
- East Pier—primarily fishing industry.
- Pierhead—historic restoration, boardwalk, restaurants, hotel, offices,
  pubs, shops, theater, maritime museum, aquarium.
- Portswood Ridge—office park of converted Victorian houses, prison
  block converted to business school.
- Old Quarry and Upper Basin—flooded to provide waterfront hous-
  ing, offices, hotels.
- Ebenezer Bluff—tree-lined canal, residences.
- Amsterdam Battery Park—canal system with tour boats to link all
  area attractions.

Essential to the plan is landscape enhancement appropriate to the proposed land uses and compatible with the surrounding scenery. Landscape
design with plantings and street furniture provide the thread that links the
many features together.

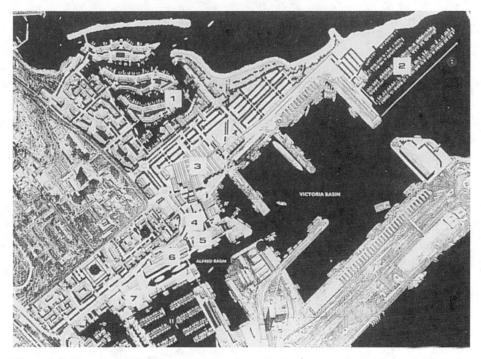

Figure 8-9. Victoria & Alfred Waterfront, Cape Town. Redevelopment of this special waterfront for tourists as well as local citizens includes: 1, residential area; 2, marina; 3, specialty market place; 4, center and services; 5, hotel; 6, maritime museum; and 7, aquarium (Victoria & Alfred 1992, 7).

The prime investor-developer is the private firm of Transnet as a subsidiary of the Victoria & Alfred Waterfront. However, planning and development have been the result of close cooperation and specific agreements with the City Council since the beginning of the project in 1988. The concept has benefited from site visits to and review of waterfront developments in Baltimore, San Francisco, Vancouver, Toronto, and Boston. Many planners/designers have collaborated on the overall concept as well as individual projects. Core planners/designers have included Gallagher Prinsloo & Associates, MLH Architects & Planners, and Waterfront Landscape Architects, as well as specialized inputs from designers of signs, transportation, hotels, hydrology, maintenance, and security.

Of the total spent on the first phase of development (R63 million), R35 million was spent on infrastructure—roads, water, stormwater, and sewage reticulation, hard and soft landscaping and street furniture and relocation of existing services. Although still under construction, Pierhead (a planned complex of Victoria Wharf & Market Square, Aquarium, hotel, and conference center) attracted 1.4 million visitors in its first four months after opening.

Special agreements with City Council have expedited progress of planning and construction. For example, the Waterfront land use control is under a new Legal Succession to South Africa Transport Services Act of 1990 rather than the former Land Use Planning Ordinance. Negotiations were complicated by the fact that the municipal boundary runs through the site. Because of this and in order to resolve the tax issue, V&AW are paying contributions in lieu of rates for the area outside the boundary.

Throughout the planning and design process, the management organization, the City Council, and the consultants have worked closely together. From initial concepts through sketch plans and final design, these group sessions were often demanding, lively, and sometimes tense. The success of the resulting development is endorsed by the many awards such as Cape Institute of Architects conservation award, Cape Times Centenary Medal for conservation and building rehabilitation, South African Institute of Civil Engineers regional award, Institute of South African Landscape Architects merit award, South African Institute of Town & Regional Planners biennial merit award, Architecture S.A. project award, Mayor's awards for greening the city and tourism, Lions International award to the Information Centre, and audio-visual award from New York Film & TV Festival (Victoria & Alfred 1992).

This case demonstrates several lessons of importance to other similar locations. Planning has combined visitor with local needs and uses. The complexity of the site required participatory planning by design professionals, citizens, and governmental agencies. And the plan integrates tourism with urban and regional plans.

THE ORKNEYS

The Orkney Tourist Board (OTB) is responsible for tourism guidance within the 67 islands that lie off the northern tip of Scotland. Warm ocean currents provide a mild climate and most soils support agriculture. In the 900s, Norse earls settled and ruled the islands. Although Scottish nobles replaced them in about 1231, the islands remained for many years under rule from Norway and Denmark. About 1468, they became a part of Scotland. It is an area of little daylight in winter and hardly any night in summer.

The Orkney Tourism Strategic Plan was prepared in 1997 and provided a set of guidelines for the future of tourism in this small (377 sq.mi., 976 sq. km.) destination. It resulted from the deliberations of a special committee, including representatives of the Orkney Islands Council (OIC), Orkney Enterprise (OE), the OTB, and the hotel and travel agency busi-

Strengths   *Weaknesses*

	Weather
	Short Searson
	Cost of travel (perception)
	Capacity of routes - Air + Ext Ferry
	Lack of integrated booking service
	Local public transport
Quality	Coordination of Attractions
Clean	Market information
Safe	Lack of plan/monitoring
Friendly	No predictive monitoring
Island - remote (exotic)	Tourist board
Culture/Heritage	Marketing budget
Products (e.g. Craft)	Training
Different wildlife	Attitude towards tourism

Electronic marketing	Competitor activity
Nature/Wildlife	Transport
Cooperation (with other sectors)	Pollution
Packages	Gov't policy - Environment/TB funding
Sport	Shift in public tastes
Farm based	Absence of strategy
Music themes/conferences	Fragmented approach
Shopping breaks to Orkney	
Cruise liners	
Niche markets	
Summer schools	
Special interest holidays	
North Isles coordination	
"Green" tourism	
Training	

Opportunities   *Threats*

Figure 8-10. SWOT Analysis, The Orkneys. An analysis of strengths, weaknesses, opportunities, and threats was the first step taken in planning future tourism for the Orkney Islands, northern Scotland (*Orkney Strategic Plan* 1997).

nesses. The plan intent was to follow the general aims of the national tourism program.

A first step in the plan preparation was a SWOT analysis, as illustrated in figure 8-10. This information together with visitor surveys and other research provided the foundation for future tourism directions.

In spite of the remoteness of this destination, high travel costs, and unpredictable weather, tourists are satisfied with their visits, primarily because of the area's very special natural and cultural assets. The area's archeology, history, and remoteness are cited by visitors as their prime reasons for visiting. Key attractions include Maes Howe World Heritage

Site (2700 BC), Broch of Gurness (Viking site, 1000 AD), and Skara Brae and Skaill House (5000-year-old Stone Age buildings and sites). Displays and interpretation help the visitor gain insight into ancient living in this far north region.

The plan states that the main objectives for their tourism include:

- entension of the visitor season
- increase visitor spending
- preservation of the Islands' culture
- preservation of the environment
- increase tourism employment
- initiate cooperation and collaboration among all sectors
- enhance quality of facilities and services.

Throughout the tourism leadership there is awareness of the importance of protecting the basic natural and cultural environment. These leaders now face the growth dilemma of many destinations today—how to maintain the environmental assets and at the same time promote more visitors. The plan asks for more cooperation among all involved in development in order to plan and manage tourism for best sustainability. Recommended is a Steering Group with diverse membership, charged with responsibility of working with subgroups, such as for archeology, heritage, sport, festivals, and other demand-supply topics.

This case illustrates the emergence of new awareness and planning concern, even in a small destination, over protecting environmental assets in balance with the economic and social returns from tourism. Commendable is the leadership recognition of linkages between promotion and delivery by the supply side. Maintaining the Islands' reputation as "Clean, Green, Serene" is a major challenge.

CUMALIKIZIK, TURKEY

The case of Cumalikizik demonstrates the planning of an important 700-year old Ottoman village adjacent to Bursa, the fourth largest city of the Turkish Republic (Oren, Woodcock, Var 2001). Already experiencing winter tourism because of the nearby Uludag, Mt. Olympus, that has been declared a national park, cultural tourism development of the area has been more difficult. Cumalikizik is the only surviving village of the original seven surrounding Bursa. Hamamlikizik, Fidyekizik, Derekizik, Degirmenlikizik, Dallikizik, and Bayindirkizik have been overrun by industrialization and population growth, completely changing their original character.

Figure 8-11. Heritage Tourism, Cumalikizik, Turkey. This view illustrates a remnant of the rare ritual architecture of the Ottoman Empire that is now being rehabilitated. Along with the structures and village pattern that have endured over centuries is a comparable life style of ceremonies, religious holidays, and neighborhood bonding. This example is typical of similar cultural restoration being developed around the world, not only for tourism but for preservation of a past culture (Bursa Local Agenda 21 Archives and Oren, Woodcock, Var 2001).

Outmigration from the village of Cumalikizik has been caused by economic decay, abandonment and massive degradation of buildings. Further decay has been halted by governmental declaration for preservation of the area (Bilenser 2000). Remaining are 270 houses, many historic sites, a mosque, a Turkish bath, cemeteries, and remains of churches (figure 8-11).

Preservation and redevelopment were stimulated by action of the Senior Committee of Monuments in 1980 and registration as an Urban and Natural Protected Area in 1981 (Kaya et al. 2000). The Bursa Local Agenda 21 Cumalikizik Conservation and Revitalization Action Plan were initiated in 1998. Local Agenda 21 is a global sustainable action plan derived from the Rio Earth Summit of 1992. The goal of this local project is not only important historic restoration but also sustainable development that will enhance the social, economic, and cultural fabric of the community as well as increase tourism.

A major impetus was provided by initiatives of the Bursa Metropolitan Municipality and the Bursa Tophane UNESCO Youth Association.

Already many improvements have taken place. Infrastructure, such as water and waste systems, electricity, and fire protection have been upgraded. Historic reuse has occurred in four houses—a Turkish restaurant; business offices, archives, cafeteria, and shop with indigenous crafts; and two residences. A basic element is participatory planning including opinions of residents. Women now operate two bed-and-breakfast facilities. Several women have been studying and replicating original village motifs in crafts, embroidery, bedspreads, and carpet weaving. Job training in agriculture, crafts, and tourism has been implemented. Public educational programs have fostered new awareness and support of community improvement. Project design and construction have been awarded through competitive bidding approved by the Bursa General Directorate of Organization of Conservation for Cultural and Natural Values, Metropolitan Municipality, and District Municipality.

Although this project is in progress, it is cited here because basic planning steps, often omitted, were put in place at the start. Public and private interests have rallied behind major restoration and development for resident value as well as for tourism. The main concern in the future will be monitoring every increment of development for negative impacts. This precautionary measure will identify capacity limits early enough to establish needed controls against oversaturation of tourism.

CAPE BYRON HEADLAND RESERVE, AUSTRALIA

The case of planning Cape Byron Headland Reserve on the northeastern shore of Australia represents an unusual mix of tourism and conservation (*Cape Byron* 2001). This headland of 95.5 hectares is managed not by federal or state government but by the Cape Byron Trust, a part of the Byron Bay Community. Membership of the Trust includes community and government representatives, as well as residents of the local and regional area.

The area is rich with natural and cultural resources and development has been minimal. It is dominated by coastal subtropical rain forest that offers animal habitat and scenic appeal. The area is ringed with beautiful beaches along Byron Bay and the Pacific Ocean that support a variety of marine activities. Aboriginal people have occupied the area for thousands of years and the Arakwal now are influential in current planning. The Mount Warning caldera in the central part of the peninsula offers opportunities for 360-degree viewing, hiking trails, and hang gliding. Figure 8-12 illustrates the regional setting and figure 8-13 shows the proposed land use.

Figure 8-12. Regional Context, Cape Byron Reserve. A growing tourist area in northeastern Australia that is now under sustainable environmental guidance by the Cape Byron Trust (*Cape Byron Headland* 2001, Map 1).

The original Cape Byron Headland Reserve was established in 1903. By 1933, portions had been excised for residential, sand mining, and grazing. The Cape Byron Trust, an independent nonprofit body, was established in 1997 to halt such environmental degradation. It is charged with management of the area but continuing under National Parks and Wildlife Service policy. Environmental and heritage planning are in accord with the New South Wales Coastal Policy of 1997 and the Australian Natural Heritage Charter of 1996. An agreement between the government and the Arakwal Aboriginal Corporation provides native title over the land.

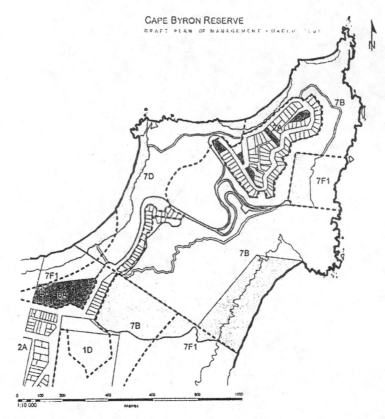

CAPE BYRON RESERVE
DRAFT PLAN OF MANAGEMENT • MARCH

Figure 8-13. Land Use, Cape Byron Reserve. Zoning categories include: 7B-coastal habitat; 7D-scenic escarpment; 7F1-coastal lands. Residential zone, 2A, was excised from the reserve because it had been established earlier. Tourists can enjoy the natural and cultural resources of the reserve whereas lodging and food services are provided in Byron Bay (*Cape Byron Headland* 2001, Map 2, by permission).

Foundation Values

This area possesses several very important foundation values for tourism and residential quality of life. Prominent are *natural values*. Trees, grasses, and other plants are of regional, state, or national conservation significance. Wildlife habitat supports over 100 species of birds, humpback whale, loggerhead turtle, greater glider, eastern grey kangaroo, and frog groups. *Cultural heritage values* include sites and artifacts of indigenous and Aboriginal peoples of thousands of years ago. European exploration, settlement, conflict, trade, commerce, and land use are cultural assets. The outstanding landscape offers considerable *scenic and aesthetic values*. This asset includes a buffer from residential uses, a 360-degree vista of the coast, viewing platforms and trails, and views of rain forest gullies, rock cliffs, beaches, marine life, and the rising and setting of the sun and moon. Among the *research and educational values* are opportunities for studies and interpretation of the many assets. These are open to school groups, researchers, and visitors. Already significant are *tourism and economic values*. Nature tourism and recreational markets stimulate business employment in accommodations, transport, tours, food services, information, and souvenirs. For the general public locally and distances away, the *existence and protection values* are symbolic of concern over the several values described above, even though some people may never visit.

Management Plan

The Management Plan encompasses four major strategies.

1. Natural Resource Management
The major intent is conservation of the rich vegetation and wildlife, already threatened by unplanned trails, unauthorized drives and parking, and unorganized recreation activities. Especially important is protection of habitat for over 100 species of birds. Removal of a herd of goats and prohibition of domestic dogs and cats are planned. Required is a revision of fire management. The natural duneland, coastal and other land forms should not be eroded by sand excavation and unregulated visitor use. Aesthetic vistas must not be disturbed by poorly designed and located facilities, signs, parking areas, and shelters. New wastewater management is needed. Sources of noise pollution, such as chartered airflights, portable vendor vehicles, generators, and motorbikes must be regulated.

2. Cultural Resource Management
Special emphasis must be placed on preservation of Aboriginal relics, sites, and ceremonial places. A conservation, management, and interpretive strategy will be prepared jointly by the Trust and Arakwal Aboriginal Corporation. Visitor access can be reduced by improved interpretive centers. Equally important is protection, restoration, and interpretation of historic resources (since European settlement). A systematic survey of significant sites is overdue. Conservation easements, appropriate historic reuse of buildings, and elimination of conflicting recreational uses are planned where needed.

3. Recreation, Tourism, and Visitor Use
The balance between resource protection and visitor use is a major challenge. With a half million population living within one hour drive, the management of visitors requires a review of carrying capacity. A starting point is a new campaign of information and public awareness of the

area's assets and need for visitor control. Planned action includes a Web
site, a new interpretive sign system, and guidelines for appropriate visitor
ethics and use. For residents, the protection of the sense of place is paramount, requiring a low-key development and use that protects this value.
The Trust intends to expand the involvement of local people in all planning and management policies and actions. An element of this cooperative relationship is the volunteer organization, Cape Byron Friends, who
already are involved in restoration of resources and their interpretation.
Among visitor control measures are road speed-bumps, speed limits, and
personal car fees to encourage bus use. New visitor waste control is
planned. New management regulations will prevent conflict of water uses
such as boat launching, swimming, and surfing. Safety and risk management require new measures. The policy of resource conservation is superior to commercial recreation development in this area. Licensing and
promotion are controlled by the Trust.

4. *Management of Adjacent Areas*

Because the use of this area is influenced by adjacent lands, integrated
planning is essential. The Trust intends to increase its cooperation with
the Byron Shire and the bordering Wakegos, and the Lighthouse Road
communities. New cooperative arrangements are being formed between
the Trust and the National Parks and Wildlife Service, the proposed
Arakwal National Park, and the Arakwal Aboriginal Corporation. A new
program between the Trust and Byron Shire Council for vegetation
integrity, water quality, visitor safety, and road access is being proposed.
Community events will be encouraged.

Even though the plan is so recent that implementation has not yet been
achieved, many aspects of a desirable plan are already in place. Cooperation between the Trust and others has already occurred. The policy of
balancing resource conservation with tourism is well known. The planning process has involved several important entities including Aboriginals and the community. Well understood are the many proposed
management strategies, based in research and challenges for sustainable
development. This project holds promise of improving Cape Byron as a
continuing conservation area and a prominent tourist destination in New
South Wales, Australia.

BALEARIC ISLES

An excellent example at the destination level of planning for sustainable
tourism is public-private accomplishment in recent years for the Balearic
Island region in the Mediterranean Sea 250 kilometers off Spain's coast

(Middleton 2001). In response to the explosive growth and consequent ills of development of the 1950–80s, tourism's quality and quantity are now under more sustainable planning guidelines.

Although beach resort development dominated past tourism, this geographic destination has great cultural visitor potential (Garcia 1972). Because of its strategic position midway in the Mediterranean Sea between Africa and Europe, much of world history was written here. Artifacts and cave paintings prove human occupation as early as 5000 BC. The cave culture was followed by centuries of the Tayalot culture, evidenced by ruins of walled settlements. Ample evidence of structures and artifacts prove occupation by Roman and Phoenician cultures. Later occupation is evidenced by remnants of early Christian churches. This rich historic background offers an abundant resource for further restoration, preservation, and interpretation for tourism.

The strategic location within easy market range for millions of tourists and its abundant natural resources combined to catapult the region into a world-class destination. Tourism has surpassed agriculture and become the major economic provider for this area of four major islands including Majorca. The early and rapid development of beach resorts, typical of similar lures worldwide, took little account of environmental impacts even though the total land consumed only about four percent of total island land area. Over 75 percent of land has been protected by law. Issues of beach pollution, aesthetic erosion, oversaturation, and social maladjustment became evident. A serious issue was the growth and dominance of foreign tour operators. Traveler volumes were more dependent upon their market manipulations than the planning, development, and management of the supply side on the islands.

As the Balearic area became an Autonomous Regional Government of Spain in 1983 it was able to establish its own tourism policy. Since then, a series of voluntary and legal steps have resulted in major environmental improvements and stabilization of the tourism economy. New sewage treatment plants prevent raw sewage from contaminating beaches. A set back regulation prevents any new structure within 500 meters of the seashore. Oversaturation of mass tourism was stopped by capacity regulation. New land use legislation guides growth and adapts tourism to other development, urban and rural. The traditional narrow focus of beach resorts has been expanded to meet interests of other market segments. The addition of heritage and cultural sites, walking and cycling trails, golf and sailing activities is directed toward leveling seasonality, so dominant of tourism in the past. For example, the ancient walled city of Alcudia has been developed and interpreted. Although ecotours are provided, a voluntary program maintains visitor ethics and environmental protection.

Linked to all the environmental and economic enhancements is an effective marketing plan. Marketing research is carried out by the Centre of Tourism Research and Technology. The University of the Balearics and especially Professor Esteban Bardolet have been instrumental in the studies of markets and resources and in the preparation of marketing plans. Marketing here is seen in its true definition of including product development as well as promotion.

By the end of the 1990s, several important planning steps had been implemented: social and political consensus on significance of tourism, privately managed tourist boards on each main island, and a Tourist Council within the Balearics Ministry of Tourism that provides a forum for planning and marketing. Local governments are now integrated with the regional government on tourism matters.

This case demonstrates how the great complexity of tourism can be integrated and managed for both environmental and economic purposes. It utilized volunteer action by developers, marketers, and tour operators combined with public policy and regulation. As with many destinations around the world, it was unfortunate that insufficient vision and plans in early development allowed so many mistakes to occur. But this case shows that many errors of the past can be rectified and that sustainability is not merely an ideal but a reachable goal. It also demonstrates that today new leaders are emerging, new management methods are being implemented, public and private interests in tourism are cooperating, and a new environmental awareness is appearing.

AILLST TOURISM STRATEGY

In an effort to decentralize tourism planning in Scotland, the several tourist boards have created their own planning and action programs. One example is the *Area Tourism Strategy* for Argyll, the Isles, Loch Lomond, Stirling and Trossachs (AILLST) destinations (*Area Tourism Strategy* 2001). This final draft plan is for the period of 2001–2006.

This plan evolved from previous plans, action achievements, and travel trends. It was the result of public-private collaboration and reflects a consensus of views and aspirations of tourism development that is productive and yet sustainable. It built upon local action plans, such as for the new Loch Lomond and Trossachs National Park and Millennium Link Canal.

This area encompasses an abundance of natural and cultural assets. Its background reaches back from the prehistoric sites of Kilmartin Glen through the birthplace of Northern European Christianity. The scenic appeal of the landscapes and diversity of wildlife are significant resources.

Strengths	Weaknesses
Scotland's first National Park Rich Wildlife/Landscape Quality Diversity of experience Built/Cultural Heritage Accessibility to key domestic markets	Mismatch between supply and demand Lack of management skills Variable public transport provision Poor basic infrastructure
National Park, Millennium Link and other flagship projects E-commerce/ICT culture Niche markets Grown in eco-tourism market Trend to short breaks/second holidays Growth of business tourism	Potential economic downturn Ongoing strength of the pound Poor perception of value for money Spiraling domestic costs High cost of transport/fuel Reluctance to respond to change Negative attitudes within the sector and the wider population
Opportunities	Threats

Figure 8-14. SWOT Analysis, AILLST. A first step in the process of developing the Area Tourism Strategy for the destination called Argyll, the Isles, Loch Lomond, Stirling & Trossachs (AILLST), was this SWOT Analysis. Opportunities do exist but planning must recognize the weaknesses and threats (*Area Tourism Strategy* 2001, 7).

Cited are several recent accomplishments including major investments in improved and new tourist information and visitor centers, and restorations of historic sites. Added was a ferry providing links between Kintyre and Northern Ireland. Introduced was a new approach to economic monitoring. However, tourism gains were tempered by a few setbacks. Among these were constraints on public sector infrastructure, lag of financial support for quality improvements, and delays in obtaining key information. Even so, the planners seek a better future.

The plan encompasses a vision statement, aims and aspiration, specific objectives, implementation action, and how progress is to be measured. A beginning point in the process of planning was an analysis of strengths, weaknesses, opportunities, and threats (SWOT) for tourism. The results are shown in figure 8-14 which stimulated the creation of the destination plan's Vision Statement:

> Working together to unlock the area's potential by enhancing its profile and the quality of the visitor experience.

The aim is to create a healthy, sustainable tourism sector that provides the social, environmental, and economic benefits desired. The aim is not merely more business but protection and enhancement of the natural and built environments. Action plan initiatives are detailed in the following categories: knowledge and market intelligence; visitor services; skills; accommodations; attractions; arts, events, entertainment; retail catering and licensed premises; environment; transport and other infrastructure; and delivery mechanisms. In each instance, the lead agencies for implementation are identified.

Commendable is the business recognition of its dependence upon the natural and cultural foundations. For example, the action plans include several specific recommendations: better planned gateway communities, new trails through natural and heritage sites, landscape enhancement of access to communities, and improved visitor corridors. A major recommendation is to establish carrying capacities and monitor environmental impacts.

Throughout, the main thrust is toward a better balance between the dynamics of supply and demand, that regular monitoring of traveler trends be made available and that the supply side maintain its place distinctiveness as adaptation is made to market shifts. Five key topics are emphasized and detailed in the plan:

- knowledge and market intelligence
- marketing and visitor services
- skills and business development
- the physical tourism product
- delivery mechanism

Finally, a detailed Local Tourism Action Plan expands on these initiatives with lists of action programs, the responsible lead agency, and a time scale of accomplishment.

This case is cited here for its several commendable elements. The plan builds upon past experience of planning and development. It is a joint public-private plan. It recognizes the need for protection and interpretation of its abundant and valuable natural and cultural resources. And it does not conclude with a vague promise of implementation but names the agencies and organizations charged with action.

NORTH CAROLINA

An important planning opportunity for tourism destinations is the use of adult education specialists. Today, community-area jurisdictions throughout the world, especially those in rural areas, could benefit from professional guidance patterned after the Cooperative Extension Service of the United States. CES was originally designed to provide technical assistance to farmers, but now many states have adapted its slogan of "helping people help themselves" to the field of tourism. Sponsorship and funding come from three sources—federal government, land-grant universities, and local county government. Extension Specialists, trained in techniques of public participation and tourism development topics, are based on the campus of a university but provide service throughout the area.

An example is the Sustainable Tourism Program, North Carolina State University, Department of Parks, Recreation and Tourism Management. This project of planning guidance is *Sustainable Tourism in the Sandhills—A Process for Community Development in Anson, Moore, Montgomery, and Richmond Counties* (Kline 1999). Over a period of nine months, Extension Specialists and local people cooperated on developing a plan for tourism development (Brothers and Kline 2000). The process included many meetings and discussions with business leaders, citizens, nonprofit organizations, civic groups, and governments at several levels. Focus groups met and created information on the following objectives:

- Inventory of the area's resources
- Suggestions for increasing community involvement
- Identification of current and potential markets
- Short and long-term marketing strategies
- Product development opportunities
- Steps toward implementation.

Based on earlier work in this region, several barriers to tourism development were discovered, as illustrated in table 8-1. These barriers are frequently encountered elsewhere and must be dealt with if the goals of tourism are to be advanced. Without proper understanding of the complexities of tourism, organizational guidance, and political coordination, any effort toward destination planning of tourism will be handicapped.

This grass-roots tourism project in four counties of North Carolina resulted in identifying ten top needs for action.

1. *A Plan for the Land.* Local citizens should prepare a land use plan for each county that includes conservation as well as development areas.

2. *Create Vision, Begin Education.* Public involvement is needed for education on principles of sustainable tourism development and creation of vision aims and objectives.

3. *Research Visitors.* Market research is needed for both present visitors and trends for the future.

4. *Inventory Resources.* Utilize existing resource information from state agencies to determine conclusions on potential.

5. *Integrate with Organizations.* Planners need to contact many organizations that influence future tourism development.

TABLE 8-1

BARRIERS TO TOURISM DEVELOPMENT AND POSSIBLE RESOLUTIONS

BARRIERS	POSSIBLE RESOLUTIONS
Lack of Vision	Facilitate focus groups with stakeholders and representatives from the community.
Lack of Education about Tourism Impacts	Hold open houses and educational assemblies where community members can learn about possible negative and positive impacts of tourism. Involve local media in development process. Involve community leaders in disseminating information: educators, religious affiliates, civic organizations, clubs, and special interest groups.
Lack of Education about Development Processes	Appoint sub-committee to research laws and regulations concerning all aspects of development: real estate, sign ordinances, DOT plans, historic preservation, ADA, etc. Involve other industries from region. Nurture their support and understanding of the big picture.
Insufficient Infrastructure	After assessing current infrastructure, prioritize those current facilities that should be updated and improved. Concentrate on bettering what already exists rather than using limited resources to add new infrastructure to an insufficient foundation. Use an innovative approach to think about new ways of using existing facilities. Incorporate efforts regionally to facilitate surrounding infrastructure.
Lack of Direction and Organization	Develop standard means of communication within the industry, such as a calendar of events, or a newsletter. include incoming groups at local hotels. Create a Community/Tourism Development Office that will centralize efforts so that the public can be informed of the common goals. Involve media in organization efforts so that the public can be informed of the common goals. Use past research and future data collection efforts to identify the target markets and to aid in product development. Maintain a healthy relationship with the Division of Tourism, Film and Sports Development as well as other essential tourism organizations, locally, regionally and nationally. Mirror local marketing efforts to state's long-range direction.
Politics	Ensure that diverse groups from the community are represented proportionately in all planning stages. Maintain the focus of all efforts as a QUALITY OF LIFE issue for the WHOLE COMMUNITY. Discourage elected officials from development positions that could be considered a conflict of interest. Employ an outside agency to aid in assessment, development and policy. Remember that tourists do not notice artificial boundaries such as county lines.
Need for Hospitality Training	Coordinate efforts with Cooperative Extension, community colleges and universities for hospitality training. Enforce training with periodic "after hours" gatherings for line staff or supervisors. Establish annual awards for line employees.

6. *Create County Cooperation.* Needed are communication systems that will foster sharing information among the four counties.

7. *Assemble a Resource Library.* All four counties need to develop a resource library on many related topics—business management, funding sources, architecture, planning, resource protection issues.

8. *Develop Interpretation Guidance.* Provide guidance on development of facilities and programs of visitor interpretation.

9. *Develop Action Timeline.* Rather than attempting too many projects at once, take action on only one at a time.

10. *Focus on Niche Markets.* Instead of generalizing development for broad markets, favor segmented market focus.

The role of the Extension Specialists is primarily that of facilitator. A minor role is that of a tourism expert but not in the sense of dictating information or action. Included in their responsibilities are: holding public workshop meetings, providing educational sessions, developing educational materials, designing research, and providing technical assistance as needed.

This project is cited here because of its strategy of participatory planning of a tourism destination. Rather than engaging an outside consultant to prepare a plan, the process began with regional conferences directed toward those who would take action in the future based on their own development plans. The objectives of these initial conferences were:

- provide the vehicle for local residents to define issues
- address these issues through training and education
- formalize partnerships for guiding development.

Throughout the process, emphasis is being placed on the stakeholders who need tourism education and will be responsible for implementation.

As community involvement focused on their own tourism planning, the following concerns were addressed:

- quality of life
- overall economic development
- desirable levels of growth
- protection and conservation of resources
- transportation issues
- land use planning and controls.

By June of 2001, progress had been made in implementing these guidelines. A Web site containing tourism information was established. Web modules include: Destination, Marketing, Tourism Research, Marketing Plan, and Public Relations. A survey of existing travel markets was initiated in all four counties, five sustainable tourism fact sheets were issued, and a teleconference program on "Issues in Tourism" took place. In addition, guideline papers on ecotourism business and agritourism potential were prepared and distributed. Several landowners are preparing tourism development project feasibilities. Names and addresses of resource organizations have been compiled and distributed. There is ample evidence that this participatory process is a desirable approach to tourism destination planning.

COROMANDEL, NEW ZEALAND

The case of tourism development in the area of Coromandel, New Zealand, represents successful participatory planning over several years. This peninsula, projecting into the Pacific Ocean along the northeastern coast of the North Island, has become the fastest growing tourist area in the nation in recent years. Fortunately, both government and private sector planning have guided tourism development in a sustainable manner.

Its natural resource features, rich history, and easy access from Aukland provide a firm foundation for tourism. Centuries ago, it was settled by a Polynesian group called the Maori. The Dutch, in an aborted attempt to land in 1642, gave the region its name, New Zealand. British settlement began following Captain Cook's arrival in 1769.

Similar to other nations during the colonization period by Europeans, there was vicious clash between the native Maoris and English conquerors. However, in recent years, a great amount of acculturation has taken place, fostered by tourism. The present Maori have retained many of their original crafts and customs and have adapted them to tourism.

Among the many natural resource features of importance to tourism are the remaining indigenous forests and replantings of pine. The surrounding coasts provide broad sandy beaches, big game fishing, diving, cruising, sailing, windsurfing, swimming, and waterskiing. The scenic beauty from the coasts to central hills is outstanding. The stamp of major historic events and artifacts from both white and Maori cultures is clear and lend themselves to greater visitor development and interpretation.

PATA Evaluation

As one of the tourism assessment projects of the Pacific Asia Tourist Association (PATA), an insightful study of the Coromandel destination was completed (*The Coromandel Experience* 1991). The Task Force included Peter James, HJM Consultants; Ian Oelrichs, landscape architect/planner; Kevin Young, tourism consultant; Bob Ashford, consultant; and was led by Ian Kennedy, Pacific PATA. A counterpart local team of nine representatives cooperated on the project. After intensive discussions locally, together with review of research materials, the Task Force reached several conclusions and offered a compendium of 43 specific recommendations for tourism improvement.

A key conclusion was the need for a central tourism organization charged with guiding future tourism development of this destination. The Task Force concluded that the area possessed an abundance of natural and cultural assets that can support year-round visitor use, contrary to a belief that seasonality is a problem. They concluded also that a thorough inventory and analysis of attraction potential would provide the foundation for creating a new and integrated tourism image. In this compact area, most of the diversity of traveler experience of the entire country can be found. It seems that this destination well fits the nation's current slogan, "New Zealand: the things that shouldn't change, haven't."

The PATA report contains an abundance of valuable recommendations regarding physical and program development if tourism is to grow into a major economic and social force. The 43 recommendation statements could be generalized into these pertinent directions for the future:

- In order to protect basic assets and guide tourism, a new organization that encompasses both political jurisdictions (districts) is needed.

- Essential to the future is stronger and better conservation of its natural and cultural resources. Such assets form the very foundation for visitor interest and essential values if the business of tourism is to grow.

- Key to such protection is perpetuation of the Coromandel life style. Any growth of tourism must respect this tradition that has stabilized the main theme of this area.

- Now lacking is a greater number and quality of attractions, especially parks, museums, trails, campgrounds, and interpretive centers. The abundance of resources supports such growth, provided that design and planning will prevent environmental abuse.

- Although Maori-white relations have improved, greater cooperation and collaboration is demanded as a future tourism development policy.

- A design principle of small and low key development is needed in order to maintain important place qualities. All expansion of accommodations should favor community locations rather than erode the special landscape and resource qualities of outside resource regions.

In response to the PATA study, a tourism plan for the Coromandel was issued (Adams 1994). This plan emerged after over 20 public meetings, review of sources of information, and collaboration with the two governing districts—Coromandel and Hauraki. The report is presented in three main sections.

The report opens with a consensus on their *tourism vision*. This statement shows an unprecedented balance of aims from economic development to protection of resources and existing life values. Concerns over too much growth were identified as: increased costs of living, environmental erosion, congestion, unbalanced economic tradeoff, and a loss of community solidarity. Eight kinds of visitor experience areas that should be retained are contained in the vision: scenic coastline, marine playground, relaxed lifestyle, special societal features, forests and volcanic hills, sense of exploration, small-scale tourism, and the Maori and heritage experience.

The second section identifies specific *tourism objectives*. These set target numbers of anticipated visitors by five and twenty-five years from the date of the plan. Included is growth of both domestic and international markets. These projected visitor numbers are then translated into anticipated visitor expenditures and employment. Mass tourism by motor coach tours is discouraged, favoring instead the smaller and independent market segments who seek slow-paced exploration, enjoyment, a mix with residents, and enrichment from the setting.

This set of objectives is then followed by *issues*. Tourism Coromandel does not have the authority to do specific development of tourism. Instead its primary role is leadership, advocacy, and guidelines and principles for fulfilling the aim of sustainable tourism growth. Action is to take place by appropriate public and private entities. Following is recognition of several important issues.

Tourism Development and Infrastructure

Any growth will require new investment in infrastructure. Recommended is new measurement of impacts of every increment of growth. Visitor growth must be balanced with the needs of resident growth. Types of preferred tourism growth must be more clearly defined. Encouragement of expanding and updating existing facilities should be favored over new growth.

Marketing of the Region

Encouragement of growth techniques needs to be defined. Marketing needs to be conditioned by the desired type of visitor, season of preference, and adaptation to the specific site situations. Long-term objectives must be defined. All marketing must be coordinated with growth of supply-side development: new interpretive centers, new facilities, new infrastructure, and new resource protection, especially along the coasts.

Transport and Roading

Because independent travelers are sought, forms of mass transit must be studied. At this time, expansion of rented car use will overburden roads, bridges, ferries, and parking. Visitor transportation issues must be coordinated with those for residents. Uniform signage for guidance needs improvement. Cycling routes should be created.

Community Support

The poor communication between tourism interests and communities must be overcome. Little of the Maori dimension is being considered. Parochialism undermines initiatives and collaboration. Community commitment to tourism is lacking in many locations. The needed new actions: building Maori involvement, honoring name "Hauraki," making an environmental audit, linkage with other industries, initiating a high school work program, and better community-tourism linkage.

Protection and Management of Natural Areas

Recognizing that the natural environment provides the greatest foundation for tourism attractions in the Coromandel, new funding and new management policies are needed. Threats to this environmental base include poorly planned coastal development, introduction of exotic flora and fauna, mining, and excessive visitor impacts on certain sites. Recommended actions include greater cooperation with the Department of Conservation (DOC), lobbying for more financial support of environmental initiatives, establish a Coromandel Conservation Trust, and establish new protective areas, especially wetlands.

Implementation

By 1997, the majority of recommendations by the PATA Task Force had been implemented (Burt 1997). Several are in process and some lack funding support. Responses to the recommendations are paraphrased as follows:

- Tourism Coromandel, Inc., an overall tourism organization, has been created. It has provided leadership, coordination, and vision. It collaborates with the New Zealand Tourist Board and is supported by both political jurisdictions.
- The DOC has created an effective plan for the protection of environmental resources. The DOC has a seat on the Tourism Coromandel Board.
- Generally, local residents have accepted tourism as not necessarily eroding their life styles. They approve particularly the sustainable directions of planning for future development.
- Several new and upgraded attractions have been developed: wildlife visitor center, new Karangahake Gorge Trail, new marina. The museums have not yet developed joint management.
- Much progress has been made between Maori populations and

whites in the planning for future tourism, especially for protection and interpretation of the aboriginal cultural background. The Maori have a seat on the Tourism Coromandel Board.

• The sustainable thrust of planning has guided accommodation growth. Community location has been accepted and new building is prohibited within remote areas. No new large resorts are planned. Much remodeling and several new bed-and-breakfast facilities have been established.

Conclusions

This case demonstrates a series of positive planning approaches to tourism development. The New Zealand government and PATA accepted an evaluation of tourism by a special task force. From the study results, the two political jurisdictions agreed to create a tourism development plan. The creation of a vision statement and detailing the many concerns, objectives, and programs of action are commendable. The goal of environmental sustainability is given generous support throughout the plan. All the desirable initial steps were taken and are now proving their value.

CHUN-CHEON RESORT AREA, KOREA

An example of destination planning is the case of the Chun-Cheon resort area, located 100 kilometers northeast of Seoul, Korea. At the request of the Korean Transportation Institute (KOTI), a consultant team was formed at Texas A&M University to evaluate existing plans and develop a revised plan concept. The team was led by Turgut Var, Texas Tourism Research and Information Program, and reported in *Chun-Cheon Plan 2001* (Var et al. 1991).

Evaluation of Existing Plan

KOTI had prepared a rough tourism development plan. Review indicated that it was superior to previous plans, identified new attraction clusters on the islands including a natural history museum, an aquaplanting garden, youth camping, a folk village, and recreational facilities. However, the plan included high-density resorts on the islands, requiring costly new infrastructure (water supply, waste disposal, police, fire protection) difficult to provide and maintain on the islands. These resorts would likely

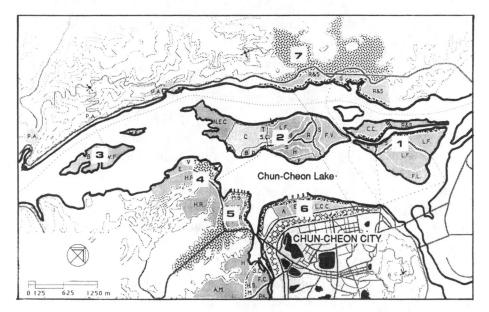

Figure 8-15. Chun-Cheon Resort Plan, Korea. Development concepts were prepared for seven areas: 1, Sangjundo Islands; 2, Hajungdo Area; 3, Bangcseom Area; 4, Samcheondong Area; 5, Toegye-Dong/Onui-Dong Area; 6, Train Station Attraction Cluster; and 7, West Shore Development Area. The main policy was to keep services on the mainland and provide visitor day use to the many attractions on the islands and far shore (Var 1991, 15).

produce sewage pollution, would present difficult access problems, would erode aesthetic appeal, and would not be economically feasible.

Recommended Plans

The consultant team visited the area, interviewed key parties, and studied pertinent documents. From this, several recommendations were made based on the assets of the area (figure 8-15). This plan illustrates the concept of considering not only the islands but the entire Chun-Cheon area as a destination zone. Such a concept incorporates the surrounding mountain region as well.

The concept puts the island development into context of the overall destination. Instead of placing high density resort development where it would be difficult to develop and service, it is recommended that it be established on the mainland adjacent to the city. The islands would be developed with day-use activities at designed and managed attractions that are indigenous to the the area. The opposite mountainside should be placed into national park status with properly designed public access appropriate to the resources.

A summary of the recommendations for seven site areas within this overall destination zone follows.

1. *Sangjundo Islands.* These two islands, separated by a canal, are well adapted to: a park area, fishing sites, reception village (vernacular architecture, shops, food services), craft center, small scale tourist services.

2. *Hajungdo Area.* Tourist activities on this major island could be expanded beyond the existing campground. Several attractions would be linked together by walkways, trains, bicycle trails, and horse-drawn carriages. Included would be: nature center complex (aquaplanting garden, nature preserve, interpretive center, trails, observation tower, museum), outdoor theater, living farm museum, folk village (vernacular architecture, cultural foods, crafts), historic museums, fishing village museum and fishing village restoration, boating clubhouse, specialty stores, and natatorium recreation center.

3. *Bangeseom Area* Because of its low elevation and special natural resources, this island would be maintained as a nature interpretive area. Included would be: botanical garden, major nature interpretive center, boardwalk trails, observation tower, new food shops, and ferry landing.

4. *Samcheondong Area.* This area southwest of Chun-Cheon city already contains the Velodrome, riding track, Memorial Hall, Children's Hall, and some resort development. Recommended: hillside resort complex, equestrian center, vista overlook parks, resort village complex, harbor, and shoreline promenade with walks and rest stops.

5. *Toegye-Dong/Onui-Dong Area.* Just east of the Samcheondong area, there is potential for an amusement theme park, new hotel cluster, new health spa and fitness center, and easy access by rail and highway.

6. *Train Station Attraction Cluster.* Located near the central business district and adjacent to rail and highway access is a site with the following potential: new "gateway" train station, lakeside convention center, exhibition conference center, educational conference center, fitness and aquarium area, shopping and entertainment, food services, and new marina and yacht service area.

7. *West Shore Development Area.* A dominant scenic resource is the mountainside across the lake from the city. This feature should be incorporated into the overall destination zone complex. Probably the best design and management policy would be to place it under national park status. Recommended: mountain scenic drives, overlook and rest stops, major nature interpretive center, children's play areas at service centers, boardwalks along lakeshore, service business expansion at connection points with scenic drives and ferry landings.

Planning Principles

In order to be of greatest assistance to KOTI, some basic principles were recommended to be applied to the physical concepts. These included:

> Accommodations should be suited to the travel market mix.
>
> All waterside development should be integrated into the natural landscape setting.
>
> A "village square" concept can foster host-guest relations.
>
> Parking should not intrude upon lakefront; multi-level structures may be needed.
>
> Landscape attractiveness must be protected and enhanced at all water and land entrance points.
>
> Pedestrian circulation must be properly planned.
>
> All development must protect the water quality of the lake.
>
> Public-private cooperation and collaboration holds promise of best support for development.
>
> Properly planned, this area has potential of becoming a major tourism destination area for both domestic and international travel markets.

As of the date of this book it is too early to anticipate implementation. When the public and private developers begin action, they now have a blueprint for directions their tourism development might take in order to attract and satisfy more domestic and international travelers and at the same time protect basic assets.

CONCLUSIONS

Although these cases are not the result of an international survey, they are from several geographic settings and contain some trends of value to planners of destination tourism. Of the three scales of planning, this level is most critical. The success of tourism as a whole depends greatly upon all the functional elements of communities and surrounding areas. Destinations dominate the supply side of tourism. Market appeal is focused mostly at the destination scale. However, in the past communities and their surrounding areas have been ill prepared for their roles and responsibilities for destination planning and management. The great majority of planning and governance of communities justifiably has been for residents. But with the great growth of visitor populations, community roles and policies must be broadened. Most of these cases demonstrate such a change in recent years.

From these tourism destination cases a few conclusions can be drawn.

Objectives have been broadened beyond economic.

To the sole economic objective of past destination tourism development is being added several others. Plans are now including environmental issues. Attempts are being made to avoid negative impacts, such as water and air pollution, vegetative and wildlife deterioration, and scenic erosion. In addition to awareness of the need for greater conservation, resources are beginning to be seen as essential to the tourism product. The new wave of "green" tourism is more that advertising hype.

Foundation analysis continues as a first planning step.

Although practiced with varying techniques, gaining an understanding of basic assets is a growing principle of destination tourism planning. Most destination planning processes include study of natural and cultural resource foundations, transportation and access, and existing tourism infrastructure. Computer GIS programs can now expedite this process. Helpful has been application of a SWOT (strengths, weaknesses, opportunities, threats) analysis.

Better market-plant match is evident.

Destinations are recognizing the importance of striving for supply development that meets the needs and interests of travelers. Greater understanding of travel market segmentation is taking place. The dynamics of change in supply and demand are being considered in destination tourism planning.

Public-private relations are improving.

Although slow to develop, the great complexity of tourism is requiring new cooperation and even collaboration on destination planning. A first step is greater communication. The many governmental agencies and private organizations are making joint progress on setting goals and taking action on new programs and plans for destination tourism.

Political jurisdiction cooperation is increasing.

In the past, joint destination planning between city and county officials has been hampered by political rivalry and apathy. Because the field of tourism bridges jurisdictional boundaries, more officials are finding the need for new communication and joint discussion on tourism development matters. Especially important is coordination of policies and regulations concerning land use, finance, transportation, health, and safety.

Participatory planning is replacing top-down planning.

The failure of many plans in the past has been due to the lack of involvement of those who will be impacted or taking action on tourism development. Plans by outside consultants and political officials are beginning to incorporate local citizens, business leaders, and investors into the early stages of the planning process. Even though this is more cumbersome and more time-consuming, it promises better implementation of desired action.

Destination tourism planning is demanding new ethnic and cultural understandings.

As communities seek new tourism growth, social issues between hosts and guests must be dealt with. Cultural conflicts can arise if host areas do not create their own plans, regulations, and management principles for hosting visitors. This challenge is especially important as aboriginal peoples seek the benefits from destination tourism development.

DISCUSSION

1. Compare the ways in which these cases show both guidance and hindrance by governmental policy and action.
2. Why does implementation of destination planning encounter so many difficulties?
3. Discuss how urban planning agencies and professionals can become involved in tourism destination planning.
4. What does participatory planning mean and how can it help integrate tourism destination planning between a city and its surrounding area?
5. Discuss the value of a SWOT analysis at the start of destination tourism planning.

6. How can host-guest relations be structured and planned for native cultural tourism destination development?

7. Discuss the reasons why an adult education leader, such as is being used in North Carolina, can provide a valuable function for destination tourism planning.

8. Why is it necessary to employ legal control measures as compared to voluntary action, such as in the case of Sanibel Island?

9. What role can tourist businesses play in determining visitor capacities at natural and cultural attraction complexes?

REFERENCES

Adams, Chris (1994). *Towards 2020*. A Strategic Plan for Tourism in the Coromandel to the Year 2020. Thames, New Zealand: Tourism Coromandel, Inc.

Area Tourism Strategy (2001). Stirling, Scotland: Argyll, The Isles, Loch Lomond, Stirling & Trossachs (AILLST) Tourist Board.

Bennett, John, Mark Smoot, and Johan van Papendorp (1995). "Cape Town: Tourism as a Catalyst for Change." Pp. 216–221. *Proceedings, the 32nd IFLA World Congress, October 21–24*. Bangkok, Thailand: Thai Institute of Landscape Architects.

Bilenser, E. (2000). Turkiye "Bursa Local Agenda 21 Cumalikizik Conservation and Revitalization 98" Project. Bursa Buyuksehir Belediyesi.

Bosselman, Fred P., Craig A. Peterson, and Claire McCarthy (1999). *Managing Tourism Growth*. Washington, DC: Island Press.

Bouctouche Bay Ecotourism Project (1996). Based on reports from Rachelle Richard-Collette, coordinator, and from www.ecotourismnb.com as of 9-22-01.

Brothers, Gene and Carol Kline (2000). "Regional Tourism Conferences: A Community Involvement Strategy for Sustainable Tourism Development." Unpublished paper. Raleigh, NC: North Carolina State University.

Burt, David (1997). An unpublished evaluation of PATA Task Force for Coromandel in correspondence from David Burt, New Zealand Tourist Board, to Dr. Graham Brown, Southern Cross University. June 4.

Cape Byron Headland Reserve (2001). Draft Plan of Management. Prepared for Cape Byron Trust by Andy Baker, Wildsite Ecological Services. Byron Bay. NSW, Australia.

Clark, John (1976). *The Sanibel Report*. Washington, DC: The Conservation Foundation.

The Coromandel Experience (1991). A PATA Task Force Study. Sydney, Australia: Pacific Area Travel Association.

Cunningham, Melissa. (Interview on May 29, 1992, now director of Main Street Program, Denison, Texas; formerly director at Mineral Wells.)

D K Shifflet & Associates (2000). *Indian Country Visitor Profile*. Phoenix: Arizona Office of Tourism.

Dowling, Ross K. (1992). "An Environmentally Compatible Approach to Tourism Development Planning." Presentation, International Seminar on Tourism Development, Shanghai, China, July 12–17.

Garcia, L. Pericot (1972). *The Balearic Islands*. London: Thames and Hudson.

Gunn, Clare A., Carson E. Watt, Sherman Frost (1985). *Mineral Wells; Opportunities for Tourism Development*. College Station, TX: Texas Agricultural Extension Service.

Hopi Tribe (1997). *Overall Economic Development Program of the Hopi Tribe for FY 1993–1995*. The Hopi Tribe.

Kaya, A. et al. (2000). Korum a Calismalari. Bursa Buyuksehir Belediyesi Etud Proje Daire Baskanligi, Tarihi Cevrekoruma ve Yenileme Sube Mudurlugu.

Kelly, Michael E. (2001). Principal Planner, Hopi Tribe. Correspondence, July 5.

Kline, Carol (1999). *Sustainable Tourism in the Sandhills—A Process for Community Development in Anson, Moore, Montgomery, and Richmond Counties*. Raleigh, NC: Parks, Recreation & Tourism Management, North Carolina State University.

McHarg, Ian L. (1969). *Design With Nature*. Garden City, NY: Natural History Press.

Middleton, Victor T.C. and Jackie Clarke (2001). "The Balearic Islands of Spain—Strategy for More Sustainable Tourism Development." In *Marketing in Travel and Tourism*. London: Butterworth Heinemann, pp. 443–448.

Midkiff, Sandra. (Interview on June 5, 1992, now director, Keep Mineral Wells Beautiful program, Mineral Wells, TX).

Oren, Ulker, David Woodcock, and Turgut Var (2000). "Sustainable Tourism. A Case of Cumalikizik, Turkey." Accepted for publication, *Tourism analysis*.

Orkney Strategic Tourism Plan (1997). Kirkwall, Orkney: Orkney Tourist Board. Also: correspondence from Elaine M. Tulloch, Orkney Tourist Board, Kirkwall, April 3, 2001.

Plan (2001). Part 3.2, Protection of Natural, Environmental, Economic, and Scenic Resources. From www.ci.sanibel.fl.us/planning/plan/article3-2.htm as of June 18.

Rogers, Bruce (2001). Interview with Bruce Rogers, director Sanibel Planning Department, June 11, and correspondence, June 20.

Schroeder, Pat (2001). *Hopi Tourism Marketing and Feasibility Study*. Phoenix: Practical Solutions.

Var, Turgut et al. (1991). *Chun-Cheon Plan 2001*. A Comprehensive Resort Development Plan for the Chun-Cheon Lake Area. Prepared for the Korean Transportation Institute by the Texas Travel Research and Information Program, Texas A&M University.

Victoria & Alfred Waterfront (1992). *1992 Review*. Cape Town, South Africa: The Victoria & Alfred Waterfront (Pty) Ltd.

Vision Statement (2001). Issued by Rob Loflin, National Resources Department, City of Sanibel. From www.ci.sanibel.fl.us/dnr.htm as of June 18, 2001.

WCED (1987). *Our Common Future*. Report of the World Commission on Environment and Development. (The Brundtland Commission) Oxford: Oxford University Press.

Wood, David S. (2001). "The Winners and Losers of Participation in Praxis: A Case Study of Strategic Tourism Planning in Australia's Northwest." Unpublished paper. Curtin University of Technology, Perth, Australia.

Chapter 9

Site Planning Concepts

INTRODUCTION

The most tangible and meaningful aspect of the tourism system is the built environment, defined here as the collection of individual sites experienced by the visitor. As one travels, a broad cross section of land uses from cities to rural areas, from domesticated to wild landscapes, and from ugliness to beauty is experienced. No other part of the economy is changing the landscape as much as tourism. The resorts, entertainment centers, hotels, motels, parks, interpretive centers, restaurants, beach development, theme parks, facilities, and land development for outdoor recreation are just a sampling of this massive land development.

In concert with regional and destination tourism planning, it is at the site scale that final development takes place. It becomes clear that this site development has resulted from a mix of amateur and professional designers/planners. As tourism has matured over many decades, the professionals have made great strides in improving the travel landscape, especially at the site scale. Even so, the challenge today is to cope with the many global changes and also make the needed design/planning changes toward the goals of better visitor satisfactions, enhanced economy, resource protection, and community integration. The purpose of this chapter is to offer guidance in the roles and processes for these aims at the site scale.

THE PROFESSIONAL DESIGNERS/PLANNERS

Very important at the site scale of tourism development are the roles of the professional designers/planners. They are the ones charged with the responsibility of not only creating concepts for built environments but also following through with specific construction drawings and specifications. Even though owners, public and private, initiate projects, such as hotels, restaurants, attractions and public parks, they turn to designers/ planners to convert these into physical realities. Generally, designers/ planners can be grouped into four categories even though they often collaborate as teams on tourism development (Cuff 2000).

Landscape Architects

Landscape architects are professional designers who bridge land and landscape characteristics with adaptation to human use. Their practice is defined as "the profession which applies artistic and scientific principles to the research, planning, design, and management of both natural and built environments" (ASLA 1992, 2). They become involved in a wide range of projects related to tourism.

For example, at a World Congress of the International Federation of Landscape Architects (*Proceedings IFLA* 1995), presentations by landscape architects worldwide focused on the theme of landscape design and in a variety of ways. Topics included: tourism impacts on indigenous cultural land use, land conservation and tourism, design of historic environments for tourism, planning and design of gateway cities and parks, community tourism design issues, and roles of landscape architects in tourism. In addition were reports on project plans, such as for a rural resort in Japan; the Parque do Flamengo, Brazil; plants suited to coastal resorts in Venezuela; redesign of St. Bavo's Square, Ghent, Belgium; indigenous land design theme at Finna, Indonesia; new tourism plan for Angkor, Cambodia; and Bintan Resort, Indonesia.

Landscape architects are involved in research, teaching, and consulting as well as professional practice. In many regions they are licensed, based on laws assuring public safety and welfare. Most tourism projects are accomplished by team action, involving other design professionals.

Architects

The creative and detailed design of buildings is the primary focus of architecture and architects. Over centuries, the most visible expression of architecture has been the evolution of style. Even today examples of Greek, Roman, Byzantine, Mayan, Moorish, Gothic, and Renaissance architecture can be found in many modern cities. Over time, building materials influenced architectural expression. Today, the uses of steel, glass, plastics, and concrete are evident in building design. In the late 1800s, the City Beautiful movement using neo-classical styles was intended to create a new social order in cities. Decades later came the Modernism movement that emphasized stark steel-glass buildings, stripped of ornamentation. Today's architecture reverses this trend of rigidity by adding greater humanism (Broto 1997).

Although technology is an important aspect of architectural design, the reliance upon creativity, artistry, and innovation remains. In spite of legal systems, zoning, and societal rules, architects cherish their individuality of expression in their designs. This creative ability provides new solutions to building exteriors, interiors, and spatial relationships. Sometimes, these innovations are poorly accepted by the public at large.

As with landscape architects, architectural professionals receive training in university programs on the many facets of architecture. Based on assuring public safety and welfare in most countries, architects must pass professional examinations and become licensed. They also follow a process of creativity through construction.

In this realm of tourism, architects continue to have an important role in the design of resorts, hotels, casinos, museums, sports arenas, theaters, and other structures. Architects today are no longer bound to past traditions and are influenced by modern social, economic, and environmental trends.

An important tourism issue today is historic adaptation (Warren, Worthington, Taylor 1998). Critics object to the juxtaposition of contrasting modern style against a building of an important historic design. At the other extreme is kitsch design that attempts to replicate an older style with a modern function. Concern over this issue, a very important one for tourism, led to *Recommendation Concerning the Safeguarding and Contemporary Role of Historic Areas* issued by UNESCO in 1976. This, together with the movement toward historicism, has stimulated architects to ameliorate their designs of new structures in historic areas.

Urban Designers

The field of urban design, a hybrid between architecture and landscape architecture, has arisen in response to dramatic changes in communities. The role of urban designers focuses on the physical, aesthetic, and functional form of cities.

Much of their concern is directed toward the trend of community globalization, the homogenizing effects upon identity. The encroachment of lookalike structures is eroding the special distinctiveness of communities.

This change is also an element of concern in tourism planning. A major reason for travel to another location is for its distinctive qualities.

Urban designers work with local politics in order to maintain individuality within the city, desired by local citizens as well as travelers.

Urban Planners

Urban planners consider it their realm to guide the social and economic policies of cities. This function includes elements such as transportation, land use, growth, and other aspects of city management. Urban planners are professionalized alongside architects and landscape architects. Basic competency exams, training, and higher education are required for licensure.

Urban planning requires close relevance to real estate, property values, and development trends. Urban planners today are challenged to include tourism in their scope of professional work. With new environmentalism this trend has begun to change.

DESIGN/PLANNING INTEGRATION

In spite of the separate professions cited here, most design and planning action today requires team effort. The complexity of actual land development, especially for tourism, requires collaborative work that calls upon the expertise of many disciplines.

For example, the design of a new hotel within or near an historic district requires joint input from an architect, landscape architect, urban designer, and urban planner as well as historians. Planning a new or modifying a major park will likely need a team of landscape architects, ecologists, sociologists, and specialists in roads, trails, wildlife management, crowd control, and interpretation. Because of increased interest by travelers, the design of public use of prehistoric sites will require several design specialists including archeologists.

The traditional pattern of a design relationship involves the designer and client (owner-developer). By means of many discussions between these two, decisions on what is to be designed and where are agreed upon. These discussions may involve many iterations in order for both to be satisfied with the proposed action. For major projects there may be several alternative plans in order to reach a final decision.

Although this relationship is still valid as applied to tourism, it requires one more dimension—the user. Generally, designers and developers do not have adequate understanding of tourist markets. Too often, designers and developers make many assumptions based on their limited experience or opinion. Unless a plan for a site development has included a thorough understanding of visitor interests and behavior at that site, the plan is likely to fail.

Needed on the design team is the input of a travel market specialist. Today, market researchers can identify travel market trends, especially market segments. The special site characteristics, such as accessibility and resource foundations, can influence the type of clientele likely to be interested in the location. Useful data on travel market trends are often available from national or local tourist agencies. More refined market segmentation sources include the several scholarly journals. However, it is unlikely that designers have the time or interest in studying travel markets to gain their own knowledge on this topic. The planning of every tourist site development requires the input of a travel market specialist.

For tourism, the designer-client-tourist trilogy faces a major complication—change. Tourism is such a dynamic phenomenon that it is difficult to plan for changes in travel markets and community variables. It is more difficult than planning for a local shopping center because local customer profiles are more easily identified. For tourism, today's clientele for a resort hotel may be lured away by a competitive development at another location and not return. Again, the input of a travel market specialist may provide a hedge against early obsolescence of a site development. Tourism mirrors societal and economic changes the world over.

SITE DESIGN INFLUENCERS

Although professional designers play a major role in finalizing plans so that they can be built, a final plan is not under their exclusive control. Several other groups can exercise direct and indirect control of a design, even to the extent that a good design may never be built. The following discussion identifies several groups that have critical roles in influencing design beyond that of professional designers.

Owners/Developers

No tourism development takes place unless land owners/developers want to do it. And, they want to do it their way. No matter whether recommendations are made at the regional or destination scales, the final decisions are played out at the site scale. As has been stated before, these decision makers for tourism are of three sectors.

In many countries, governments have delegated development of certain aspects of tourism to their agencies—parks, recreation, streets, highways, air travel, historic preservation. This delegation is bound by certain legal mandates that define the powers and activities of each agency. Generally, however, most of these agencies were established for purposes other than tourism. For future planning of tourism at any scale, government agencies will need greater understandings of tourism, input from other sectors involved, and even changes in policies.

Nonprofit organizations, by definition, are not engaged in activities to accumulate wealth but to fulfill other purposes of value to each organization and to society. But, for each development activity related to tourism, the decision will be based primarily on their goals. For example, restoration of an historic building may be a prime objective of an historic society. The tourism implications, however, such as parking, toileting, and interpreting, would be secondary. So final action hinges more on how well each project fits the aims of the organization.

Private sector commercial owners usually make land use decisions primarily on economic feasibility. But, experience demonstrates that for small business there may be many other factors that influence final decisions. Continuing family ownership, interests of heirs, love of location, security, and other external factors often have more bearing on development than profit-making. Certainly, entrepreneurship in other cultures may not be as strong as in industrialized nations, tending to inhibit development of tourism. Final plans and implementation of development vary greatly no matter how designers may have conceived of plans.

Moneylenders

Critical to all implementation are the decisions of the financial sector. Financial institutions have their own policies and practices regarding funding. Both public and private organizations have precise criteria for financing projects. Many excellent plans have never reached fruition because either judgment or regulations prevented financing. On publicly-supported projects, such as parks, arenas, and recreation complexes, critical to implementation will be the public vote to issue bonds or increase taxes. Often, plans for projects are modified following first estimates, usually downscaled. Financing has a major influence by limiting or fostering what is designed and built.

Construction Industry

All projects that require land modification and construction require bids from construction firms. Often, the designer's estimates and inflation do not reflect true building costs, causing major modification of plans.

Labor strikes, supplier business failures, shortages of supplies, material, bad weather, and many other contingencies can often delay or even deny completion of projects. Until costs are known, plans are only ideas, excellent as they may be. Construction costs can either halt or permit the implementation of plans.

Managers

Unfortunately, future managers are seldom brought into planning negotiations until after projects have been approved. Especially important for hotels, restaurants, marinas, entertainment centers, and resorts is the input of experienced managers. Details of functional operation are much more easily changed at the planning stage than after construction. Often owners and developers do not have sufficient experience and knowledge of these details to make sure problems are avoided.

Publics

Several public groups can influence greatly whether a plan is executed. For any governmental expenditure, the voting public would have supported the agency's budget with taxes. In a democracy, if this public is dissatisfied with an agency's management, change can come via the next vote for officials. Or, as American politicians are learning, the electronic age allows direct feedback via radio and television without waiting for the next election.

Increasingly, public organizations are voicing advocacy on many issues relating to tourism. Hotel associations, tourism associations, tour organizations, airline associations are more active than in the past. Many environmental organizations are bringing political pressure to bear on public and private tourism developers. Health and safety organizations often foster new regulations that tourist businesses must comply with. In areas where planning is becoming more active, nearly all plans require several stages of public input and approval.

Public Involvement

Several techniques are being utilized to obtain collective public opinion on tourism projects. The Nominal Group Technique has been applied effectively in many instances. Ritchie (1987) used this technique in Alberta with the public and private sector groups—Tourism Alberta (the provincial public agency) and the Tourism Industry Association of Alberta (the provincial private sector). This study was carried through in three phases: definition of priority issues, identification of needed initiatives and action, and monitoring recommendations. This resulted in identifying 15 themes of highest priority and 22 themes of second level priority. This technique was used in Fredericksburg, Texas, to determine information needed for more effective operation of tourism (Watt et al. 1991). Workshops of a broad cross-section of community residents yielded needs of market profile, advertising effectiveness, and economic impact. These have led to new plans for development and promotion.

Another technique for obtaining public input has been devised in Canada, called the "Co-Design Approach" (Callaway et al. 1990). This was applied to harborfront planning for tourism development in British Columbia. Through a focus-group session, waterfront images and themes were identified. The design features derived from this process included:

- use of natural landscaping with indigenous species;
- a marine architectural theme for all commercial buildings, and whenever possible, street furniture;
- uniform structural theme of round posts and square or rectangular beam pedestrian features;
- water theme areas—fountains, children's play area;
- all "urban-native" walkways, to connect the downtown to the wharf via Harbourfront Village's commercial core.

This technique utilizes a seven-step process of public participation and features artists who capture the themes with graphics. For this waterfront plan, this technique is credited with helping to implement the plan into reality.

There is little question about the power of public involvement in all tourism planning. Although the use of these techniques is no substitute for utilizing the talent and experience of professional designers and planners, there is strong symbiotic value in combining public input with that of professionals.

A classic example is one of the most popular tourist attractions in the United States—the San Antonio River Walk (Gunn et al. 1972). Over its history, a great many individuals and groups have been involved. Perhaps this diversity is responsible for the resulting amalgam of characteristics that make it so dynamic and popular. Following a flood in 1921, a bypass canal was built, providing a stabilized water level for the horseshoe bend. A local architect, Robert H. Hugman, developed a plan for a major visitor attraction. But only after 1938, when a local bond issue provided some funding, supplemented by the Works Progress Administration of the fed-

eral government, was major work begun—rock retaining walls, pictur-
esque foot bridges, rock-surfaced walks, and landscape enhancement of
river edges. In 1961, a chamber of commerce plan for converting it to an
amusement area was rejected. Since then, under the umbrella supervision
of a River Walk Commission, the parks department has continued landscape beautification of the river edges; private owners have developed
nightclubs, restaurants, and shops; the San Antonio River Authority has
controlled water volume and quality; and the Conservation Society has protected historic structures and plants. No one individual or group has been
fully responsible for this outstanding, enriching, and popular travel and
local attraction.

As another case, a different but comprehensive waterfront redevelopment plan for the Baltimore harbor has met with equal success. It is
now one of the most popular gathering places on the U.S. east coast.
Redevelopment of the derelict waterfront was sparked by a local water
resource engineer upon returning from a visit to the impressive harborfront of Stockholm (Billing 1987). The challenge was picked up by
Mayor Theodore R. McKeldin who then authorized the City Planning
Commission, the Greater Baltimore Committee and the Committee for
Downtown to work with planners Wallace, McHarg & Associates. Representatives of many publics were involved throughout the process. A
quasi-governmental agency, The Neighborhood Progress Administration,
formed in 1984, manages 283 acres that include Center, Inner Harbor,
Inner Harbor East, and Inner Harbor West. Since then, new parks, marinas, 90 major new buildings (restaurants, shops, museums, aquarium)
valued at over $2 billion have been established. As a result, not only has
a major tourist attraction been created but also the surrounding area has
been stimulated into complete renovation. This demonstrates that there is
no single most effective category of planner type to do tourism planning
and design.

The Western Australian Tourism Commission has produced the guide,
*Public Involvement in Tourism Development: What Does it Mean for the
Developer?* This publication firmly recommends that all tourism developers have the obligation of letting the public know about plans because
it can be in their best interest. Local people have the right to know how a
project will affect them. Several mechanisms can be utilized before and
after sketch plans have been prepared: personal contact, open houses,
workshops, community liaison group, media releases. As the involvement
progresses, monitoring is suggested, for the following objectives:

To determine public preferences.

To identify outstanding issues of concern.

To incorporate public input into the planning process.

To determine if the public is receiving and understanding the information.

To identify areas which may not have been covered by the program.

(Western, n.d., 20)

Even though public involvement will entail some cost and perhaps delay, it can avoid the need for weathering protest too late and often provides constructive recommendations for improving the project. Public involvement today is an essential element of tourism planning and design.

Governments and Regulations

Governments, from federal to local, establish policies and enact legislation that often impinge on tourism site scale development. In many instances tax abatement and other incentives are offered to tourism developers. Cities have building inspectors who make sure that building codes have been followed concerning issues such as materials of construction, fire resistance and control, moisture resistance, sound control, energy conservation, ventilation, hygiene, air quality, and elimination of safety hazards. All of these will have an influence on the design of buildings and landscapes. These challenge the designer by making the task more difficult, often restricting creativity. Regular review of these regulations is necessary to meet new needs and eliminate those that are obsolete.

Zoning is a legal control of land use. Unless it is coupled with planning it often degenerates into a bureaucratic handicap. When properly used and monitored by competent zoning boards, zoning can guide land uses in better social and economic ways. For tourism site development zoning can prevent use of fragile and rare environmental situations. It can guide the use of signage and layout design, such as building lot setback and coastal area setback.

With the growth of cultural tourism and increased visitor interest in historic sites, special guidelines are often enacted into regulations. Historic districts are designated, within which special controls are created in order to protect the values of one or more historic periods. For example the coastal city of Newport, Oregon, has enacted an overlay ordinance in the historic Nye Beach area, important in fishing a century ago (*Historic Nye Beach* n.d.). A design review process assures landscape beauty, strengthens historic character, improves vehicular and pedestrian access and safety, and other historic qualities. Building orientation, parking, lighting, building height and setback, and architectural styling are also part of the permitting process, especially pertaining to all new development within the zone. This governmental intervention has stimulated both commercial historic reuse and tourist appeal as an attraction. Care needs to be exercised in overdevelopment of adjacent sites.

PLACES AND PLANNING

Place Meanings

As has been described, regions and destinations of tourist development are collections of individual sites. In a democratic society, these sites are owned, developed, and managed by individuals, governments, organizations, and firms. But, even in a democracy, social values are ascribed to sites, such as health, safety, and general welfare. And, all the attributes of places, as described in chapter 7, are equally important for site and building design. Relevance to place is a basic principle of all tourism design.

Places have a variety of meanings, important to all who now own or wish to develop sites. Perhaps a better descriptor is the landscape—land with all its collective attributes. Over time, and in all cultures, landscapes have accumulated characteristics that must be dealt with for any development, including tourism. Motloch (1991) has paraphrased Meinig's (1979) classification of how landscapes are perceived:

    as nature—unspoiled, deserving of conservation

    as habitat—supportive of man, animals, vegetation

    as artifact—to be subdued, conquered by man

    as system—holistic, human-nature as one

    as problem—all is in disarray, needs solving

    as wealth—a commodity to be owned, sold, used

    as ideology—holds ideals, cultural meaning

    as history—cumulative record of man's use

    as place—visual and spatial geography

    as aesthetic—intrinsic beauty, visual value.

In the design of hotels, restaurants, museums, parks, and other tourist attractions and services, the decisionmaker and designer must recognize these many perceptions of place. It is not enough to design the basic functions solely for the tourist service or activity. All attributes of the site and surrounding area must be taken into consideration.

Many times efforts toward increased economic development of communities are so misdirected that qualities of place are eroded and even destroyed. For manufacturing plants and other industries these qualities are not as important as for tourism. Essential to most tourism products are the unique place characteristics that set a community and a site apart from others. Designers of tourist attractions and businesses have the obligation to maintain major qualities of place in their land and building plans. Garnham (1985, 9) has identified several items that give a site special sense of place:

Architectural style.

Climate, particularly the quality and quantity of light, amount of rainfall, and variations in temperature.

Unique natural setting.

Memory and metaphor, what the place means to people who experience it.

The use of local materials.

Craftsmanship.

Sensitivity in the siting of important buildings and bridges.

Cultural diversity and history.

People's values.

High quality public environments which are visible and accessible.

Townwide activities, daily and seasonal.

What is called for today in design for tourism is a shift from the traditional architectural and landscape project emphasis to *placemaking*. Placemaking is not merely the manipulation of materials of architecture and the landscape. Rather, it is the creative adaptation of given site characteristics to new uses, such as for visitors. This adaptation is in sharp contrast to the bulldozer-scalping mindset that destroys the meaning of place that a site once possessed. Placemaking is the retention of the essence of place while giving it new physical and psychological meaning.

For tourism, new placemaking is being exemplified in renewal of deteriorated urban waterfronts. Owners, designers, local citizens, and developers are now incorporating new project designs into a larger context of human use and enjoyment. For coastal planning and design this larger whole encompasses retention of historic values, retention of viable economic uses (shipping, fishing), adaptation of historic structures, retention and restoration of land-water interface values, as well as linkage with land uses surrounding the waterfront site. While new uses for visitors—boardwalks, marinas, shops, restaurants, entertainment—are incorporated into projects, the overriding essence of place is retained. This is place-making for tourism.

New Paradigms

As observed by Motloch (1991), important changes are taking place in how sites and buildings are designed. The traditional approach has been oversimplified and narrowly focused. As a consequence, both sites and their surroundings suffer from lack of understanding interrelationships. The former mindset of designer-client agreement on project objectives is giving way to the reality that sites are integral parts of a larger whole. For tourism, the context of destination and region must be incorporated into every site design decision.

As has been discussed earlier, sustainable development is more than a catch phrase and slogan. It is a principle of significance to every site design. Environments surrounding sites impact upon them as well as the impact of sites on external environment. Even though sites and buildings will continue to require working drawings in order for them to be built, every internal feature and specification must have been influenced by many external factors.

An essential part of this issue arises from the traditional "great man" and "creative genius" philosophies held by some designers in the past. When designers are carried away with their own egos and argue that they know best what good design really is, the finished project is likely to be no more than a monument to the designer's ego.

Designer Bob Scarfo (1992, 3) admonishes his professional colleagues to be more concerned about the quality of the eventual users of projects than winning awards from peers. He states:

> Do we have the scientific knowledge to understand the dynamics of a site within its larger ecosystem, or do we simply apply technical formulas within limited site boundaries? Do we stop to identify the political, economic, and spiritual dynamics that will sustain our project? Do we outline management strategies that contribute to its maintenance? Do we really know how to identify the expressive appreciations of a user group, and transform indigenous aesthetics and historical character into new landscapes?

This is the paradox of design. It is essential that the special talent, insight, and creative intuition of the designer be unleashed as driving forces for every design project. But good design is more than this. The final and enduring test will be its value not only to the designer and the new investor but to all people who will view and use the site as an integral part of human experience in the environment.

This new design paradigm requires a new kind of designer. It requires architects, landscape architects, and urban designers with new sensitivity to human behavior and the science of the environment. For tourism, it is essential that designers cooperate with "clients" to focus more on the characteristics and interests of the travelers than only on the financial "bottom line" of the project's feasibility. It is equally essential that designers and clients turn more attention than in the past to the causes of environmental degradation now being proven not by sensational journalism but by scientific fact.

An increasing number of designers today are proving the value of implementing this new paradigm by accepting the new role as catalyst or "facilitator." Marshall (1983, 88) states that the designer is

> ... one whose role is to bring participants and processes into sync and insure progression toward common quality-of-life goals. The design facilitator would be a prime mover, a mediator, a conciliator, and a communicator of ideas, concepts, and viable alternative solutions to design issues.

DESIGN CRITERIA

Functional Criteria

Although no one would deny that all tourism plans and designs should produce development that functions, it is not always clear just how it should function. Certainly, one can readily document many monuments to plans that didn't work—many communities still have relics of bad urban renewal plans of the 1960s. Tourism leaders of site plans could benefit by testing their plans against the following criteria.

First of all, tourism site plans (structures and landscapes) must provide for *tourist flow* to and within the site. Such a human use function includes vehicular and pedestrian forms of circulation. Today's site plan must provide for easy entrance, movement, and parking of personal automobiles, buses, and RVs. Important also is movement of service vehicles, such as for food supplies, waste removal, fire trucks, and emergency vehicles. All anticipated pedestrian movement within the site must be planned for all subsite activities and include access by handicapped individuals.

All building plans must include comfortable and efficient movement and use by anticipated market segments. Designs at resorts, interpretive centers, entertainment complexes, and campgrounds will require different functional layouts. Special features will be required for elderly, handicapped, families, and special youth groups. Especially critical are historical buildings refitted for visitor use without damage to interiors.

Decisions must be made on the estimated capacity of roads, walks, decks, and building spaces. These decisions usually cannot be based on peak use because of overdesign and excessive costs for all other times of use. For natural resource-based attractions, aesthetics and limits of environmental stress may dictate that management control use, setting arbitrary standards of capacity. People-use also involves highway directional and informative signage that is legible and not misleading. Functional building design requires cooling and heating easily manipulated, interior design that provides for proper reaching, walking, and seating, especially for children and tall or short adults.

All these functional criteria require not only thorough design conceptualizing but also close coordination among owners, developers, landscape architects, and architects.

Designed and built development must also meet the very critical criterion of *structural stability.* With the availability of today's engineering technology, there should be little excuse for structures that are weak and possibly even dangerous for people to use. Yet sometimes standards are overlooked and incompetent builders take short cuts. Most cities have building codes that require adequate standards that must be demonstrated on all plans and specifications. Most business managers of amusement and theme parks enforce their own rigid standards of inspection and maintenance for public safety. Buildings, drives, walks, drainage systems, and all other development of the land must be designed and maintained to withstand use as well as weather conditions that could shorten the life of the structure or threaten the safety of users.

Throughout the world, as tourism expands into a wide range of environments, histories of earthquakes, floods, hurricanes, and other hazardous threats should be reviewed. Architects and engineers are regularly upgrading structural standards to withstand such disasters. Again, team design and planning is required for the creation of site development plans that withstand potential structural threats.

Equally difficult but necessary is function that is a*esthetically pleasing and appropriate.* Here, standards are not as easily defined as for engineering standards of structures. And, the traveling market includes people of a great variety of tastes. But with greater sophistication of travelers, they no longer will patronize ugly and inappropriate landscape and structural planning and design. Some cities and some attraction complexes have provided for legal permitting of the appearance of landscapes and buildings based on the opinion of a panel of specialists of planning and design. Some developments are opened to planning and design competitions. The choice of a winner may be made by government officials, a panel of planners/designers, or public opinion workshops. It should be emphasized that for all planning and design, there must be adequate allowance for creativity and innovation in order to avoid repetition and stagnation.

For the business person and developer, the concept of beauty may seem remote from success factors. For tourism, it is essential to success. Although principles and standards of beauty are less scientific or technically precise, experience has demonstrated its value. Even though travelers may not realize the details of why some establishments are more satisfying than others, they can discriminate between ugly and pleasing. It is the responsibility of designers to incorporate principles of aesthetic expression into even the smallest detail of tourism development. Professional designers generally have the talent, training, and experience to create aesthetically satisfying development.

Among the several principles of beauty in tourism design a designer should seek order. The positioning of buildings on the site and the order in architecture make all development easier for the visitor to grasp. Disorganized sites and building interiors cause confusion and disorientation. Orderly development need not be rigidly oriented but should make the flow of walking, viewing, and experiencing places a pleasurable and unobstructed event.

Especially important in the design of tourist sites and structures is continuity. By design, each part of the landscape and buildings logically flows from another. Too often, especially in places that have been modified over the years, tourist masses block movement. In other instances, the detailing of architecture and landscape features is so mixed up that the visitor sees a jumble of disconnected parts. Good design would engage repetition of style and theme to provide continuity throughout the site.

An aspect of continuity is the balance between harmony and contrast. Too much repetition of the same design can be monotonous. Some resorts are monotonous because all bungalows were designed exactly alike. However, a high degree of repetition can create harmony in both settings and buildings. The same sign format and style, light fixtures, benches, and walk materials throughout a landscape can produce harmony. One way of relieving monotony is to introduce a contrasting element. In an otherwise flat and repetitious landscape, the introduction of a tower or tree cluster can provide interest and focus.

Visitor interest and feelings of beauty can be created by a designer's tasteful use of color, texture, and ornamentation. In both architecture and landscapes, visitors are increasingly weary of bland and sterile settings for their activities. How these elements are expressed in the design of tourist sites varies from urban to rural settings. Generally, more formal and man-controlled styling are most appropriate for urban sites. But, for outlying areas, natural resource settings are dominant and site and building design should be compatible with these settings. In all cases building and site design need to be in harmony, requiring close collaboration between developers, landscape architects, and architects.

These are just a few of the design principles to create beauty available to building and landscape designers. They have become an integral part of their training and creative talent. But, fundamental to all design is first to recognize the intrinsic beauty of a place. The given beauty of a piece of property—trees, slopes, distant vistas—is too often destroyed with a "slash-bulldoze" policy. Landscapes already have character and it is the designer's duty to retain as much of the natural beauty as possible.

Design Aids

Although there is no substitute for the artistry and creativity of the designer, there are many new tools that assist in the design process.

For example, the Environmental Systems Research Institute, Inc. (ESRI) in 2001 donated one million dollars' worth of geographic information system (GIS) software to the American Society of Landscape Architects (ASLA) which, in turn, offers members equipment at a reduced rate and provides new funding for the association. This offer includes: ArcView GIS (spatial information and analysis), ArcView Spatial Analyst (spatial modeling), ArcView 30 Analyst (creation of views of three dimensions), and ArcView Network Analyst (finding preferred routes).

A recent program of technical assistance to landscape architects and others involved in environmental design, and applicable to tourism, was produced by the U.S. Green Building Council (USGBC) and is called Leadership in Energy and Environmental Design (LEED) (Calkins 2001). This program offers guidance and a scheme of credits for building design in several categories—sustainable sites, water efficiency, energy, atmosphere, and indoor environmental quality.

Training workshops and examinations are offered for designers to become a "LEED-certified designer/LEED-accredited professional" (Boryslawski 2000). This exam certifies proficiency in use of the LEED Green Building Rating System, standards and credits for green and sustainable design strategies and resources. Projects that meet these standards can be certified by the USGBC LEED technical committee. A detailed Reference Guide offers details of specifications. Although the program is new, its application is demonstrating worthwhile communication and understanding between designers and clients concerning environmental sensitivity of designs, especially for structures.

Landscape architects creating projects related to tourism can apply credits for many aspects of the design, including site selection, redevelopment, brownfield development, alternative transportation, reduced site disturbance, water efficiency, wastewater control, and recycling. As this system becomes applied more extensively its value for tourism planning will become more evident.

An important Web-based innovation for the coordination of design management is available (Sipes 2001). It is capable not only of tracking, scheduling, and sequencing but also of communication among several design team members located some distance away from each other.

OTHER DESIGN RELATIONSHIPS

In addition to functional criteria for the design of tourism sites, several important relationships need to be considered. Experience has demonstrated that the simple owner-designer contract is not sufficient in today's tourism development. Although this is an important step, it is not sufficient for creating a resort complex, a theme park, a hotel, or other tourist facilities. This statement is not arbitrarily intended to complicate the design process. It is simply a reflection of the complex nature of tourism development. Following are some design relationships at the site scale of tourism that are essential to today's success.

Integration with Other Plans

All tourism planning must be integrated with other plans. New tourism plans may run counter to official city plans. City plans may not allow for the traffic expansion, increased demands on water and other utilities, and the extra burden on police and fire protection required by a site project. Rural and extensive area development planning for tourism must be integrated with plans and policies of major resource managers, such as for forests, hunting and fishing areas, or historic and archeological sites. Again, because many jurisdictions are involved, the task of fulfilling this criterion is not easy but necessary.

Such integration begins with the first designer-client contact. When an entrepreneur, investor, or developer first contacts a professional design firm, the first item on the agenda is communication. Designers will need to obtain information about the proposed project. But equally important for successful tourism is the need to raise questions about the relevance to all other factors that will influence success of the project.

Individuality

In the past there has been a tendency of financial institutions to favor support of new projects that are similar to existing establishments. Apparently this policy suggests less financial risk. Perhaps this has some merit regarding some types of development. Franchise lodging and food services standardize their designs on the basis that the traveling public becomes "brand" conscious and are assured of the same services and quality wherever they go.

However, when attractions and some services are being designed, there is merit in strong design creativity. Creative design is sensitive to the unique characteristics of sites and travelers and produces new and different solutions. The seeming risk of being different is offset by a better match between market and supply, a basic principle of tourism development. Design professionals are not mere technicians, although technical accuracy is essential. They possess the mental capacity and special talent to be creative. Their intuitive powers must be allowed to flourish in order to avoid the sameness that is encroaching upon the landscape. Creativity is essential if the visitor is to be enriched by the travel experience.

Authenticity

To strive for authenticity is a desirable design goal. Travelers resent being promised attractions, services, and facilities only to be disappointed upon arrival. If ethnic dances are promoted, for example, they must be delivered by the site attraction. If historic architecture is promised, it should be genuinely available upon reaching the destination site. These qualities of authenticity again require close communication between the decision maker, promoter, designer, and especially tour operator.

However, increased volumes of visitors and their subsequent wear and tear on sites sometimes demand a modified design policy. For an historic building, for example, five or ten thousand visitors a day could damage rare carpeting and historic artifacts if the restoration design allowed close visitor contact. Channeling visitors over durable walkways and controlling access so that views can be obtained without close contact are often design requirements. Other design modifications, such as air conditioning, new electrical systems for lighting and exhibits, and installation of public toilet facilities, can be tastefully accomplished in historic buildings with care in design.

Authenticity as a principle again endorses the need for sensitivity to place and special environmental characteristics. The more that a designer can emulate the special place attributes, the more competitive will be the business establishment.

Marketability

Although the artistic and creative talents of designers are essential for attractive tourism development, the finished project must be marketable. Tourism facilities and development must meet the test of matching traveler demand. In order to meet this objective, close collaboration between designer, developer, and marketer is required. Innovation may gain favorable response from a new or untapped market segment but cannot be so extreme that no one would wish to visit and use the finished project. As stated in *Successful Tourism Design* (WATC 1990), "Ultimately, developments which have a strong, unique sense of place become quality marketable products."

SITE SUSTAINABILITY

The concept of sustainable development, introduced in chapter 4, has greatest application at the site scale. It is here that the specific issue of balancing resource protection with development moves from policy to action. As stated in a guide to tourism developers, "In tourism, perhaps more than in most industrial developments, there is an opportunity to blend conservation and development in a continuing, lasting and sustainable marriage" (WATC 1989, 5).

Implicit in every design for a tourism project must be a clear identification of potential negative environmental impacts. Developers need not consider this action a constraint but rather a necessity for doing business. The more that harmony can be struck between the natural setting and new development, the more appealing a project can be for visitors. If the scale and intensity of a project is so massive that all the original site conditions are drastically modified, one may question why travelers would go there. Following are some concepts directed toward achieving sustainable tourism development.

Eco-Design Ethics

A major concern relates to irreversible actions and the possibility of regenerative design. A landscape architect may be caught in a power struggle regarding a plan demanded by a client that removes most site vegetation and flattens site undulations. Rather than withdraw from the project, an astute designer may be able to offer an alternative concept that retains original landscape values and yet allows desired development. Both client and designer must take a strong ethical stance against an irreversible deterioration of a landscape's indigenous values. In instances where a site has already been eroded by bad land use decisions, it may be feasible for a landscape architect to employ regenerative design—reinstating desirable environmental characteristics.

If tourism developers and designers were to adhere to a set of "eco-design ethics," environmental protection could become an integral part of project success, either public or private. The following is a capsule of some important eco-design fundamentals.

1. Resort design shall be in scale with the setting and not dominate the natural resources. All waste shall be disposed of without polluting air, water, or soil.

2. Waterfront development design shall not be separated from the water's edge by a road. Access shall be provided back of the development.

3. The location of tourist facilities shall be separated from major cultural and natural resources. Access to these resources shall be planned and controlled so as to not exceed the capacity without environmental damage.

4. Public agency-held parks and preserves shall zone these lands to identify those locations deserving protection as well as locations of "hardened sites" suited to commercial development.

5. All project design shall be acceptable to nearby community residents.

6. Attraction development dependent upon natural resources shall be designed with overlook towers, boardwalks, trails, and other features in accord with protecting wildlife habitat and the native flora.

7. Cultural attraction development (historic and archeological sites) shall be planned ethically so that visitors are given a rich experience without eroding cultural values.

8. Major resort and tourist projects should not be developed in remote areas where the infrastructure (water supply, waste disposal, police and fire protection, and road access) are unavailable.

9. The immediate area within and surrounding important cultural or natural resource attractions must not be developed with incompatible uses that diminish the quality of visitors' experiences.

10. All shoreline development, such as for marinas, residences, and shops, shall respect the natural erosive forces of waves, storms,

and flooding and be placed far enough back from the water's edge so that development will not be damaged.

11. Major land use projects, such as golf courses, shall not be developed in pristine natural areas but rather at locations where minimum stress upon the environment will be made.

12. In the siting of buildings as much of the natural topography, trees, and other plant materials shall be retained as possible in order to provide maximum visitor enjoyment and also protect the resources.

Adoption of these basic environmental ethics by all designers and developers can help build and protect the very resource foundations so important for tourism to succeed—for visitor satisfaction and for business. A step in this direction is the guide, *The Eco Ethics of Tourism Development*, produced by the Western Australian Tourism Commission (1989).

Cultural Sites

The increased travel demand places much greater responsibility on planners and designers to create attractions that protect cultural resources and yet open them for visitor enrichment. Regarding the designer's new role, Antoniades (1986, 398) states, "We could say that 'this discipline' is a most refined one regarding civilizations; it concerns itself with the continuity of civilization, the values and physical testimonials of the past."

The first responsibility of the designer, but not easily accomplished, is cultural preservation. Today, the field encompasses prehistoric sites, historic districts, cultural landscapes, ethnic centers, and places of significant historic events. Prevention from demolition is a first step. But many structures will need restoration and even complete replication. The designer must work closely with owners and cultural specialists to maintain the patina of an earlier era without creating an artificial appearance. Addition of heat, air conditioning, fire control, new utilities, and toilet facilities must be done with great design care.

Equally important is the role of designers for integrating the site into the community. Too often, citizens view restoration as an unnecessary and non-economic act. For tourism, cultural development is a part of attraction development, an essential part of new economic and social input to a community. Stokvis (1984, 180) has identified several key steps for cultural adaptation to tourism:

> Establish a responsible organization.
>
> Establish public-private input for decision making.

Coordinate site and regional cultural planning

Return a portion of tourism revenues for cultural development.

Design sites as integral parts of the community.

He emphasizes the need to coordinate the local groups involved in the arts, economic development, design, and planning, as well as tourism. McNulty (1984) further emphasizes the need in cultural tourism to relate site design to public policy, educate for public understanding of cultural tourism, design facilities and restoration with environmental care, and manage sites with regular monitoring of visitor impact.

Another important aspect of cultural tourism design is the creation of plans for festivals and events related to culture. Museums, exhibition halls, and outdoor exhibit areas require specialized design for display, interpretation, and especially the physical handling of masses of visitors. The relationship of these designed places to support tourism facilities is particularly important. These cultural attractions not only stimulate local interest and pride but provide strong economic impact. For example, the Rameses the Great exhibition featuring 74 Egyptian antiquities, including a 47-ton, 25-foot statue of Rameses the Great, required special design for housing and interpreting this important exhibit in the city of Memphis, Tennessee (Kyle 1992). It attracted over 675,000 visitors, including 110,000 schoolchildren. The economic impact is estimated at $85 million.

In addition to historians, archeologists increasingly play an important role in tourism planning. To protect rare artifacts and also offer worthwhile visitor interpretation requires close collaboration between designers and archeologists, both terrestrial and nautical. Some time ago, King et al. (1977, 193) admonished archeologists to "serve as critics, advisors, and watchdogs over both statewide survey and planning and over the execution of particular projects."

Cultural site design promises to take a major place in all tourism planning and development in the future.

SPECIAL SITE INTERESTS

Interpretive Centers

An excellent design and development solution to many of the environmental concerns at the site scale is the establishment of visitor interpretive centers. As defined here, *a visitor interpretive center is a facility and program designed to provide a rich, accurate, and entertaining visitor understanding of natural and cultural resources.* These installations are proving to be successful in providing a fascinating and memorable visitor experience, and preventing degradation of the resource because masses of tourists are accommodated outside the protected resource zone. If created by a joint public-private venture, it could be less costly for the public agency. The keys to success for visitor interpretive centers are as follows.

The purpose should be clearly defined as:

- to interpret the area's natural and cultural resources;
- to provide an enjoyable and stimulating visitor experience;
- to provide a setting for environmental education;
- to provide a design alternative to mass erosion of natural and cultural resources; and
- to add an important tourist attraction complex to others within a destination.

Interpretive visitor centers should be located on "hardened" sites, away from rare and fragile resources that would be damaged by excessive human intrusion. These locations should be accessible by automobile and tour bus. Adequate parking around the facility would allow for volume use.

The interpretive center would include a lobby, receptionist, restrooms, exhibits, demonstrations, and video room. Optional features would include food service, souvenir and gift sales, and lecture room. Exhibits and dioramas, as used in museums, would provide resource information in themes for best enrichment, education, and enjoyment by visitors. The facility would often have outdoor exhibits of features important to the resource setting. These exhibits would have descriptive signs and/or video presentations for visitor understanding. Appropriate would be a living museum nearby that would mix cultural with natural resource interpretation.

The landscape and building design styling should not be extreme but rather should be a timeless expression of purpose as well as a themed adaptation to the site. All structures should be designed in harmony and provide the visitor with a sense of place. The visitor should be more aware of the content and purpose of structures than of the architectural styling of the interior. The center could also serve as a staging area for guided or self-guided tours to nearby natural or cultural resource features. Such tour walkways would be designed with proper surfacing, stairways, and sizes to accommodate a maximum capacity of visitor use.

Management and service facilities should be located off-site where their appearance and functions will not intrude upon the aesthetics and dominant theme of the center. Only sufficient office space for minimum operational activities should be included in the main interpretive center.

When properly located, designed, and managed, such visitor interpretive centers can function as a "surrogate attraction." In other words,

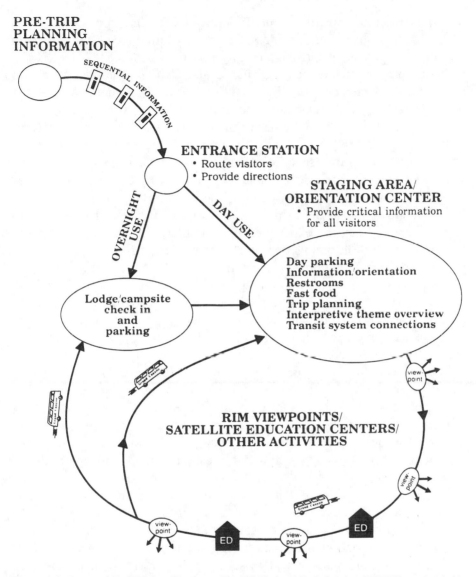

PRE-TRIP PLANNING INFORMATION

SEQUENTIAL INFORMATION

ENTRANCE STATION
• Route visitors
• Provide directions

STAGING AREA/ ORIENTATION CENTER
• Provide critical information for all visitors

OVERNIGHT USE

DAY USE

Lodge/campsite check in and parking

Day parking
Information/orientation
Restrooms
Fast food
Trip planning
Interpretive theme overview
Transit system connections

RIM VIEWPOINTS/ SATELLITE EDUCATION CENTERS/ OTHER ACTIVITIES

view-point

ED

Figure 9-1. Proposed Park-Visitor Sequence. The US National Park Service (USNPS) prepared this model visitor sequence for Grand Canyon National Park. This plan is equally applicable to other natural and cultural resource areas, placing lodging and other services just outside the park (USNPS 1992, 12).

instead of allowing visitors to wander over the resource base promiscuously, threatening the resource elements and learning little about them, visitors gain great insight and leave without damaging the environment. For the purists who wish a more intimate natural resource experience, a quota system by management can provide a limited number of properly trained individuals to penetrate the resource to a greater depth along designated trails. (See Figure 9-1 for proposed visitor interpretive sequence.)

For example the results of a public-private workshop sponsored by the U.S. National Park Service included "the absolute need for adequate funding for state-of-the art, high quality interpretation" (NPS 1992,19) to meet the present visitor needs of the Grand Canyon National Park and other areas managed by NPS. Among the recommendations was the redesign of the park orientation into a visitor center that would provide introduction to the several themes of the park. Existing historic structures should be studied in terms of what they represent regarding the area, natural or cultural history, specific resources, and what the structure represents in terms of park development or design style. These structures should be converted into museums. Where shops are appropriate, sale items should relate to the exhibits and theme of the site. The workshop results included also an expansion of park interpreter responsibility and function, such as initiating impromptu comments, presenting brief programs at noon in restaurants, and relating factual information in an entertaining manner. The intent is to give visitors not only an enriching experience but greater pride and respect for the resources and facilities provided by NPS. Volunteers and employees of concessions should also be trained as park interpreters. Especially important are programs in other languages to assist visitors from other countries.

When located at the edge of protected resource areas, such as parks and preserves, visitor interpretive centers could be augmented by other supplementary services, such as an outdoor theater for appropriate drama production, a specialty restaurant featuring foods of the area, base for motorcoach tours, and possible car service. The private sector could provide valuable input to government for establishment and operation of these facilities.

Modern-day aquariums, museums, zoos, often located in urban settings, are proving extremely popular. Similarly, visitor interpretive centers for natural and cultural resource enjoyment and education are becoming equally popular tourism design features.

The design and management of cultural centers need to be directed toward desired objectives for both visitors and local people. Six objectives were identified by those sponsoring the design and development of a model center in New Mexico—Pueblo of Pojoaque (Warren 1991, 4):

1. Involve 800 local artisans—dance, pottery, crafts.

2. House headquarters of the Eight Northern Indian Pueblos Art Council.

3. Retrieve significant local artifacts from the Smithsonian Institution, Washington, D.C.

4. Expand the visitor center—greater interpretation, cultural exhibits, crafts, shops, traveler services.

5. Create 50 jobs locally.

6. Distribute report on this model to other Indian tribes.

This project is partially sponsored by the Administration for Native Americans, U.S. Department of Health and Human Services. The overall goal is to increase native self-sufficiency as well as educate the public on this interesting cultural era, ethnic group, and locale.

There is little question about the greatly increased demand world-wide for traveler visits to cultural sites. Critical then is the increased application of creativity, skill and cultural sensitivity by designers and developers of these sites in order to enhance visitor satisfactions and perpetuate cultural qualities that are being threatened.

Again referring to Figure 5-4 (page 137), this model of attraction design suggests a planning concept whereby new and contrasting development does not completely smother its cultural sense of place. However, historic small towns often face great controversy in their plans to preserve or "become modern." The case of the historic mining town of Central City, Colorado, provides a typical example.

From the perspective of today's visitor, two images of the community are projected. New gambling establishments within and beside historic structures of this small town (population 335) have engulfed the community. Raw land grading, a cacophony of gaudy signs, and voluble hucksters at high-priced parking lots dominate the landscape. For the gaming travel market segment, this new image is an ideal match. For the cultural tourist segment, however, it is a shock. It is extremely difficult for a traveler to imagine the character of this important jewel of Americana as it appeared in the 1870s. Even as late as 1964, it was such an excellent example that the entire canyon was designated a National Historic Landmark.

This dichotomy of tourism planning is symptomatic of longstanding community divisiveness. Stokowski's study (1992) of Central City concludes that planning has not yet resolved the stress between the preservation of an historic spirit of place and externally imposed gambling. Her challenge to planners is paraphrased as follows:

> Tourism development should be concerned with producing and creating sustainable futures, not only in economic terms, but also in terms that are human, and on very personal and collective scales.

Xeriscape

A major issue of sustainable site development is the large quantities of irrigation water needed to maintain landscape improvements that have been popular in the last few decades. In many regions, water availability has been reduced greatly, humidity levels have escalated and costs have risen.

The preferred solution is to design landscape changes at tourist developments so that a minimum of new irrigation is required. In many instances, large lawn and ornamental plant beds can be avoided with studied design. Designers, owners, and maintenance staffs need to collaborate on design in the early stages to minimize the demand for water. Today, recycling of wastewater from the facilities is becoming a practical solution in many locations. Many resort golf courses are now irrigated with such waters, reducing the demand on local aquifers.

These trends have given rise also to the concept of "Xeriscape," the selection of plant materials requiring much less water and yet offering desirable aesthetic functions. Landscape architect Ueker (1992) has summarized this concept with the following seven principles.

> *Planning and Design.* Plan and zone areas according to water needs and microclimatic conditions.

> *Soil Analysis.* Results can assist in reshaping land form for greater water retention.

> *Plant Selection.* Select plants adapted to area that require minimum amounts of supplemental irrigation.

> *Practical Turf.* Place turf in areas to be irrigated separately and only where absolutely necessary for effect.

> *Efficient Irrigation.* Select system that can provide irrigation only when and where needed.

> *Use of Mulches.* Mulched planting beds can retain moisture, reduce weeds, and prevent erosion.

> *Appropriate Maintenance.* If these recommendations are followed, less fertilizer and less water and chemicals will be needed, reducing maintenance costs.

Cultural Landscapes

A major tourism development thrust today is historic restoration and interpretation, focused primarily on structures. Important is the adaptation of historic structures to visitor use, retaining the original architectural elements but modifying for visitors to observe and understand the structure's role in history.

Equally important but receiving much less attention are the sites and landscapes of history. The land forms, vegetation, wildlife, waters, and weather, for example were powerful factors in historic migration into America's western frontier. Throughout the world, the land characteristics have had major influence on settlement and economic development.

When historic redevelopment is made for tourism, landscape restoration is as important as structural preservation. For example, throughout the route of the Lewis & Clark exploration in 1804–06 across America new interpretive centers are being established. Few artifacts remain as clues to structural development, but the log of naturalist Meriwether Lewis offers fertile documentation of the diversity of plant and animal life along the trail. Landscape architect Don Brigham (2001) has researched this issue and recommended the use of historically accurate plant materials during site restoration, a lesson for historic redevelopment design for tourism everywhere.

The National Park Service (NPS) of the United States has responded to the need for guidance on protection and restoration of cultural landscapes by providing its *Guide to Cultural Landscape Reports* (Page, Gilbert, and Dolan 1998). Although directed primarily to designers and managers within NPS it is available and equally applicable to others. The term cultural landscapes encompasses four categories of geographic areas.

Historic Site—a landscape significant for an event, activity, or person.

Historic Designed Landscape—designed building or site noted for its historic value.

Historic Vernacular Landscape—site reflecting endemic conditions of culture, customs, land use.

Ethnographic Landscape—heritage sites, massive geological formations, special plant and animal grounds, contemporary settlements.

Required for each project is a Cultural Landscape Report (CLR) that defines the issues, tasks required, management objectives, and project agreement. The "Process for Preparing a Project Agreement" includes three major steps: project initiation, preliminary research, and site visit. This begins the identification of design needs—restoration and design, new treatment needed, site access, land use, visitor services, operations, and interpretive programs.

There are thirteen descriptive documents within the Cultural Landscape Reports that are available to landscape architects, architects, and private developers of cultural sites as well as within NPS. Included are topics such as landscape characteristics, geophysical survey techniques, geographic information systems, and treatment of plant features. Particularly

valuable for both designers and developers is the extensive listing of refer-
ences. Implementation is required by NPS but would also be worthwhile if
used by the private sector. Essential to better cultural tourism planning and
development universally are these process recommendations.

Site Continuity

One of the first to publicize the virtues of making an environmental evaluation of land before development was landscape architect Ian McHarg
(1969). In 1971, resort developer Charles Fraser engaged McHarg to
apply such an evaluation to his proposed resort development of Amelia
Island off the coast of Florida. Almost thirty years later, landscape architect Jonathan Sutton (2000) returned to the site to analyze how well the
plan recommendations had been followed.

The results were mixed, demonstrating the reality that plans are guidelines and subject to many influences. Although the first buildings built followed the design recommendations and were low profile and sited behind
the dune, later expansion allowed taller and wider structures to be built
even closer to the shore. A large hotel has violated the plan's emphasis on
protecting shoreline landscape aesthetics. Instead of rigid protection of
marshland, encroachment is degrading this important resource. Long
Point, an area originally planned to remain open, has been built upon. If
these trends continue, environmental sustainability will be severely threatened, in spite of many aspects of the development that have shown respect
for the original plans.

This case is typical of the difficulty of following through on protecting
the integrity of landscape values against the competitive pressures of
market demand for less sustainable projects.

New Design Philosophies

Designers need to be aware of two very important contemporary philosophies of design.

An ancient but newly applied site design philosophy being used for
tourism as well as other development is *Feng Shui*. Originating in China
some 3,000 years ago, this philosophy is based on the belief that everything on earth, such as landscape, water, wildlife, and mountains, has a
spirit and exists best in certain environments (Dankittikul 1993, Lip
1986, Too 1994). It is applied today in the siting and design of hotels
and resorts in Asia and is spreading elsewhere. Testimony to the effec-

tiveness of Feng Shui design principles is the overwhelming success of these businesses. The practice is based in an animistic spiritual tradition of relevance among geographical, topographical, and climatic conditions as well as powerful spirits of a locality. A designer will need the trained input of a specialist, a *geomancer.*

Feng Shui is based on the belief that energy flow, the *chi*, from slopes to valleys and water should not be interrupted. In Asia, one can see new hotels with an opening several stories high and few units in width to allow *chi* to flow from background hills to the river in front. When *chi* is respected, the owners and guests will be favored by success and personal satisfaction. The best sites are sloping and undulating, not at the peak of hills or mountains. Landscape vistas and soils suited to vegetation are most auspicious. Paramount is keeping nature in balance.

Many of the principles of *Feng Shui* are similar to those of landscape architectural training. Before designers discount this design approach as fanciful and impractical, the literature should be reviewed and examples of *Feng Shui* application in the real built environment should be examined.

These Chinese land design principles were part of early beliefs in biophilia, that when properly respected, nature is beneficial to humankind. Recently, researchers have proven scientifically that exposure to natural landscapes is not only pleasing but also a force in reducing stress (Ulrich 1979, Ulrich and Parsons 1992). This belief, now proven, was fundamental to all pioneer works in landscape architecture.

Much of this research is being performed at the Visualization Laboratory at Texas A&M University. Physiological techniques are used to measure responses to landscape scenes that nominally have been called beautiful or repulsive. Methods of nonverbal response included the use of sensors on respondents to record muscle tension, electrical conductance of the skin, and pulse transit time, a noninvasive method that correlates highly with blood pressure. When combined with verbal responses, the research demonstrated that beautiful landscape settings reduced tension and offered pleasing sensations.

Much of Dr. Ulrich's work has been applied to hospital layout, landscape setting, and interior design with remarkable success. The research offers many implications for parks and tourism as well. The effects of landscape qualities can now be attributed to positive tourist reward from a great diversity of activities based on natural resource factors.

Landscape architects, in their site analysis process, can determine characteristics of sites that should be retained and would allow a minimum of modification for tourism site development. Important is the siting of buildings and layout of other features, such as circulation. This step fosters visitor enjoyment of landscape beauty at the same time it promotes sustainability. Architects, in their design of buildings, can place windows and decks to gain best landscape view from interiors. Collaboration between architects and landscape architects can utilize new philosophies to advance tourism site design to new levels.

Resource Protection Controls

Many developers, especially business interests, resent the concept of controls. There is no doubt that in many instances, governmental controls have become excessive, conflicting, and obsolete. But even accepting a degree of truth to this connotation, there is another, and very constructive, side to controls. In a modern society, rules and regulations are necessary in order to accomplish objectives and pursue goals.

The major focus of this book is the absolute necessity of resource protection for tourism's survival. Owners, developers, and designers can *enhance* success of establishments by utilizing the land use tools available to them. For example, the New Castle County Department of Planning, Delaware (1989), has identified several strategies for resource protection available in the United States that could be utilized to the betterment of tourism land development. These are paraphrased as follows.

Government Regulation

County zoning and land use controls may be of value. These can place limits on disturbance of identified resource areas. A "delay-of-demolition" process can foster the preservation of historic sites. Flood plains, steep slopes, and certain resource/open space areas could be given resource protection. Controls exercised by governmental agencies at the local, state, and federal level could be effective in protecting important resources.

Some other strategies that have potential: prohibition of demolition of selected historic structures, performance zoning, mandatory clustering, scenic road overlays, improved flood and drainage controls, density transfer, conservation easements, acceptance of donated lands for protection, tax abatement, and recreation dedication of subdivision lands.

Site Planning Techniques

Increased dialog between land developers and governmental planners early in the planning of projects can increase understanding of the benefit of resource protection. The maps and reports prepared by professional planners can help explain the rationale of balancing resource protection with development. Site design processes that include resource analysis can heighten awareness of protecting fragile sites rather than using them for development. A site plan review process between developer and government planner can foster mutual agreement on common goals of development and resource protection.

Private Land Stewardship Actions

Where land is owned primarily by a few large landholders, cooperation on resource protection issues is more readily obtained than among a great number of owners. A smaller constituency can more readily reach consensus on principles of land development and protection. When put into practice it can be used to leverage owners of smaller properties. Donation of tracts of special resource assets can often be to the advantage of landowners, such as: gifts of land with retained life estates; bargain sales of land in which part of the value of the land is given and the remainder is purchased by an organization or government; and donation of conservation easements which allow some uses but restrict others. The publishing of planning documents can serve as a stimulus to private landowners to initiate conservation easements. Because this is interpreted as a legal donation, there may be tax reduction advantages.

Historic Resource Protection

Official state and national designation of historic sites, following inventory recommendations, assists greatly in resource protection. Tax abatement, revolving fund, and general public education can aid in the application of this strategy. Important as a foundation is the cultural resource inventory that must be extensive and regularly updated. Efforts need to be increased to encourage owners of potentially worthy historic redevelopment to nominate them to federal and state historic protection authorities. For example, owners of income-producing National Register properties, rehabilitated to National Park Service standards, can receive a 20 percent tax credit on the project cost. It may be necessary to revise historic zoning regulations to allow a greater number of properties to obtain historic protection status and to exercise more realistic controls. For example, legislation may be needed to increase fines for unauthorized demolition of historic sites. There needs to be a program of increased incentives for adaptive reuse of historic resources. New zoning stipulations could increase identification of historic resources during all land planning processes.

Highway Planning

A major need as related to tourism is to encourage and assist highway authorities to consider the needed changes in standards required for scenic highway planning development. A scenic road could be defined as one having a high degree of natural beauty and historic or cultural value. The criteria of the Federal Highway Administration were modified slightly for this area. The results of scenic landscape inventory were also used. Positive and negative attributes of roads were identified leading to designation of highway segments with scenic potential. Highway planning policies may need revision to favor less scenic routes for expansion as traffic increases. Especially important will be collaborative planning between highway officials and land owners adjacent to scenic routes. The county officials, responding to their constituencies, may be the best catalysts for accomplishing scenic highway protection.

Public Agency Coordination

A major need identified in this study was increased cooperation among all governmental agencies if better resource protection is to be accomplished. For some issues, such as water, interstate cooperation is needed. Intergovernmental cooperation will not come about without better communication on environmental issues. Perhaps a joint advisory group is needed.

This example demonstrates some of the environmental control problems and techniques for solution that are applicable to tourism planning and development. Without clean water, abundant wildlife, attractive vegetation, reduction of soil erosion, reduction of air pollution, and protected cultural resources, tourism cannot thrive. Resource protection is an essential part of doing business.

A MODEL SITE DESIGN PROCESS

Even though the design steps for tourism sites are similar in principle to other applications, there are many special needs of tourism that must be recognized by planners and designers. For example, for virtually all other site and building design, the user of the finished project is more readily

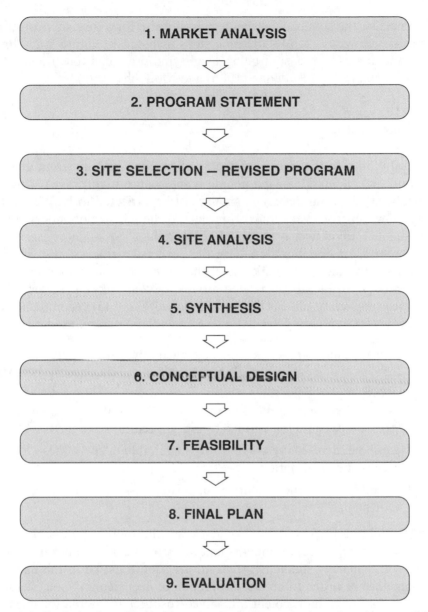

Figure 9-2. Site Design Steps. Sustainable tourism development at the site scale can be accomplished by team planning that includes professional designers, property owners, developers, and involved publics. Such a sequence is desirable for attractions, services, and interpretive centers, as well as for other specific tourism site developments.

visible. Second, a great amount of tourism design is for rural and small community settings, often outside the normal professional work load of designers. Finally, no other site design demands resource protection to the extent required by tourism.

Generally, the sequence in the design of a site for a tourism project put forward here would follow the nine steps as listed in Figure 9-2. This is an adaptation of site design being taught and practiced today and advocated by Molnar and Rutledge (1986) and Motloch (2001).

1. Market Analysis

Generally outside the designers realm of training and experience is the field of tourism markets—the people who do and might travel. This is a specialized topic and designers should not be expected to have expertise in it. However, for every tourism project at the site scale, understanding the potential travel user is absolutely essential.

Designers, in collaboration with the owner-developers, will need to engage the services of a market analyst. Today, the total travel market is segmented into several specialized groups. Few attractions, services, and facilities can satisfy all. Forbes and Forbes (1992, 143) recommend several points for evaluating special interest travelers, paraphrased here:

> Identify and articulate specific, significant profiles of potential travel groups that contain detailed characteristics, purchasing behavior, needs, and expectations.
>
> Develop clear management goals as a foundation to conceive, organize, and execute the various roles required to serve the traveler.
>
> Determine the existing competencies available to serve these traveler needs and expectations.
>
> Carefully assess new market opportunities—areas of significant potential.

Although the responsibility for assessing the market potential for the proposed project to be designed is that of the owner-developer in cooperation with a market specialist, such information must be in the hands of the designer in order to create the proper project design. Consideration should be given to both potential domestic as well as international travelers. Sources of help in assessing market demand include government agencies, nonprofit organizations, and private consultants.

2. Program Definition

Based on the developer's intended market and management goals it is essential for the developer in collaboration with the designer to reach consensus on program definition. This definition encompasses every planned structure and land use intended for the site design project. Motloch (2001) has observed that more projects fail because of incomplete or erroneous program definition than from any other cause.

In order to avoid this risk, a great amount of time for study and discussion between designer and developer should be scheduled. A designer cannot create concepts of site designs without a clearly stated program definition. Many trial drafts will be needed in order to reach consensus on what the owner seeks, what uses are to be made of the site by visitors, and what the designer's intuition suggests is feasible on the site.

The outcome of this step is an estimated program statement, not a fixed list or pattern. Subsequent steps, such as a detailed site analysis, likely will suggest needed revision of the program.

3. Site Selection/Revised Program

Ideally, designers and developers collaborate on the site selection process. Landscape and building architects are experienced in qualities of suitability of land for project development. Often, a preliminary study is made comparing the suitability of three or four prospective sites for the project. Even without detailed analysis and estimating, rough feasibility can be determined for each, resulting in one being selected as superior to all others. Many factors need to be considered such as access, extent of land preparation, size, location, cost, availability of land, land use regulation, relationship to competition. Often, property that may appear relatively costly may be favored because it has a better feasibility for financial success.

After narrowing the choice to one site, cooperative discussions between designers and owners may result in modifying the program. For example, because of the social and physical environment, the program for a resort hotel may have to be changed from the original highrise to a series of separate lowrise units.

4. Site Analysis

Detailed analysis of the piece of land selected for the proposed project is a two-part investigation. Very important is a detailed description of the several internal characteristics of the site. But, equally valuable in the design process is assessment of external characteristics. Molnar and Rutledge (1986, 93) have identified several factors needed in the analysis of a site for the design of site projects based on experience with parks.

ON-SITE FACTORS:

1. Constructed elements
   - Legal and physical boundaries, public easements
   - Existing buildings, bridges, and other structures including those of historical and archeological interest
   - Roads, walks, and other transportation ways
   - Electric lines, gas mains, and other utilities
   - Existing land uses
   - Applicable ordinances such as zoning regulations and health codes.

2. Natural resources
   - Topography, including high and low points, gradients and drainage patterns.
   - Soil types, for clues regarding permeability, stability, and fertility.
   - Water bodies, including permanence, fluctuations, and other habits
   - Subsurface matter: geology and functionally valuable material such as sand, gravel, water.
   - Vegetation types—trees and other plant materials and individual specimens of design importance
   - Wildlife habitat and species of importance
   - Climate and weather: temperature, days of sunshine per year, sun exposure, precipitation, wind and storm frequency and directions.

3. Perceptual characteristics
   - Aesthetic characteristics including views, features
   - Smells and sounds within the site
   - Spatial patterns, such as distribution of wooded areas, open spaces, water features.
   - Design character as derived from lines, forms, textures, colors, size relationship.
   - Overall impressions derived by the designers.

OFF-SITE FACTORS

1. Surrounding land uses and their characteristics, especially compatibility with intended site program, trends in land use change; shading from high-rise development.

2. Stream and drainage sources and their characteristics such as stability, flooding, and potential pollution.

3. Negative influences of sounds and smells generated nearby, such as from agriculture, industry, and entertainment.

4. Aesthetic influences, such as views from site outward. If landscape factors are beautiful in surrounding area, are protective measures in place?

5. External utilities—their accessibility and capacities.

6. Transportation and access, new and future trends.

It must be emphasized that these qualities are identified only for one reason—to assist in design decisions. Every factor may have a bearing on what is to be placed where and how the layout will function best. Sometimes designers become so caught up in details of analysis they fail to relate these facts to their design objectives. Close communication between designers and clients can avert this problem.

The documentation and presentation of results from this analysis process often utilize maps, descriptive narratives, and verbal presentations. Computer GIS overlays and other graphic programs may be utilized to simplify complicated analysis tasks, make rapid visual images, and reduce costs.

5. Synthesis

Although intuitively, thus far in the process, the owner-developers and designers have been considering how well the proposed site and its development will meet the needs of visitors, there must be a time for double-checking these factors. Too often, designers are tempted to skip this step and go directly into the preparation of plans.

This step of synthesis is intended to derive meaning from the great amount of facts that have been gathered. Throughout the study of travel market trends, program definition, and site analysis, the designers and developers should have been getting ideas for appropriate development. Perhaps the earliest concept of a high density resort has now been modified to serve a different market segment. Perhaps this modification has completely changed the potential sponsorship from a purely commercial operation to a joint venture with a nonprofit organization. Perhaps the site analysis revealed new opportunities to develop facilities for a market segment not considered possible in earlier stages.

An important part of synthesis is experimenting with functional relationships between the several elements of the project. Illustrated in figure 9-3

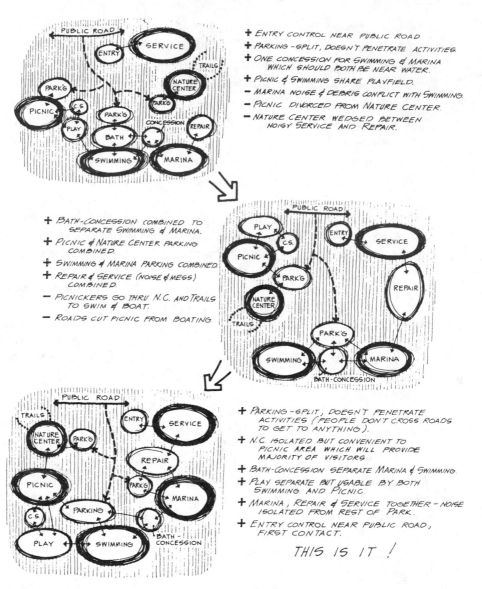

Figure 9-3. Bubble Diagrams of Functional Relationships. It is desirable for the professional designer of site projects to create alternative site development concepts for discussion. Such graphics illustrate potential functional relationships, roughly adapted to the site. Subsequent sketch plans have much greater chance of success when preceded by this step (Molnar and Rutledge 1986, 99).

are "bubble" diagrams showing rough relationships between the program elements required by a park agency (Molnar and Rutledge 1986, 99). Although this effort is roughly related to the site, it is primarily for the purpose of establishing plan policies and spatial linkages or barriers. Such experimentation by the designer can raise many questions of the owner. The owner can react by visualizing how well different arrangements will suit visitor needs, fit the general site conditions, meet feasibility requirements and provide for efficient management and maintenance.

Within the synthesis step should be two very important actions—public participation and sustainable assessment.

Conflict with residents can be avoided by revealing to them the results of site studies and program information. Tourist businesses are tempted to maintain secrecy within these early steps. The public, however, deserves to know what is being developed and how the site will be used. Although this step is sometimes forced by community regulation, there is greater publicity value if it is initiated voluntarily by the developer/designer team. Properly planned and managed workshop meetings can provide valuable public participation at this point.

Because of the groundswell of environmentalism, the developer/designer as well as the local public need to define how the intentions of this project encompass environmental sustainability. This point includes social and economic potential impact as well as impacts associated with natural and cultural resources.

These actions may seem time-consuming and delaying but can prevent major roadblocks as the planning process continues.

6. Conceptual Design

If all the preceding steps have been taken and considered thoroughly by the developers and designers, the design team can now engage in the creative thinking and conceptualization to give form to a plan. This results in what is often called a sketch plan or a preliminary plan. It is at this stage that the value of the first five steps becomes clear. The many hours of discussion among the designer, developer, and public concerning the problem statement, site selection and analysis, and synthesis now begin to pay off.

Recommended are two or three alternative sketch plans based upon different assumptions. Alternatives allow discussion and debate upon the relative merits of each. These are based on facts but also incorporate the imaginative thinking and artistic creativity of the designers.

These sketch plans of both site and structures give form to the project for the first time. The site plan shows protected areas, all program use areas, circulation systems (drives, walks, parking), entrances—all logically and creatively related to program and site conditions. The creation of sketch plans reveals the bulk, character, functions, and spatial relationships of all structures.

7. Feasibility

With a clearer vision of what is to be developed and where, it is now possible to come to conclusions on feasibility. In the past, feasibility has meant primarily *financial feasibility*. Today equally important are short and long-range sustainability.

The first step in preparation of financial feasibility is an estimate of total capital costs—site, site development, architectural development. Designers, in collaboration with landscape and building contractors can make estimates of total costs of construction for each of the alternative sketch plan solutions.

A *physical environmental feasibility* refers to the negative and positive environmental impacts of each of the designs. Even if environmental impact statements are not a legal requirement, an accurate assessment of how the project will influence all existing natural and cultural resource factors is needed.

Social environmental feasibility is an increasingly important step. All new tourism projects will exert some impact on the social, economic, and life style factors of an area. Too often in the past these considerations were not made early in the planning process, resulting in strong local opposition. The local people of an area have the right to pass upon changes that will affect their lives. Social feasibility includes also meeting all legal requirements of land use and buildings.

Taking all these feasibility factors into consideration, the developer can now make revenue cost-return estimates for each of the alternative plans. Perhaps one shows a favorable profit conclusion but will cause great social and physical stress in the area. That alternative should be passed over in favor of the alternative that places least stress on the area and also shows profitability.

8. Final Plan

After the most feasible alternative sketch plan has been selected, final plans for development can be prepared. The sketch plan generalities now need detailed refinement so that the landscape and building contractors will have clear instructions on exactly what is to be built and how. These final plans include three main documents: *construction drawings* (working plans), *specifications,* and *contract.*

Construction drawings show precise dimensions for every aspect of development. On layout plans, all drives, walks, and structures are defined. Often, separate drawings with details are made for grading, drainage, plumbing, heating, air conditioning, and plantings. It is at this stage that earlier steps prove their worth. Changes at this stage are confusing and costly. These drawings are prepared by licensed specialists in all related fields of design in order to meet requirements of health and safety.

Specifications provide detail descriptions of materials and how they are to be used. Again, these details are exacting and meet all the needs of regulations as well as the designed development. Specifications make sure that no matter the final contracting firm, the same standards are maintained.

The contract is the final agreement between contractors and owners/designers. Contracts define the complete and itemized costs of construction as well as schedule of work to be performed. Methods of payment, insurance, bonding, and other legal requirements are included.

With the completion of these documents the project is advertised for construction bids. Implementing the policies of the owner (public or private), selection of the best firm is made.

The project is then developed and ready for staffing and management so that visitors can be received.

9. Evaluation

Often in the past, no mechanism was put in place for evaluation of the success of the designed project after completion. Needed is an assessment by both owners and designers of how well the design met its objectives. Because the entire project was based on *estimates* of market interest and quality of management, there may be need for modified design after the experience of a year or two. Tourism is extremely dynamic making it difficult to make precise forecasts of success no matter how well designed a project may be. Feedback from visitors, managers, and local people can provide needed enlightenment.

CONCLUSIONS

Site scale design is the culmination of all tourism planning.

It is at the site scale that regional and destination planning yield concrete results. Regional and destination tourism planning lay the foundation for the best general areas and types of development that have potential. But, it is only when the three sectors of decision makers—governments, non-profit organizations, commercial enterprise—actually create the supply side development that all planning efforts bear fruit. This fact endorses the need for great care in all site scale design for tourism.

Today's site development success demands professional design.

Although decision makers and others contribute greatly to site development, land and building design requires input from professional designers. Today, most projects require team collaboration among several professionals—architects, landscape architects, urban designers, engineers, and often several other specialists. They translate concepts of decision makers into buildable and manageable development for visitor fulfillment.

Tourism design is greatly influenced by groups other than designers.

Often overlooked is the great influence imposed by moneylenders, owners-developers, construction industry, managers, publics, governments and regulations, and public involvement. The final design of hotels, resorts, restaurants, shops, travelways, and other facilities and services is the result of their input as well as that of professional designers. This fact requires close collaboration among all for best design results.

The new design paradigm requires new kinds of designers.

No longer can the design professionals operate as narrowly as in the past. Their role includes several new dimensions, especially for tourism design. Today, they need greater insight into travel markets as well as the social, behavioral, economic, and environmental aspects of every project. Integration of the many constituencies and influences on tourism design problems requires a new catalytic and facilitating role and new ethics of design.

Placemaking is essentially the creative process of tourism design.

Although resource characteristics are critical foundation factors, they are not truly available to visitors unless so designed, built and managed. Natural and cultural resources are not attractions until decision makers, designers, and others develop these resources for visitor use. A major responsibility is maintaining the integrity of places, especially their social, environmental, and economic distinctiveness.

Public involvement is essential throughout the process of all tourism design.

Unlike other economic development, tourism has impact upon many public groups. They have the right to know about a tourism project planned for their area. Their input can provide constructive recommendations and raise issues that will need to be resolved for best public acceptance. For all sectors of developers, public involvement is an essential part of tourism design and development.

Tourism project design must meet several criteria.

Although most designers practice their profession to meet these criteria, they are especially important for tourism projects. Criteria such as functionality, visitor needs, integration with other plans, individuality, authenticity, aesthetics, and marketability are critical to tourism project success. Best results are derived from designer-developer collaboration toward meeting these criteria.

Tourism project design is the final expression of sustainability.

Theories and philosophies of sustainability remain as postulates until given substance in actual project development. Sensitivity to environmental factors is a basic rule of all planning and design today, but has special meaning for tourism. Tourism depends heavily on the perpetuation of resource qualities. All techniques should be employed, such as interpretive centers, land design details, xeriscape, and protection controls so that every design for tourism can provide maximum visitor satisfaction and environmental enrichment.

Tourism project design requires a special process.

In addition to traditional design process steps, tourism projects have some special steps of importance. Information on travel markets, clarity of program statement, care in site selection, special understanding of on-site and off-site characteristics, and feasibility are important for tourism. Follow up evaluation is especially needed to determine how valid were all of the design decisions.

DISCUSSION

1. How has designer and design specialization helped tourism site design?
2. Discuss the need for greater collaboration among the several designer specialists.
3. Discuss how the important input from those who influence design (moneylenders, builders, etc.) can be incorporated into a tourism site project.
4. In what ways have past tourism site developments failed to function properly?
5. How can design continuity be retained at the same time that individuality and creativity are encouraged?
6. Name the best actions needed today in order to gain better design ethics by both public and private sectors.
7. If installation of interpretive centers is beneficial to both visitor satisfaction and resource protection, what are the obstacles preventing greater use?
8. Throughout tourism site design, how can science and art be combined in order to guide tourism toward the goals of better visitor satisfaction, improved economy, greater sustainability, and acceptable integration between tourism and communities?

REFERENCES

American Society of Landscape Architects (1992). *1992 Members Handbook.* Washington, DC: ASLA.

Antoniades, Anthony C. (1986). *Architecture and Allied Design,* 2nd ed. Dubuque, IA: Kendall/Hunt.

Billing, John C. (1987). "Baltimore's Past Harbors Its Future." *Landscape Architecture,* 77 (5) pp. 68–73.

Boryslawski, Mieczyslaw (2000). From www.usgbc.org/programs/leedrsc as of May 1, 2001.

Brigham, Don (2001). "Lewis & Clark Plants—Commemorating Their Expedition." *Landscape Architecture* 91 (1): pp. 24–26.

Broto, Carles (1997). *Architecture: An Overview.* Barcelona: Links International.

Calkins, Meg (2001). "LEEDing the Way." *Landscape Architecture* 91 (5): pp. 36ff.

Callaway, Clive et al. (1990). "The Salmon Arm Waterfront—A Win-Win Project." Proceedings, "Planning for Special Places Conference," Banff, Alberta, May 13–16, pp. 107–110.

Cuff, Dana (2000). "The Design Professions." In *Design Professionals and the Built Environment*. P. Knox and P. Ozolins, eds. New York: John Wiley & Sons, pp. 31–40.

Dankittikul, Chaiyasit (1993). The Influence of the Spirit World on Landscape Design and Planning in Thailand. (Unpublished dissertation), Texas A&M University, College Station, TX.

"ESRI Donates Software to ASLA" (2001). A newsletter insert published by the American Society of Landscape Architects, Washington, DC.

Forbes, Robert J. and Maree S. Forbes (1992). "Special Interest Travel." In *World Travel and Tourism Review*. Ritchie and Hawkins, eds. Oxon, U.K.: C.A.B. International, pp. 141–144.

Garnham, Harry L. (1985). *Maintaining the Spirit of Place*. Mesa, AZ: PDA Pub.

Gunn, Clare A. et al (1972). *Cultural Benefits from Metropolitan River Recreation— San Antonio Prototype,* TP-43. Texas Water Resources Institute. College Station, TX: Texas A&M University.

Historic Nye Beach Overlay Zone Handout (n.d.) Planning Department, City of Newport, OR.

King, Thomas F. et al.(1977). *Anthropology in Historic Preservation*. New York: Academic Press.

Kyle, Jack (1992). Personal correspondence from Manager Communications & Public Relations, for Wonders: The Memphis International Cultural Series, letter October 22, 1992.

Lip, Evelyn (1986). *Feng Shui*. Singapore: Times Book International.

Marshall, Lane L. (1983). *Action by Design.*. Washington, DC: American Society of Landscape Architects.

McHarg, Ian L. (1969). *Design with Nature*. Garden City, NY: The National History Press.

McNulty, Robert H. (1984). "Tourism Development and Cultural Conservation: Ways to Coordinate Heritage with Economic Development." In *International Perspectives on Cultural Parks,* proceedings, First World Conference, Boulder, CO, pp. 183–187.

Meinig, D.W., ed. (1979). *The Interpretation of Ordinary Landscapes*. New York: Oxford University Press.

Molnar, Donald J. and Albert J. Rutledge (1986). 2nd ed. *Anatomy of a Park*. New York: McGraw-Hill.

Motloch, John L. (1991). *Introduction to Landscape Design*. New York: Van Nostrand Reinhold.

Motloch, John L. (2001). *Introduction to Landscape Design,* 2nd ed. New York: John Wiley & Sons.

National Park Service (1992). *Visitor Use Management Workshop Findings and Recommendations*: *Grand Canyon National Park*. Washington, DC: USNPS.

New Castle County Department of Planning (1989). *The Red Clay Valley Scenic River and Highway Study*. Newark, DE: NCCDP.

Page, Robert R., Cathy A. Gilbert and Susan A. Dolan (1998). *A Guide to Cultural Landscape Reports: Content, Process, and Techniques*. Washington, DC: National Park Service, USGPO.

Proceedings, IFLA World Congress (1995). International Federation of Landscape Architects, 32nd Congress, Bangkok, Thailand, October 21–24. Thai Association of Landscape Architects.

Recommendation Concerning the Safeguarding and Contemporary Role of Historic Areas (1976). UNESCO.

Ritchie, J.R. Brent (1987). "The Nominal Group Technique—Application in Tourism Research." Chp. 37, pp. 439–448. In *Travel, Tourism and Hospitality Research,* Ritchie and Goeldner, eds. New York: John Wiley & Sons.

Scarfo, Bob (1992). "Ethics: What We Say and What We Do." *L.A. Letter,* 3 (3) pp. 2–3.

Sipes, James L. (2001). "Web-Based Project Management." *Landscape Architecture* 91 (10): pp. 46–50.

Stokowski, Patricia A. (1992). "Place, Meaning and Structure in Community Tourism Development: A Case Study from Central City, Colorado." In *Mountain Resort Development,* proceedings, conference in Vail, Colorado, April 18–21, Alison Gill and Rudi Hartmann, eds. Burnaby, BC: Simon Fraser University, Centre for Tourism Policy and Research.

Stokvis, Jack R. (1984). "Utilizing Tourism, Both as an Economic Stimulus for Community Development and to Improve the Quality of Life for Residents." In *International Perspectives on Cultural Parks,* proceedings, First World Conference, Colorado, pp. 179–182.

Sutton, Jonathan Stone (2000). "Amelia Island Revisited." *Landscape Architecture* 90 (1): p. 112.

Too, Lillian (1994). *Practical Applications of Feng Shui.* Kuala Lumpur, Malaysia: Konsep Books.

Ueker, Raymond L., Jr. (1992). "Water Conserving Landscapes: Focus on Xeriscape." *Landscape Design,* 5 (7), pp. 22–24.

Ulrich, Roger S. (1979). "Visual Landscapes and Psychological Well-Being." *Landscape Research* 4 (1): pp. 17–23.

Ulrich, Roger S. and Russ Parsons (1992). "Influences of Passive Experiences with Plants on Individual Well-Being and Health." In *The Role of Horticulture in Human Well-Being and Social Development,* D. Relf, ed. Portland, OR: Timber Press.

Warren, John, John Worthington and Sue Taylor, eds. (1998). *Context: New Buildings in Historic Settings.* Oxford: Architectural Press.

Warren, Winonah (1991). "A Model Cultural Center at Pejoaque Pueblo." *Cultural Resources Management,* 14 (5), pp. 4–6.

Watt, Carson et al (1991). "Rural Tourism development Case Study: Fredricksburg, Texas." Texas Tourism and Recreation Information Program. College Station, TX: Texas A&M University.

Western Australian Tourism Commission (n.d.). *Public Involvement in Tourism Development: What Does It Mean for the Developer?* Perth, Australia: WATC.

Western Australian Tourism Commission (1989). *The Eco Ethics of Tourism Development.* Perth, Australia: WATC.

Western Australian Tourism Commission (1990). *Successful Tourism Design.* Perth, Australia: WATC.

Chapter 10

Site Planning Cases

INTRODUCTION

In all of tourism development, the real brick-and-mortar action takes place on sites. Although regional and destination planning provide important policies, guidelines, and stimulation for development, it is at the site scale that development materializes. Tangible development occurs within all five functioning components of the tourism system—attractions, transportation, services, information, promotion.

Worldwide, casebooks and files of tourist-oriented site planning and design are much more plentiful than destination or regional plans. It is more commonly accepted and often a legal requirement that the design and planning professions are more often engaged by owners to develop properties at this scale, such as for attractions, parks, hotels, resorts, and restaurants. In recent years, with the shopping mall concept, more tourism site clusters, such as theme parks, have appeared.

It is at the site scale that the many characteristics of the land area become very important. The physical elements of topography, soils, drainage, wind flow, and plant materials provide both opportunities and limitations for design. External factors of adjacent land use and access are also important in guiding design solutions. A major challenge to the designer is to balance protection of resource assets and fulfillment of the owner's and user's objectives.

As in the last chapter, the term site planning encompasses all aspects of both building and land design of specific sites for tourism and all those entities involved in the process. The search for examples of good design of sites revealed a great diversity of approaches and kinds of development for tourist use. It became clear that today, whether sites are designed by professionals or amateurs, tourists are offered many more places that are environmentally sustainable, attractive, functionable, and well suited to market needs. Although the design of many of the current chain and franchise motels, hotels, and restaurants are stereotyped and tend to homogenize the travel landscape, individually they serve tourists better than in the past.

The cases offered in this chapter exhibit many ways that site design has been applied and in different settings. Each one demonstrates that in spite of the complexities of tourism, some valuable accomplishments have taken place. They vary from a renovated city square to a major parkway and from an Indian riparian site to an Italian resort complex. But for all, the basic site design framework described in the last chapter applies.

Many more examples of exemplary site planning for tourism are briefly described in *Vacationscape: Developing Tourist Areas* (Gunn 1997). Included are over 40 cases including an ecolodge, a visitor center, a coastal resort, a Spanish parador, and a canopy walkway in Ghana.

BOSTON HARBOR ISLANDS

An important planning and management case that demonstrates collaboration of many constituencies, often in conflict, is the resolution of joint management of the 30 islands in Boston Harbor (figure 10-1). This is the first case in the United States whereby a federal agency, the National Park Service, went beyond its traditional scope of ownership and management of an area and acted as a catalyst for recreation and tourism development. After many years of fragmented ownership and control, these islands are now managed by a consortium of partners.

These islands, ranging in size from less than an acre to 214 acres, have experienced misuse and some habitation for over 8,000 years. Multiple phases of history have included coastal defense of the nation, a prison, almshouses, forts, lighthouses, American Indian settlement, archeological remains, collections, and archives. In the 1960s, one island burst into flame from spontaneous combustion of buried garbage. Several remain attractive landscapes with sumac, aspen, pine, birch, and white poplar as vegetative cover. The islands provide wildlife shelter for marine mammals and fishes.

Following many years of impasse, an effort to identify issues and begin dialog was the Boston Harbor Islands Roundtable Discussion (Paolini 1982). The twenty participants included representatives from attorneys, planners, the National Park Service, tourism specialists, recreation and park managers, developers, arts agency, educators, and two key Massachusetts state environmental and park agencies. Michael J. Pitas of the National Endowment for the Arts acted as facilitator. The discussion

Figure 10-1. Setting of Boston Harbor Islands. These 30 islands that have been subject to misuse and mismanagement for decades are now under new policy and integrated control for visitors as well as residents. The National Park Service now has the role of collaborative planning, management, and development (*Boston Harbor Islands* 1994, 3).

was sponsored by the Office of Coastal Management, National Oceanic and Atmospheric Administration.

The charge given the group was to define the issues, public and private roles, and offer planning opportunities. This forum resulted in new understandings of the environmental characteristics of the islands and the need for better collaboration on management. A consensus was reached on several conclusions: the islands have special environmental and historic resources; they deserve better collaborative management; and, future facilities for recreation and tourism users should be minimal on the islands and located primarily on the mainland where they can be serviced. Although it took several years for further action, progress had begun.

Unique Approach

After much negotiation over a decade, an unprecedented action gave the National Park Service (NPS) the role of collaborative planning, management, and development of the islands on October 26, 1992, as a result of Public Law 102–525 (*Boston Harbor Islands* 1994). This legislation stipulated that the NPS explore expanded cooperation among all public and private agencies to conserve the islands and provide appropriate public use. The planning process began with a detailed resources study to evaluate the natural, cultural, and recreational values of the islands. This was followed by the preparation of alternative plans for development and management. A small startup grant from the National Fish and Wildlife Foundation was then augmented by an appropriation from Congress of $250,000.

The significance of the case lies in the unusual mix of agencies and organizations that were brought together by law in 1996 to create the Boston Harbor Islands National Recreation Area and the Boston Harbor Islands Partnership. The purpose "shall be to coordinate the activities of the federal, state, and local authorities and the private sector in the development and implementation of an integrated resource management plan for the recreation area" (Boston Harbor Islands 2000, 9). This 13-member partnership includes 12 members appointed by the Secretary of the Interior and one by the Secretary of Transportation. Two very important and unique aspects of the partnership were the establishment of the Boston Harbor Islands Advisory Council and the Islands Alliance.

The Advisory Council's purpose is to advise the Partnership on the development and implementation of the general management plan. This Council has 28 members, including representatives from municipalities, educational and cultural institutions, environmental organizations, businesses and commercial interests, tourism, transportation, advocacy organizations, Native American interests, and community groups. It elects two of its members to seats on the Partnership and is mandated to obtain advice from citizens. The Partnership is directed toward the program areas of operations, education, marketing, planning, and finance.

The Alliance is a nonprofit group charged with generating private funds.

The Plan

Of three proposals, the one preferred directed its policies toward preservation of the island's resources and to the design and management of designated visitor areas. The purpose statement follows:

- to preserve and protect a drumlin island system within Boston Harbor along with associated natural, cultural, and historic resources;
- to tell the islands' individual stories and to enhance public understanding and appreciation of the island system as a whole, including the history of American Indian use and involvement;
- to provide public access where appropriate to the islands and surrounding waters for the education, enjoyment, and scientific and scholarly research by future generations (*Boston Harbor Islands* 2000, 26).

Specific planning of future land use areas used the following criteria:

- mainland gateway areas
- visitor services and park facilities
- historic preservation
- managed landscape
- natural features
- special uses (existing sewage treatment, health facilities).

Because of the newness of this unprecedented approach, details of accomplishment are not yet available. It is cited here because it demonstrates a first step that too often is not made because of turf protection, narrow perspectives, and the tremendous array of actors in the planning and development of tourism and recreation resources.

Special Study

Because of this unique approach to planning and management, this project became the focus of a special study for two main purposes: to identify factors that stimulated the creation of such a partnership and to conceptualize a partnership operating model (Makopondo 2001). This work was founded on past studies of collaborative theories by Reed (1997), Hall (1996), Jamal and Getz (1995), and Wood and Gray (1991). The study utilized methods of qualitative research including participant observation, document analysis, and personal interviews of 21 key individuals representing the Partnership, Advisory Council, and NPS staff.

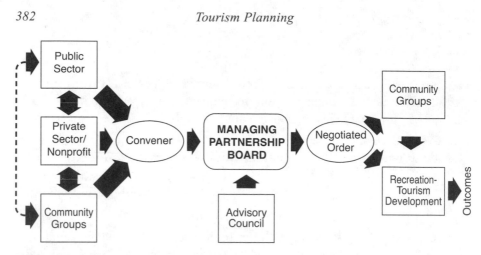

Figure 10-2. Collaborative Managing Partnership Model. An elaboration of the new collaborative approach to planning and managing the Boston Harbor Islands resulted in this generic model. It represents a participatory planning and management approach especially adapted to conflicting jurisdictions (Adapted from Makopondo 2001, 106).

Results help explain the success of this type of planning. It revealed that the major motivators for consensus on collaboration were common stakeholder interest, resource dependency, national pride (and shame), the urgent need for resource control, distrust of external authorities, local and national politics, and the vision of desirable outcomes. Additional factors were skillful and committed administrators and negotiators, visionaries, and political leadership. Important was the collective recognition of the need for collaboration that would be inclusive. Emphasis was placed on a flexible process that could respond to changes in the future. Equally significant was not placing management under one agency but rather a consortium. The study concluded that such a partnership empowers residents to be effectively involved as equals in the planning, management, and use of recreation and tourism resources.

Emerging from the study was the conception of a collaborative managing partnership model, figure 10-2. This "Collaborative Managing Partnership Model" was conceived as a participatory planning and management approach. It would be legislated and composed of stakeholders representing a cross section of a community. It could be called an inclusive and empowering partnership that is more effective than its separate parts, such as a marketing group or trading alliance, and can address common goals with broad social, economic, environmental, cultural, and political dimensions. It would be composed of a team of Managing Partners with requisite decision making authority and a Perpetual Advisory Council.

Two major requirements were revealed during the structuring phase: to collaboratively plan, develop, and manage recreation resources, and to involve local residents in the process. To fulfill these objectives, the convener must insure the following:

- open meetings at venues that are easily accessible by the public
- interested citizens involved throughout
- a diversity of representatives is included.

Such a model with open meetings and all-inclusive participation provides for a public-private interface that generally is not employed in resolving the complexity of issues associated with tourism and recreation development of resources. It is dynamic and practicable, as demonstrated by the Boston Harbor Islands case. The professional designers of the NPS will now carry out the collective planning wisdom of all participants.

BLUE RIDGE PARKWAY

One of the most outstanding and popular attractions in the world is the Blue Ridge Parkway, nearly 500 miles of uninterrupted scenic drive in the eastern United States, linking Shenandoah National Park on the north with Great Smoky Mountain National Park on the south. As a restored scenic landscape, this parkway is visited by over 20 million people a year, the most popular natural resource attraction in the United States (Stahlecker 2001). It proves the value of professional design input by landscape architects to replace cutover forests, muddy streams, and eroded land with outstanding scenic beauty. The economic spinoff of tourism for nearby communities is substantial. Although its size might suggest a more appropriate classification as regional or destination, its scope and planning process are more closely those at the site scale. The following description is paraphrased from *Blue Ridge Parkway: The First 50 Years* (Jolley 1985).

Political/Economic Origin

What today is a preeminent example of landscape design and resource protection grew primarily from an economic need. In the 1930s the United States experienced unprecedented hard times with mass unemployment. One solution, led by President Franklin D. Roosevelt, was the National Industrial Recovery Act of 1933, including a Public Works administration (PWA) with a $3.3 billion budget to provide relief employment.

Several political leaders are credited with proposing to the president that the building of a scenic highway was an appropriate project for this federal program. Senator Harry F. Byrd (formerly governor of Virginia) had for some time recommended the extension southward of the Skyline Drive of Shenandoah National Park and recommended this to the PWA. Theodore E. Strauss, then advisor for District 10, PWA, claims his suggestion initiated the parkway concept. Governor John G. Pollard of Virginia, on September 23, 1933, appointed Byrd chairman of a Virginia Committee to seek federal aid for such a project. A meeting on October 17, 1933, brought together representatives of the Bureau of Public Roads, the National Park Service, and PWA, resulting in the following basic policies that have guided development ever since:

1. The project would be totally funded with Public Works Administration money, thereby eliminating tolls.

2. The scenic highway was officially designated a *parkway.*

3. It would be a unit in the National Park System.

4. The states would acquire and donate the required right-of-way. In return, the federal government would design, construct and maintain the Parkway.

5. $16,000,000 would be requested to finance a projected 414 miles of road and provide employment of four thousand men.

But, even as the first $4,000,000 was appropriated from PWA, considerable controversy surrounded siting of the project. Public hearings generated much acrimonious wrangling regarding the location and states involved. Although Tennessee's representatives fought for a route in their state, they lost on the basis that they were already benefiting from federal funding of the Tennessee Valley Authority development. The final routing decision, giving it its present route, was made by the then Secretary of the Interior, Harold L. Ickes.

Design Concept

Because the National Park Service had no staff available to design such an unprecedented project, it turned to the private landscape architectural firm of Gilmore D. Clark. His staff member, Stanley W. Abbott, was appointed as Resident Landscape Architect on December 26, 1933, as director of all design and construction of the Parkway.

The first step was landscape analysis but it was handicapped by "unfamiliarity of the region, rural isolation, lack of roads, sparse food and lodging accommodations, a rugged terrain, and pathetic map resources (Jolley 1985, 16)." Even so, Abbott worked with state engineers to reconnoiter and locate the route, determine the right-of-way, donate the land to the federal government, design a plan to fit the terrain, and put the designed project up for bids by construction contractors.

By 1934, Abbott had been assigned two assistants—Edward H. Abbuehl and Hendrik E. Van Gelder. Together they conceived of their mission as *preservation, reclamation, and vistas.* For their design policies, they agreed upon the following:

Utilize that which exists.

Carve and save, not cut and gut.

Preserve the lived-in look.

Keep a managed landscape in mind.

Preserve nature and history.

Marry beauty with utility.

Emphasize simplicity and naturalistic.

The horizon is the boundary-line.

These policies dominate all design decisions on routing, varieties of elevation, frequent parking overlooks, short hiking trails, wayside museums, and roadside parks. Funds from another federal program, the Resettlement Administration, were used to acquire adjacent worn out submarginal lands and resettle the owners.

In order to facilitate planning and design, Abbott divided the route into segments. Borrowing from railroad practices, the decision was made to set mileposts throughout. A logo, representing open sky, mountain peaks, a windswept white pine tree, and a swath of the road, was designed and is still in use today. A major innovation for its time was Abbott's policy of communication. He established a regularly issued bulletin, the *Blue Ridge Parkway News*, that published information on work as it progressed and why the several design decisions were being made. This became a dynamic public relations piece, establishing community support.

Two years after the Parkway was initiated, sufficient right-of-way and design plans had been completed to invite bids for the first stage of construction. Today's travelers are not aware of the thousands of legal transactions that accompanied the process from design through construction of this major project. Most of the construction labor came from members of the Civilian Conservation Corps, another federal employment program. They were involved in final adaptation of the project to the original landscape, such as grading and planting slopes, constructing native rail fences, clearing and erosion control, salvaging and rehabilitating historic buildings, constructing utility lines and substations, and constructing park facilities.

A major setback came with the advent of World War II. After two-thirds of the project had been developed, funding and support came to a halt. The major sponsor, the National Park Service, found itself with many new responsibilities nationwide but with much reduced moneys and staff. However, following public outcry regarding the status of parks, by 1956, National Park Service Director Conrad L. Wirth was able to initiate a new program of development labeled "Mission 66." This program gained enough support from Congress to renew Parkway development, especially for more construction, new visitor centers, new campground facilities, new campfire and outdoor amphitheaters, historic restoration, and establishment of the Museum of North Carolina Minerals. A unique part of the mission was retraining of neighboring farmers in skills of land conservation to enhance the beauty of the Parkway and protect resources.

A Public/Private Conflict

In spite of its many successes, the issue of service facilities along the Parkway became critical. Abbott had envisioned the need for some lodging and food services directly along the Parkway, sensing a market demand from the millions of visitors who began to use this attraction. However, he was confronted by strong protest of these concession proposals, especially from older resort areas around Blowing Rock, North Carolina. The argument was that the federal government had no right to use taxpayer's money to establish competition. Over the years since, a balance has been struck, allowing only a few concession operations in the most remote locations of the Parkway and supporting private sector businesses in communities nearby.

Another major conflict occurred with the Parkway section around Grandfather Mountain, North Carolina, toward the southern end of the project. Right-of-way had already been purchased from landowner Hugh Morton. But, the National Park Service had new plans for a route at a higher elevation, based on environmental reviews. Morton refused to accept the new plan and a stalemate ensued. A final compromise included a new alignment between the "high" and "low" road, Morton's exchanging right-of-way for other state land, and implementation of environmental impact assessment not available in the 1930s. An innovative solution for an especially sensitive one-fifth mile segment was the building of the Linn Cove Viaduct at a cost of twelve million dollars in 1983. It is structurally supported independently alongside the mountain to avoid environmental damage to the slope.

Figure 10-3. Community Cooperation along Blue Ridge Parkway. Tourist services are provided primarily in communities alongside the Blue Ridge Parkway. Most of these communities maintain their own land use in a manner compatible with the resource protection and aesthetics of the parkway.

Planning Success

Today, the Parkway regularly receives worldwide acclaim as an environmentally sound natural and cultural resource visitor attraction. The outstanding scenery, natural resources, and many cultural sites now are protected for the future. At the same time, adjacent communities have received a substantial economic boost from tourism and millions of visitors are educated, enriched, and fulfilled by their experience. Superintendent of the Parkway Gary Everhardt, a former NPS Director, developed expanded liaison with nearby communities and political jurisdictions (figure 10-3). Testimony to its success is a prayer uttered by the Reverend Arsene Thompson, full-blooded Cherokee Indian:

> Where once there was only a buffalo trail, where Indian campfires once blazed . . . where once the red man and the white man fought . . . there is a road of peace and we are thankful (Jolley 1985, 44).

This planned natural resource feature is proof of the long-term value of perpetuating environmental values at the same time visitors are given an enriched experience. Conceived many years ago, the design is as valid today as when first built. It is a major U.S. tourist attraction with economic spinoff to many communities along the way.

FRANCONIA NOTCH

The Challenge

An unusual travelway has now been established in New Hampshire as the result of design and planning effort combined with cooperation from agencies and organizations. Completed in 1988, the project resolved conflicts of opinion that had held up action for many years (figure 10-4) (Rigterink 1992).

Environmental groups had resisted standard Interstate Highway specifications that, in their opinion, would damage the environmental character of the landscape. The project involved an eleven-mile segment of Interstate 93, which also traverses several miles of Franconia Notch State Park. Already the park attracts over two million visitors a year and at all seasons. A natural rock profile, "The Old Man of the Mountain," is a special visual landmark, the state symbol of New Hampshire since 1945.

Chronology

Because of the unusual circumstances surrounding this special site problem, the project had an equally unusual chronology of development. This aspect is related here to demonstrate the need for the planning process to include the diversity of opinion of both proponents and critics.

Figure 10-4. Franconia Notch Scenic Freeway. After many years of conflict between highway planners for Interstate Highway 93 and environmentalists, catalytic action by outside planning consultants resolved the iussue. Both transportation and scenic preservation are now the result (Green 1986).

For many years, a single highway had provided north-south transportation through Franconia Notch, a pass in the White Mountains. When the original system of Interstate and Defense Highways was planned in the 1940s, this stretch of highway was scheduled to become part of this system. The rigid high-speed nationwide standards immediately became the core of controversy over future plans. At the center of the controversy were environmental and social issues which delayed construction for over three decades. A major delay was caused by U.S. Congressional passage of the National Environmental Protection Act of 1970 that required greater public involvement.

A breakthrough came in 1973 with a Congressional amendment to the Federal Highway Act providing for a "parkway-type highway" through Franconia Notch. This allowed specifications less than normal Interstate standards. Soon thereafter, several setbacks occurred, including an injunction against construction. In all, seven different alternative routes were proposed but none was acceptable to all parties. Several major environmental groups had allied themselves to form the White Mountain Environmental Committee (WMEC). It seemed that the only solution was to create a compromise acceptable to both the state agencies and WMEC. In order to work toward this objective a memorandum of agreement was signed in 1977.

In 1981, a Design Team was assembled and charged with close cooperative planning among the several state and federal agencies and the WMEC.

Process

Of special significance for resolving conflict and planning a solution was a three-part catalytic and review process. Even though the designers were free to develop a creative and aesthetic design solution, an important public input process was established at the outset.

1. *Design Team.* Because of the diversity of needed site development, four design and planning firms made up the Design Team: DeLeuw, Cather & Company, responsible for project management, planning and engineering; Roy Mann Associates, Inc., responsible for park master planning; Johnson Johnson & Roy Inc., responsible for landscape and site design concepts and details; and Gruen Associates, responsible for architecture. The Design Team met monthly with the principal environmental group, (WMEC) for the following purposes: monitor progress, schedule upcoming work, present and critique alternatives, and identify problems, coordination and data requirements. No work was to be taken to the main client, the New Hampshire Department of Public Works and Highways or Division of Parks and Recreation without having first been reviewed by the Design Team.

2. *Staff Review.* After work proposals had undergone interdisciplinary review of the Design Team, they were submitted to the official staffs of the two key state agencies. Comments and recommendations were made concerning the acceptability of the sometimes innovative design solutions. Especially important were the aesthetic and environmental characteristics of the concepts.

3. *Technical Review.* Senior technical staff members of the Department of Public Works and Highways and the Division of Parks and Recreation then made more critical review of each step in the planning process. Each month, the Design Team's proposals and modifications were examined to determine how they had been changed based on comments of the month before. At each month's meeting, the following were discussed: progress from the month before; work schedule for the next month; planning and design alternatives; and recommendations requiring policy review and comment.

In addition to these three elements of the review process, all planning recommendations were reviewed once a month by an Overview Committee. This committee included representatives of White Mountain Environmental Committee (WMEC), the Department of Public Works and Highways, the Division of Parks and Recreation, and the Federal Highway Administration. The purpose was to regularly monitor planning progress in terms of a WMEC agreement.

Design Issues and Analysis

Because there was no precedent for this special highway plan and its unusual corridor, many environmental issues had to be studied throughout the process. Following are some of the highlights of these concerns.

The Notch posed unusual landscape problems because of high winds and intense cold. A twelve-year-old spruce tree is typically only two feet tall. Modifying the land forced the decision to obtain new specially adaptable plantings. Interstate standards prohibited bicycle paths. They also demanded divided highway construction that would have destroyed much of the setting. Road shoulder widths were excessive for the conditions. Concerns over the state park landscape and features required special design consideration. Water runoff was a major issue that had to be resolved.

A great amount of time and effort went into studies of all the physical and aesthetic resources. Because it was a special case, common practices, familiar to designers and highway builders, had to be modified, one item at a time.

Solution

Today, a pleasing and innovative highway corridor provides for multiple functions in a manner now acceptable to travelers, environmentalists and traffic engineers. The solution was not a divided highway but rather a four-lane design with "granite rumble strip" three feet wide, separating opposing traffic. In order to blend the highway color with the surroundings a special asphalt mix was used producing a permanently dark surface. Native rock was often used on severe cuts. But workers had to avoid using traditional techniques of orderly stone courses. Landscape architect Richard Rigterink stated, "We wanted stones and boulders to look as if they fell down from the mountain. In fact, we spent more time with boulders than with trees." Because bicycles are prohibited from Interstate highways and their use is popular in the adjacent state park, an entirely new bike path was created. Specially designed shoulders and right-of-way widths were used to adapt construction to the setting.

Park and recreation facilities were relocated and redesigned. Newly developed features included a visitor center and park interpretation facility, park headquarters building, special interpretive and observation facilities and trails, walk-in and drive-in camping areas, boat launch maintenance complex, and trailhead facilities. A special nine-mile bikeway through the Notch was designed. The route parallels the parkway through the forest and meadows of the corridor, crossing from side to side through underpasses and over bridges, connecting various park facilities along the valley.

Conclusions

This special site planning project is proof that, with the right goals, an adequate mechanism for cooperation among affected parties, and application of professional design/planning talent, environmental assets can be protected at the same time tourism is developed. It proves that seemingly insurmountable conflict can be resolved. In the words of one member of the Design Team, " ... My reaction was one of respect and amazement that these folks who had talked to one another primarily through attorneys for so many years could calmly discuss pros and cons of various alternatives ... I think one of the strengths of the consultant team was that each of us could establish a relationship with our counterparts within the client group and appreciate their perspective and work to convince them the proposed alternatives were responsive to their project objectives" (Rigterink 1992). The results are obvious to the travelers. Response from all sides has been positive and the design has received widespread recognition and many awards—from the federal highway agency, the American Society of Landscape Architects, and a regional engineering organization.

In this case, the major role of the designer was a catalyst to bring conflicting groups together for a final design solution.

WHITEMAN PARK

Illustrated here is description of how a special Task Force can guide the planning and development of a site for tourism. The location is a 5,000-acre area of native Australian bushland called Whiteman Park (*Whiteman Park* 1989). It was named after Lewis Whiteman, who purchased it in 1936 for cattle breeding. It contains typical bushland vegetation and wildlife, the only such site remaining in the region, and it is within 30 minutes of Perth.

The State Planning Commission and the Western Australian Tourism Commission were of the opinion in 1989 that an outside evaluation of this interesting site for its tourism potential was required. Because the Pacific Area Tourist Association (PATA) had been involved in such projects it was asked to perform this task. Ian L. Kennedy, vice president Pacific Division, accepted the assignment and began to build a Task Force.

As was typical of such projects, he sought a diversity of reputable specialists especially suited to the challenge. The selected Task Force included a specialist in commercial tourism-park operations with background in zoology and biology (Terence Beckett), an expert in heritage conservation and tourism (Robertson Collins), a landscape architect specializing in tourism design and planning (Clare A. Gunn), an expert in land and water resources and geologist (Robert Priest), and a team leader (Ian Kennedy). The format was to spend a week of intensive inspection of the site and interviews and conferences with a great many organizations and agencies involved in the park. On the final day both an oral and rough draft of a report provided findings and recommendations of the Task Force. In spite of the team members coming from different geographical locations (New Zealand, Singapore, the United States) and from several disciplines, consensus on these final conclusions was readily reached.

Objectives

The purposes of this study, as specified by the client were to:
>   Prepare an action plan for future management.
>   Comment on the "Maunsell Report" (1978).

Identify the park's tourism potential.

Prepare a long-term "vision" for the park.

Examine financial considerations.

Process

The methods used included:

Inspection of site.

Reconnaissance of area context.

Extensive interviews with agencies, organizations.

Review of documents.

Interviews with park users.

Study of cultural and natural resources.

Results

1. *Assessment of the Maunsell Report.* The Task Force determined that this report addressed recreational issues but not plans for tourism adaptation. The technical description was helpful as a reference. The plan concepts did not apply to travel markets.

2. *Present Operation.* The Task Force concluded that existing operational management had no stated policies to guide it. There appeared to be safety hazards, lack of an interpretive program, little or no formal tourism planning or marketing, inadequate sewage disposal, erratic land use patterns, and absence of natural and cultural resource management. At issue was the random acceptance of several land use claims and developments within the park—a hobby railroad organization, a hobby trolley car organization, a firing range, an equestrian park, and a youth camp. For political reasons, Aboriginal heritage, important in the park, was omitted.

3. *Natural Resource Protection.* There appeared to be complete lack of protection of wildlife and plant materials. All development and management are contrary to principles set forth in Western Australia's "Draft Environmental Guidelines for Tourism Development," 1988.

Recommendations

1. *Goals.* Recommended by the Task Force were the following goals for adapting the park to tourism development:

Protect the natural/cultural resource base.

Provide visitor satisfactions for domestic and international travel markets.

Promote economic development.

Integrate Whiteman Park into the Swan Valley tourism plan.

2. *Tourism Potential.* It was the consensus of the Task Force that the park did have potential for becoming a major tourist attraction, but only with significant change in policies, plans, organization, and management. The cultural/natural resource assets could provide the dominant theme to meet travel market interests. New policies should call a halt to more non-conforming and erosive use areas and support stronger resource protection and interpretation.

3. *Physical Plan.* Using a new theme of "natural/cultural heritage park," several physical plan recommendations were made. The railroad and trolley operations could continue if adapted to visitor use. An outstanding interpretive visitor center, located on the south of the property, would enrich the visitor experience without damage to the environment. It should contain indoor and outdoor exhibits, a museum of agriculture, crafts, auditorium, classrooms, a heritage and natural resource library, and a heritage food center. Demolition or re-adaptation of existing structures should be considered. This new visitor interpretation complex should also serve as an educational and research center for study of the important natural and cultural history of the area, especially early Aboriginal impact on the site.

Several other basic recommendations were made. Safety hazards of the rail line must be eliminated. Well-designed on-site commercial facilities could include a restaurant, souvenir shop, photo shop, bike hire, theater, and an aviary and reptile house. The proposed golf course should be denied on the grounds of non-compatible use and poor feasibility because another is being established nearby. All design should be aesthetically appropriate to the natural/cultural heritage theme. No accommodations are recommended on-site but rather in the nearby community of Guildford. A new sewage treatment plant must be established. New plans for visitor access and transportation need to be created and implemented. Adaptation of the park to tourism should include integration into an overall Swan Valley Tourism plan as shown in figure 10-5.

4. *Financial Considerations.* Recommended is strong public-private cooperative leadership and investment. Resource protection requires governmental support. Economic impact can come from two sources: revenues from commercial operations on site and revenues and employment in new restaurants, hotels, and shops in Guildford after the park is reno-

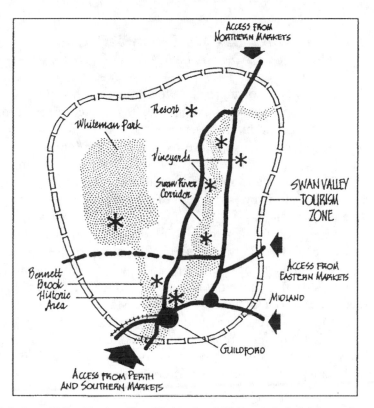

Figure 10-5. Swan Valley Tourism Plan. The study of Whiteman Park for its tourism potential revealed the need for new policy and management that would reduce obtrusive development and emulate the special natural and cultural assets. A major recommendation was to incorporate the park into a Swan Valley zone with tourist services in Guildford rather than in the park (*Whiteman Park* 1989, 19).

vated for tourism. The Interim Board of Management should be abolished and a new organizational structure created—one dedicated to the goals of tourism and resource protection. The park should be developed only in context of the area and its tourism development.

This project illustrated a form of planning that involves several experts, local input, and a brief but intensive investigation. Because the experts volunteered their time and the time-frame was only one week, the cost is relatively low. Yet the Task Force process allows a full grasp of the existing situation and major issues. Pertinent recommendations were published and widely distributed for greater understanding of the potential and stimulation of action. Implementation of this plan has not been reported. Even so, this task force process is a viable method for identifying site planning needs for tourism.

GREAT LAKES CROSSROADS
MUSEUM AND WELCOME CENTER

Included in the concept for tourism development of the Upper Peninsula of Michigan in 1966 (Blank, Gunn, JJR) was the recommendation for a special "gateway" at the highway entrance to the region from the south and across the Mackinac Bridge. An example of the long-term nature of regional planning is the fact that it took over twenty years for detailed initiative and project plans for such a gateway center to be completed (*Bridging* 1988). Following is a sketch of this example of planning for an interpretive visitor center.

In 1986, Governor Blanchard appointed a task force to analyze the potential for more effectively developing the northern Mackinac Bridge base for the benefit of local citizens and visitors. The administrative agency for much of this land (as well as Mackinac Island and Fort Michilimackinac at Mackinaw City) was the Mackinac Island State Park Commission. The task force engaged consultants in marketing, planning, landscape architecture, and architecture to produce a feasibility study and sketch plans for a specially designed facility at this important entrance to the Upper Peninsula from its largest travel market source.

Market Research

A first step in the planning of this interpretive museum and visitor center was a travel market survey, performed by Market Opinion Research. A representative sample of 1987 travelers through the Straits of Mackinac area revealed that 53 percent believed they would visit such a site and pay $2.50 for the museum visit—a total of 950,000 to 1,300,000 visitors a year. The market researchers, however, based on past experience of respondents promising more than they would actually do, revised this estimate to approximately 300,000 a year. Even so, this was evidence that a significant travel segment was interested in visiting a museum and interpretive center at this important gateway location.

Site and Building Program

The firms of Quinn/Evans, architects, and landscape architects, Johnson, Johnson & Roy/Inc., collaborated on site and building concepts and designs. This was in response to the development program articulated by the Mackinac Island State Park Commission. This program included a Michigan Department of Transportation (MDOT) Welcome Center and a transportation museum focusing on the Mackinac Bridge and the Straits of Mackinac. The Welcome Center and museum were to be contained in one structure and the parking, service, and other site development were to be shared. The building program included preliminary space allocations and space descriptions. The interpretive concept was refined with the assistance of a museum consultant and was named "Bridging the Straits." Two major themes were defined: "The Bridge" and "Straits Maritime History," with subthemes of "Transportation" and "Tourism."

Site Design Concept

Review of market findings, the planning program, and the proposed site revealed conclusions pertinent to site design. Figures 10-6, 7, 8 illustrate the site context plan, the site plan, and a conceptual aerial view.

Because the majority of travelers come from the south and arrive at the Upper Peninsula via the Mackinac Bridge, one of the largest suspension bridges in the world, the vista toward the land-water edge is particularly important. It is here that the Center/Museum is to be established, offering a visual welcome, as well as easy access from the highway, immediately adjacent to the toll plaza. The site context plan shows relationship to Highway 2, the trunk line to western Upper Peninsula, the Father Marquette National Historic Site, Straits State Park, Memorial, and Museum, as well as to the main north-south highway, Interstate 75. For southbound traffic, a parking area on the west side of the highway and a pedestrian bridge would provide access to the Center and Museum.

One goal of the site plan was maximum preservation of the native landscape vista including existing vegetation. The upper and lower parking lots are adjusted to existing wooded edges. New planting would use the same kinds as found locally and be placed in a naturalistic form. The plan is simple and strongly ordered. First, it presents the arriving visitor with a clear sequence of experiences. Second, it makes efficient use of a compact site, preserving Bridge Authority functions, fragile woods and wetlands, as well as a considerable amount of existing mature landscape vegetation. And, finally, "Bridging the Straits" placement provides maximum opportunities for fine views of the Mackinac Bridge and the panorama of the Straits of Mackinac and Mackinac Island, an important historic and resort area.

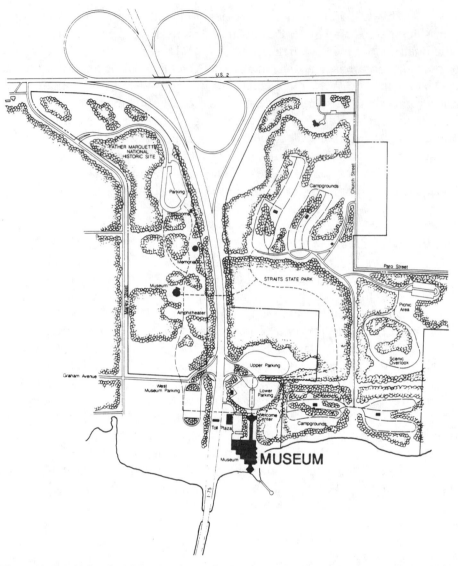

Figure 10-6. Bridging the Straits Site Plan. Following the concept provided in 1966 (Figure 6-8), this site plan includes proposed circulation, historic site, camping, landscape integration, and location of major interpretive center and museum (*Bridging the Straits* 1988, 21).

Building Design Concept

The site concept with its north-south orientation strongly influenced the building design, providing the visitor with a smooth transition from site to building functions.

Upon entering the building, one continues to sense the axis moving southward toward views of the bridge and water. A skylit orientation area

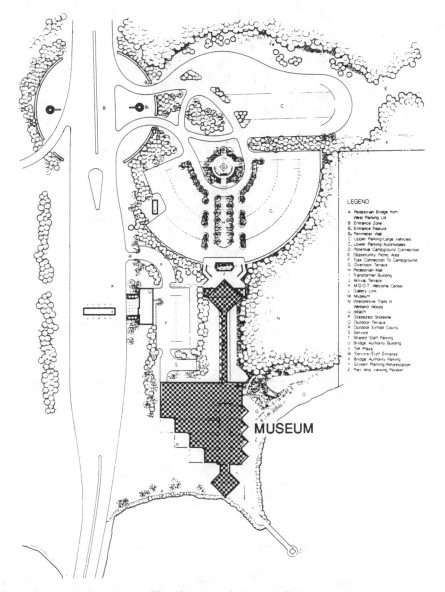

LEGEND

A Pedestrian Bridge from
   West Parking Lot
B Entrance Zone
B₁ Entrance Feature
B₂ Perimeter Wall
C Upper Parking/Large Vehicles
C₁ Lower Parking/Automobiles
D Potential Campground Connection
E Opportunity Picnic Area
F Trail Connection To Campground
G Overlook Terrace
H Pedestrian Mall
I Transformer Building
J Arrival Terrace
K M.D.O.T. Welcome Center
L Gallery Link
M Museum
N Interpretive Trails In
   Wetland Woods
O Beach
P Stabilized Shoreline
Q Outdoor Terrace
R Outdoor Exhibit Courts
S Service
T Shared Staff Parking
U Bridge Authority Building
V Toll Plaza
W Service/Staff Entrance
X Bridge Authority Parking
Y Screen Planting/Reforestation
Z Pier And Viewing Pavilion

MUSEUM

Figure 10-7. Bridging Museum Site Plan. An enlargement of the Museum area of Figure 10-6. Important is providing ease of access and capitalizing on scenic views of the surrounding area, especially toward the Straits and the Mackinac Bridge (*Bridging the Straits* 1988, 20).

offers a dramatic introduction to the museum and its setting. Upper and lower exhibits carry out the several historic themes and subthemes. The visitor may also continue on to an observation deck that offers a panoramic view of the Straits and access to a multimedia theater.

Special services within the structure include tourist information and guidance, gift shop, food service, and lounge areas.

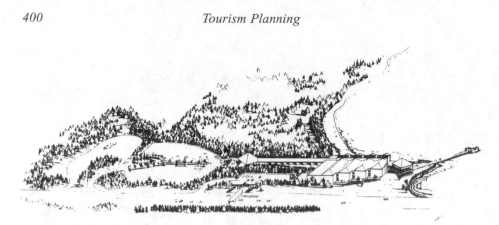

Figure 10-8. Bridging the Straits Aerial View. This sketch by the landscape designer shows the setting for the Museum and Welcome Center. The low-rise architecture is well adapted to the site (*Bridging the Straits* 1988, 19).

The guiding concept is the creation of a strong relationship between major interior spaces and the unique setting. The design thereby heightens the visitor's experience of both the museum and the surrounding natural features. Drama and anticipation are created by this special interaction between building and site. External to the building are outdoor exhibits as well as dining and viewing areas. Movement from inside to outside within the exhibit and public spaces of the museum is encouraged. The overall facility is proposed as an exciting and informative attraction and one that complements the quality of its dramatic surroundings. Its purpose goes way beyond the site, providing an introduction to the many features of the Upper Peninsula, close relationship to the tourist services of nearby St. Ignace community, and linkage with the Lower Peninsula and dominant market access.

This tourism plan illustrates a basic planning principle—implementation often takes many years. Regional and destination plans may provide recommendations for strategies and specific projects. But action must wait until a mixture of needed factors is just right—a willing investor, commitment to design and build, feasibility of success, and competent management for operation.

ST. BAVO'S SQUARE, GHENT

An example of the restoration of a medieval urban square for tourists as well as residents is the case of St. Bavo's Square, Ghent, Belgium (DeRoo 1995). Centrally located in Europe, Ghent occupies a coastal plain and is connected to the North Sea by a canal built in 1886.

Background

In the Middle Ages, Ghent was a center of economic activity of the region. Its own and surrounding economy was based mainly on textiles and agriculture. City market squares were the focus of the purchase of goods and produce. Government buildings surrounded the squares. With the industrial revolution, the city grew and obtained new access by train and automobiles. In recent years the city squares became overrun by traffic even though the monumental architecture of an older era remained. Tourist traffic, seeking enrichment from the historic buildings, stimulated touristy shops alongside.

Renovation of Squares

City planners have succeeded in making major renovations in several of the downtown squares., Following are highlights of these improvements.

- St. Michael's Square, an important landmark, was renovated in 1985.
- Former medieval port area has been modified for visitors.
- The Vegetable and Fruit Market and St. Veerle Square dominated by Castle of the Counts has been improved.
- The Corn Market has become a favored spot with cafes and shops.
- Gradually, the sequence of squares has been converted into a pedestrian mall.
- In the Town Hall Square, culture, protocol, and tourism have become the main functions.

Urban redesign criteria followed for these renovations include orientation to sun and shade, key land use, and relevance to lodging and food service. But all changes have enhanced the older architectural aura and provided greater visitor opportunity to learn of the area's rich past.

St. Bavo's Square

St. Bavo's Square is a larger open space in the heart of the medieval core of the city. The plan purpose is not merely nostalgic but rather to retain its historic quality and adapt to more public enjoyment. By removing streets and parking, pedestrianism has been restored. The square is flanked by the Municipal Theater and St. Bavo's Cathedral with its tower of 85 meters in height. Other ancient towers give meaning to the site. But in niches wherever possible, trees, benches, and other plantings humanize the space.

This case, even though in progress, is presented here to again demonstrate a trend toward cultural retention and restoration everywhere. Its adaptation to tourist and resident use is commendable.

ORANGE VALLEY RESORT

On October 12, 1990, Lester Bird, Deputy Prime Minister and Minister of Tourism and Economic Development for the island of Antigua, signed an agreement with planners to develop concepts for a new national park and resort (*Orange Valley* 1991). The objectives were to create:

> A national park which will cater to the needs of islanders and the desires of the visitors.
>
> A destination resort which will support the government's future economic plans for tourism in the 1990s.
>
> A development to be constructed in "Caribbean style" which respects and complements the vernacular architectural style of St. Johns, the island's capital.
>
> An improved landscape through the introduction of new plants and trees to enhance the existing setting.

The consultant team included developers, development managers, landscape architects, golf course architect, environmental analyst, and cable car firm. This team was guided by two basic principles: "(1) to ensure the permanent preservation of the wonderful natural beauty of Orange Valley, Ffryes Mill and Beach, and (2) to make that beauty accessible without spoiling the natural scenery in the process" (*Orange Valley* 1991, 1).

Existing Conditions

The analysis of existing conditions (Figure 10-9) and master plan (Figure 10-10) encompasses both the proposed national park and Orange Valley Resort. Following are summary descriptions of elements of the entire project.

Orange Valley, a portion of land on the southwestern coast of Antigua, offers tourism opportunities year round. The valley floor is quite spacious and views toward the sea are spectacular. It provides wildlife habitat, especially in the salt marsh and lagoon. Waterfront is dominated by Darkwood Beach.

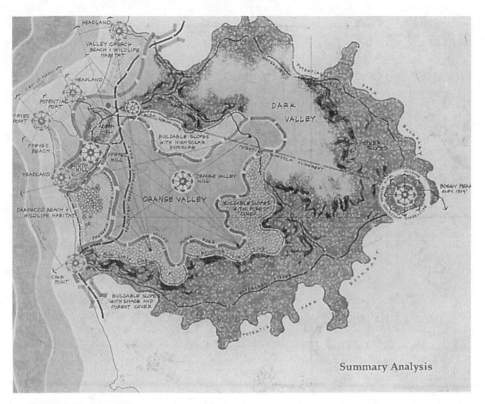

Figure 10-9. Orange Valley Resource Analysis. A first step in creating the development plan for Orange Valley Resort, Antigua, was study of the slopes, forests, beach areas, and buildable sites. This study, together with development proposals, laid the foundation for the site plan, Figure 10-10 *(Orange Valley* 1991, 4).

Dark Valley, a watershed of 785 acres, lies to the northeast of Orange Valley. It is an upland area with limited views to the sea. The pristine beauty of this valley suggests a dominant policy of resource protection rather than intensive development. Some public use could be allowed, such as for trails, picnic areas, and campgrounds.

Boggy peak is the highest point on the island—1,319-foot elevation—and is located toward the east of Orange Valley. It has opportunity for 360-degree vistas if made accessible by cable car.

Valley Church offers an idyllic approach to the national park but the beach requires restoration.

Ffryes Bay. Ffryes Mill and Beach with its magnificent beach are premier natural settings for the entire project. With salt pond enhancement, small shore habitat can be improved. Ffryes point should remain undeveloped. The remains of a windmill punctuate the northern portion of the hill.

Environmental Considerations. Each of the four watersheds is unique. Managing the levels and quality of water of Orange Valley presents a

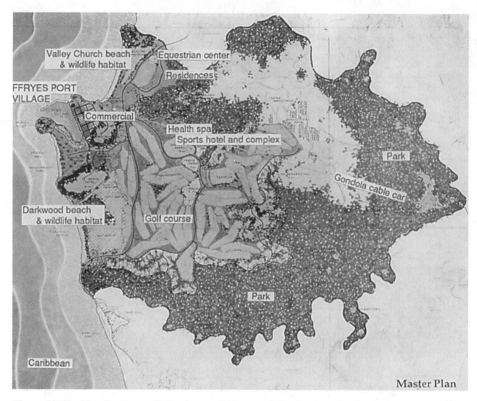

Figure 10-10. Plan for Orange Valley Resort. This master plan is adapted to the basic resources of the area and market potential. It includes location of the golf course, residential villages, road circulation, commercial development, and national park (*Orange Valley* 1991, 7).

challenge. New wetlands and mangrove swamps have potential but are contingent on use of safe and biodegradable fertilizers and pesticides. Present grazing and agricultural uses of Dark Valley should continue. Constructing a port and village at Ffryes Bay will demand great care to minimize disturbance of natural assets. Proposed separation of wetlands from intense park use includes the establishment of botanical gardens. Development of this area is an ecologically sensitive challenge.

National Park Zones

Proposed is dividing the area into three major zones.

Natural Reserve Zones total 1,565 acres (70 percent of the park). These are typified by steep slopes, quality vegetative cover, scenic resources, aquatic and terrestrial wildlife habitats, and historical sites.

Recreation Zones total 413 acres (18 percent of park). Because resource protection is critical here, these areas require special planning care. The golf courses must follow principles of adaptation, enhancement, and resource protection.

Development Zones include 272 acres (12 percent of total) and offer the setting for most intensive visitor use. Economic and cultural benefits generated from this area would support scenic and other resource protection.

Master Plan

The dominant theme is to capture the beauty and ecological integrity of the seascape and landscape. The intent is to demonstrate that with proper design the public and private sectors can co-exist as resource protection becomes a paramount policy.

Several characteristics of *Natural Reserve Zones* are emphasized in the plan. These zones would allow only swimming, sunning, hiking, equestrian trails, picnicking, camping, wildlife observation, photography, and similar activities. Environmental tours, educational, and invitational use will be encouraged. Minimum facilities will be provided. The landscape concept includes not only resource protection but rehabilitation for improved wildlife habitat, stimulation of native landscape character, and enhanced aesthetics. Beach use will be improved by removal of the beach road, providing access at either end. Replenishment of Valley Church Beach is needed. Plans include reintroduction of commercial citrus fruit agriculture. This will not only help supply local markets but also allow export and reinforcement of the Orange Valley theme.

Orange Valley plans call for both intensive and extensive development and recreational use. A new sports hotel, health spa, sports complex, and residences would be placed on the foothills of surrounding mountains, affording views of magnificent scenery toward the sea. Golf course design policy is *least change* to the landscape. It is planned to blend into the natural setting and conserve water use. A gondola cable car system, from Valley Church to Boggy Peak will provide ecologically sound access and 360-degree vistas of the surrounding Caribbean destination. Trails will provide alternate access.

Ffryes Bay plans include forest protection, a museum and park orientation center, a botanical garden, and Ffryes Port and Village. The port is planned to be a functioning fish landing area, marina, and new village designed to emulate vernacular architecture and provide many visitor services. New housing and hotels are planned.

For *Valley Church Bay,* several functional areas are planned. An equestrian center, multi-family dwellings, and preservation of open space are key themes for this area.

Conclusion

The planners, designers, and developers stated the following as their plan conclusion:

> Our objective is to create a world-class tourist resort, fully integrated into Antigua and its way of life. The resort will be promoted world-wide to attract a growing number of visitors, contributing to the island's tourist economy and providing long-term employment for present and future generations of Antiguans.
>
> (Orange Valley 1991, 10)

This plan illustrates a national park-public use concept proposed for other locations many years ago but seldom implemented. The fundamental land use policy of this concept is to concentrate public use facilities and services, thereby protecting the remaining extensive resource qualities. It accepts volume public use in specific zones that can be designed and managed for capacity functions. At the same time, rare and fragile resource zones are preserved from degradation. It demonstrates that environmentalism and development are compatible rather than antagonistic ideologies if properly planned, designed, and managed.

FRAMURA, ITALY

An important planning/design case in northern Italy involves restoration of landscape degradation and concepts for new tourism opportunities (Avagliano 1995). Four sites were considered, all located within the Cinque Terra-Punta Mesco Natural Park (figure 10-11). The area is flanked by the Tirreno Sea on one side, backed by mountains on the other. Included are the communities of Framura, Dieva Marina, Bonassola, Carro Carrodano, Levanto, and Monterosso. Resource features include:

- historical trails, of environmental, archeological, geological, and vegetative significance,
- ancient rural and vernacular agricultural structures,
- climatic support for verdant cover and bird migration.

Because of the limitation on public funding for development, planning favored private investment.

The challenge before the designers was reclamation of degraded site development, a first step for future tourism planning. Flooding of the Deiva River has eroded banks and even changed its course. Remedial

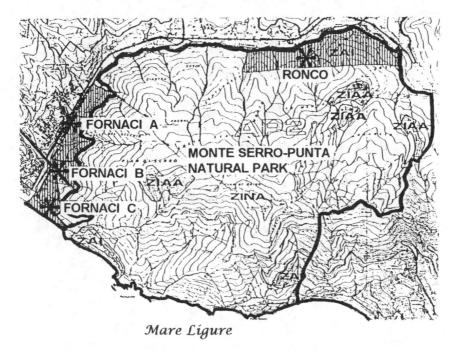

Mare Ligure

Figure 10-11. Monte Serro-Punta, Mesco Natural Park. An area not far from Genova, Italy, that was redesigned for both restoration and new development. Identified is the location of the four sites that were chosen for new planning, Figures 10-12 through 10-15 (Avagliano 1995, 62).

methods have been inadequate. Piecemeal development of vacation homes and subdivision of estates have not been planned in unity.

Main planning/design objectives included:

- new concepts for tourism facilities and services
- updating and refurbishing existing facilities and services
- new gateway visitor development for the park
- new and improved road access and parking
- new recreational opportunities, such as trails
- riverbank stabilization
- public site as focal visitor point on the river.

The national park law, established in 1991, identified several major goals: protection of natural, cultural, and important landscapes; promoting use and education for these resources; and contributing to local social and economic development by means of agricultural and tourism enterprises, better service levels, and controlled soil erosion. Different policies for four zones were created. Most of the park is designated as "partial

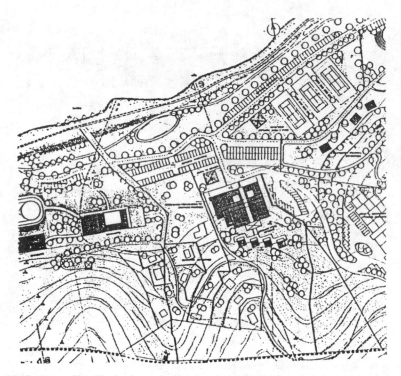

Figure 10-12. Ronco Site. The landscape architects redesigned this degraded area to take advantage of the existing natural assets and provide for new facilities. Included are sites for an upgraded village, new lodging, horseback riding, and visits into the natural park (Avagliano 1995, 62).

preserve" (RP), dominated by a policy of resource protection. The "natural environment" zone (ZINA), allows a minimum of development but fosters resource protection. An "agricultural-environmental" zone (ZIAA) is primarily an agricultural area. The ZAI zone includes areas of settlement and infrastructure in support of recreational and tourist use of the other zones, labeled "infrastructure zone."

The landscape architectural/planning firm of Avagliano Gambardella, located in Genoa, contracted for planning of four sites in this area.

Ronco Site

The Ronco Site, located in the ZAI zone, was dominated by rundown structures and poor protection of the landscape assets. Their plan (figure 10-12) recommended a higher quality level of development, favoring upgrading of a public park along the Deiva River, a new youth hostel, a horseback riding facility, and landscape restoration using indigenous plant materials. New architectural regulations were proposed for better

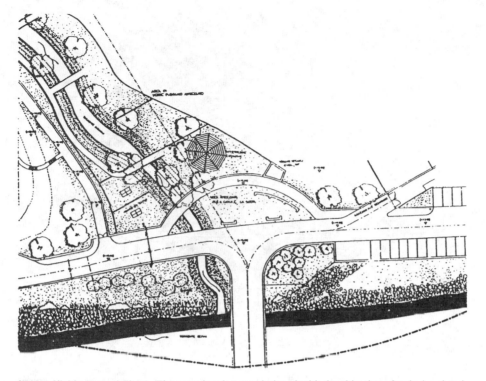

Figure 10-13. Fornaci Site A. The new site plan was designed with the objective of replacing deteriorated areas with new landscape renovation and upgraded vacation homes. Along the river are designed new park facilities and gardens. Included is a new tourist information center and several connecting trails (Avagliano 1995, 63).

adaptation to the site and enhanced beauty of the area. Recommended were new lodging facilities and vacation homes.

Fornaci, Site A.

The area, once an olive grove, had deteriorated and many buildings were abandoned. The plan (figure 10-13) includes expanded road and bridge access, removal of decaying structures, new vacation homes and residences, and protection of existing landscapes of special quality. Included is a tourist information facility and a public garden.

Fornaci, Site B.

This site borders Site A on one side and on the other the railroad between Genoa and LeSpezia. A few trees remain from an old olive grove on the

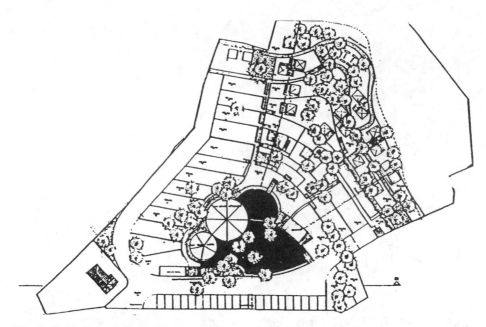

Figure 10-14. Fornaci Site B. Bordering on Fornaci Site A, this area, once an olive grove, is planned primarily for camping. Facilities are located centrally (the dark area). The site overlooks the Deiva River to the west (Avagliano 1995, 63).

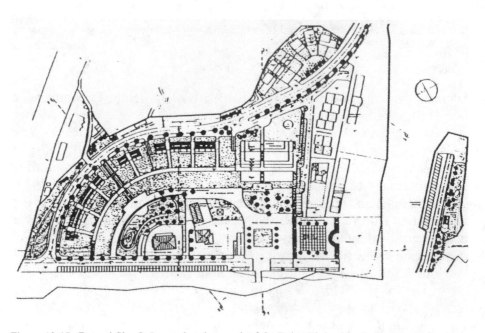

Figure 10-15. Fornaci Site C. Located at the mouth of the Deiva River, this six-hectare site has been replanned to remove older hotels that obstruct views to the sea and to integrate new lodging and vacation homes with the attractive site features. The centerpoint is a redesigned public square (Avagliano 1995, 64).

flat area backed by an incline. The designers' plan (figure 10-14) kept all major features of the landscape, adding a new camping area. This development is flanked by new and remodeled buildings appropriate to low-scale utilization of the site.

Formaci, Site C.

Located at the mouth of the Deiva River, the site is generally flat and supports a few vacation homes and a poorly integrated overall landscape. Two hotels and several houses have spoiled the original beauty of the seacoast. The plan (figure 10-15) features a balance of protected open space and new housing. A focal point is a new public square with terraced gardens sloping to the sea. From this, several hiking trails radiate. The design integrates new development with existing structures and the many landscape factors.

These projects and their design are cited because they illustrate park-tourism relevance, protection of all natural and cultural assets, carefully located new development, and creative planning within the context of governmental policies.

THE IRONBRIDGE GORGE MUSEUM

Description

Declared a World Heritage Site by UNESCO in 1986, the Ironbridge Gorge Museum, has been organized, planned and developed as a major attraction site, drawing 690,000 visitors in 1991 (Ironbridge 1991). The museum is made up of seven major features covering six square miles at Telford, Shropshire, England (figures 10-16, 17). The Ironbridge Museum Trust was established in 1967 and the first site opened in 1973. The theme, "the birthplace of industry," commemorates the source of the world's first iron rails, wheels, boats, the first cast iron bridge, and over two centuries of industrial development. The key features of the Museum complex include the following.

Museum of Iron and the Darby Furnace. Visitors can view many displays, models, and exhibits and experience the sound, light, and smoke display of the original Coalbrookdale smelter established by Abraham Darby in 1709. The museum's main collection is housed in an 1818 warehouse nearby. Original Coalbrookdale products—iron rails, boilers, saddle tank locomotive—are on display.

Museum of the River and Visitor Centre. Here, in the Severn Warehouse,

Figure 10-16. Ironbridge Gorge Museum. Located at Telford, Shropshire, England is this historic complex including the 1779 iron bridge. This setting on the River Severn is the focus for the Birthplace of Industry (Ironbridge Gorge 1991,1).

built in the 1840s, is contained an interpretive introduction to the early history of the Industrial Revolution. The significance of the River Severn is explained by means of a 40 foot model of the gorge as it was in 1796. Changes in management of the river over two centuries are described.

Rosehill House. Approximately 100 yards from the Darby Furnace are Dale House and Rosehill House. Dale House, originally commissioned by Abraham Darby I, who did not live to see it completed, is now being restored. Rosehill House was built for a son-in-law and now displays artifacts of the way of life of a Quaker ironmaster in the early nine-teenth century.

Iron Bridge and Toll House. This centerpiece of the museum complex was designed by Shrewsbury architect Thomas Pritchard in 1775 and work began in 1777. It was the first such cast iron bridge in the world with a single span of 100 feet, six inches. It became the prototype for such construction elsewhere. It was in constant use until the 1930s, when vehicular traffic was prohibited. Inside the Toll House is an exhibition about the bridge. On the north bank the town of Ironbridge developed from the 1780s.

The Tar Tunnel. Visitors are able to go underground to observe natural bitumen that has been oozing from the walls since discovered in 1785.

Blists Hill Open Air Museum. On a 50-acre site, a turn-of-the-century living community has been redeveloped and opened to visitors. Travelers can walk along gas-lit streets of this Victorian town, past railway sidings, yards and pigsties, shops and offices. The sounds of machinery in action add to the realism. A candle factory, pub, and butcher shop add to the atmosphere. Skilled tradesmen in costume assist in interpretation.

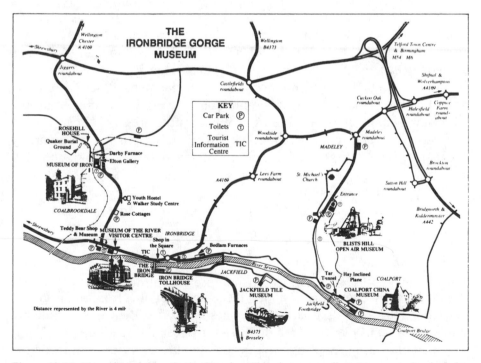

Figure 10-17. Plan of Ironbridge Gorge Complex. This six-square mile museum complex includes Blists Hill Museum, Museum of Iron, the Iron Bridge, Coalport China Museum, Jackfield Tile Museum, Rosehill House, and Museum of the River. It represents a well-planned and managed cultural attraction that also serves local needs of cultural protection and education (Ironbridge Gorge 1992, 3).

Coalport China Museum. An old china works producing Coalport ware has been restored as a museum of china. Techniques of manufacture are demonstrated and products are displayed.

Jackfield Tile Museum. Two of the largest decorative tile works in the world—Maws and Craven Dunhill—were thriving in nearby Jackfield in the 1880s. Today, the original Craven Dunhill works have been restored and the manufacturing process can be observed. On display is an abundance of decorative wall and floor tiles produced from the 1850s to the 1960s.

Planning and Development

What is now an attraction of world-class stature nearly missed this opportunity as the Shropshire Coalfield and the area was in rapid decline (Smith 1989). In 1950, an ironfounder firm made the first excavation to reveal the Old Furnace at Coalbrookdale. By chance, the area became the locale for the development of the New Town of Telford, based on the New Towns Act of 1946, stimulating interest in restoration. A solicitor in the New Town encouraged the establishment of a Museum Trust in 1967. Meanwhile, architects from Birmingham had been commissioned to prepare a feasibility study for the Museum. Basic development principles included: that the Museum would be a charitable trust; that operations would be funded from entrance fees; development funds would be solicited; a community support group would be established; and that the Museum would include a complex of related structures and sites.

Restoration and redevelopment have been accomplished by a variety of planners and designers over a period of several years. Although much of this work is now done by staff, outside professional designers, such as landscape architects, architects, engineers, exhibit designers, archeologists, and historians are engaged as consultants from time to time.

Organization and Management

This nonprofit operation is under the control of the Ironbridge Gorge Museum Trust with its Board of Trustees Executive Board, an Academic and Curatorial Committee, and the Rosehill Trust. All retailing, wholesaling, manufacturing, catering, and accommodation services are managed by a Trading Company, also under the control of the Trust.

A Monuments Management Team has been established to supervise all physical maintenance and is setting high standards for this work. It is responsible for maintaining a large collection of archeological material, once handled by a contractor. A senior archeologist assisted by two other archeologists and an illustrator have been appointed on contract to record major monuments, monitor excavations and repair works, and produce reports for English Heritage.

The Ironbridge Institute has programs of postgraduate courses, short courses, research and consultancies, and maintains close liaison with the University of Birmingham. Special emphasis recently has been placed on training for museum practice, heritage tourism, and international relations. Many foreign students have enrolled in these programs. The Institute has sponsored special studies and plans for evaluating archeological importance, surveys of buildings, and recommendations for interpretive exhibits and programs (Ironbridge 1992).

Close linkage between visitors and exhibit demonstrators is a basic planning and management policy. Although a limited number of explanatory panels are used, the managers depend greatly on staff input to explain and interpret the specific objects as well as the fascinating story of this site's role in the beginning of the Industrial Revolution.

The Museum maintains a library that provides service to the staff, students of the Ironbridge Institute, and the general public. It derives its operating and capital improvement revenues from several sources: admission fees, a capital endowment fund, commissions from retail sales, grants, and donations.

A Tourist Information Centre provides tourist guidance and information about the Museum and surrounding area. It is part of the overall information network of the English Tourist Board.

Future plans include restoration and redevelopment of several sites within the area, such as a Museum of Industrialization, Social History Gallery, a PipeWorks Museum and a Geology Gallery.

A major merit of this attraction is the exponential value to visitors as a result of creating an overall complex, linking several related historic sites together. This integration principle is carried out not only by well-designed and managed physical development but also by attractive, readable, and informative literature, such as *The Visitor's Guide 1992* (Wrekin Council 1992). This planning vision of the responsible developers enhances all tourism goals—greater economic impact, expanded visitor satisfactions, protection of significant resources, and integration with local and area economic and social life.

WETLANDS PARK, NEVADA

An unlikely case of park-tourism design is the planning of Wetlands Park, dependent on stormwater runoff from the gaming mecca of Las

Vegas (France 2001). Collaboration of environmental consultants, landscape architects and officials of Clark County has resulted in an award-winning plan and beginning phases of the 2,900-acre Wetlands Park.

The location is a grassland meadow that became a dump of trash until proponents persuaded the Nevada legislature to gain citizen approval of a $13.3-million bond issue in 1991 for restoration and redesign of this area for a park. Landscape architect Dirick Van Gorp of the county parks department is credited with the initial proposal. This plan had three purposes: recreational and visitor use, erosion control, and policies of resource management.

Several physical elements were included in this plan that received a Planning and Urban Design Merit Award from the American Society of Landscape Architects in 1997. Essential was the construction of erosion control measures. Although intermittent, intensive storms can create major erosion, so a $4.4–million project has resulted in major control structures. Ponds and channels add aesthetic landscape beauty as well as habitat for wetland wildlife. Water sources include runoff and effluent from Las Vegas. The centerpiece is a 100-acre Nature Center providing interpretation and nature education. An extensive trail system has been built. Plans include also an Arrival Plaza and Shelter, a Visitors' Center, amphitheater, water sampling area, and other visitor facilities for viewing and understanding local wildlife.

This example demonstrates several significant actions. First, it was the vision of a landscape architect that initiated advocacy for this unusual natural resource project. Many park-tourism projects have never utilized local opportunities due to a lack of such vision. Second, this vision had sufficient public appeal to secure a vote for financial support. Third, in order to plan such a complex project, a team of specialists was created—public park agency, landscape architects, wetland rehabilitation consultants, recreation and tourism representatives, a consortium for implementation, Las Vegas schools, engineers, and a local chapter of the Audubon Society. This case exemplifies a true planning search for sustainable development. These are important ingredients for others to consider in future tourism site developments.

WILPENA STATION RESORT

A continuing planning challenge is integrating tourism with national parks. Although national parks around the world vary greatly in size and administrative policy, resource protection is a dominant theme. But increasingly there is recognition of the important visitor function in national parks. New planning policies permit only a minimum of visitor services—lodging,

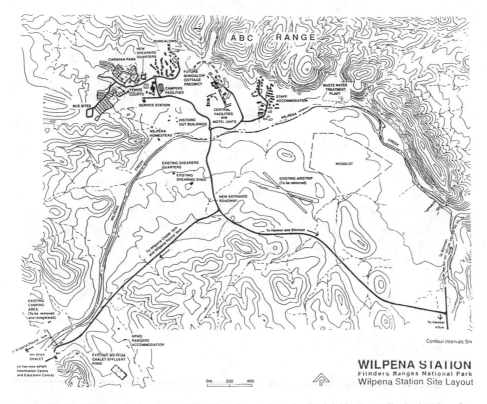

Figure 10-18. Wilpena Station Resort Site Plan. Located north of Adelaide, Australia, is the site of a resort at the edge of Flinders Ranges National Park. Design of site and structures is integrated with the dominant landscape aesthetics of the area (Williams & Associates 1988).

food—within the park, encouraging major traveler service development in nearby communities. Because most of these services are provided by the private sector, new cooperation and collaboration are required between national park officials and commercial enterprise developers.

A special design case within a national park as a private concession is the proposed Wilpena Station Resort within Flinders National Park, 430 kilometers north of Adelaide, South Australia (Williams and Blake 1990). Figure 10-18 illustrates the general site layout of this project. Figure 10-19 illustrates the relationship between the major feature of the park, Wilpena Pound, a giant oval shaped basin within great serrated walls, and the proposed resort. This is a plan for visitor use, but in a managed manner.

Objectives

Required was an Environmental Impact Statement (Williams & Assoc. 1988). This report stated project objectives as follows:

Figure 10-19. Wilpena Pound, Flinders Ranges National Park. This perspective view illustrates the intimate relationship between the dominant features of the park and integrated services for visitors (Williams & Associates 1988).

- Facilities that will expand opportunities for visitors to Flinders National Park to use and enjoy the park.

- Facilities that will increase opportunities for visitors to recognize and understand the natural and cultural features of the Park.

- Facilities in a manner that offers alternatives, and minimizes the impact that existing numbers of visitors are having, and that predicted numbers of visitors may have on the Park.

- A site specific framework through municipal services and other lessee obligations (sewerage, water, power, wood supply) to manage the consequences of visitors staying in the Park.

- A variety of facilities that satisfies almost all of the accommodation needs of the range of visitors that visit the Park.

- Facilities that enable the Flinders Ranges to realize its potential to attract state, national, and international visitors, thereby adding to the value of the tourist industry's contribution to South Australian economy.

- And for the development of a site that the South Australian government has identified as being the most appropriate site in the region for the development of a large-scale integrated resort.

Solution

As occurs in many similar situations, controversy began to surround this proposal. This was based primarily on the generalized assumption that environmental degradation is a foregone conclusion from increased volumes of visitors. Part of the objection arose from the proposed hotel that would accommodate a travel segment contrary to more traditional camping and recreation travelers. However, the proposed design and management plan demonstrate that both of these assumptions are false. In fact, the State encourages a diversity of visitor types, endorsing the variety of accommodations. Those who visit will leave with new information about environmental values, producing a public constituency with a strong resource protection ethic. By recognizing several visitor segments rather than just one, there is less likelihood of the design and management being as exclusionary as in the past.

This is an example of a well-designed visitor center within a national park that promises to satisfy criteria of resource protection and tourism. The plan encompasses issues of visitor behavior control, interpretation, use of "hardened" site for major services and facilities, tour guide training, zoning of uses to avoid conflict, marketing linkage with resource capabilities, and aesthetically appropriate design of all structures.

LOWER COLORADO RESTORATION

An excellent example of how site planning can convert an abused riparian desert landscape into a multipurpose attraction is the case of the Ahakhav Tribal Preserve Plan (Phillips 1998, Thompson 2000). Remarkable renovation of over 1,000 acres of desert land in the Colorado River Indian Reservation in Arizona west of Phoenix is the result of an innovative landscape plan and implementation. Although many people and

Figure 10-20. Ecologically Restored Ahakhav Nature Park. An example of the many restorations of a deteriorated landscape. The project encompasses redevelopment of over 1,000 acres of abused riparian land into restored endemic vegation and new visitor interpretation and services. It represents innovative collaboration among many entities including Indian, governmental, and professional landscape architectural input (Phillips 1998, 141).

agencies have been involved, this development likely would not have happened without two key leaders. For some time, the Mohave Indian, Dennis Patch, had visualized returning his homeland area to the verdant past known by his ancestors. His concern was that modern children knew nothing about this early landscape. Meanwhile, a major in landscape architecture at Purdue university, Fred Phillips, traveled to the Indian site in 1994 in the hope of finding a job after graduation. His meeting and tour of the area with Patch stirred his creative design imagination, the start of his becoming project director later on.

Phillips soon learned how the area had once provided basic resources for the livelihood of local Mohave and Chemehuevi tribes for generations. Not only a water resource but also materials for shelter, baskets, medicines, dyes, food, clothing, and paints came from the land. In the early days true sustainability had been accomplished. Then came white settlement that changed all this. Agricultural development, river dams, and introduction of exotic plants completely changed the site, generally depleting its original resources.

Following basic professional landscape architectural practice, Phillips began detailed analysis of the site characteristics. With Patch's vision in mind he began to discover zones that held best promise of vegetative rejuvenation. With no office or staff except a couple of student aides, he forged ahead with concepts for development. Well into the design process he realized his role encompassed obtaining financial backing and involvement of several specialists, such as ecologists, hydrologists, and plant experts.

The ambitious plan included changing a watercourse, restoring a wetland, rejuvenating riparian vegetation, creating a nature park (Figure 10-20), establishing an indigenous plant nursery, providing an interpretive visitor center, and beginning a cross-cultural educational program. In spite of some obstacles along the way, most of these objectives have been met, producing an outstanding area for visitors as well as residents. Over $5 million in grants have been received from various state and federal agencies. The Bureau of Land Management dredged a 2.5-mile channel back to its original depth. Marsh vegetation is now returning. A 225-acre tree plantation has been established and a four-acre day use area with a visitor center, picnic tables, barbecue pits, canoe rentals and extensive nature trails are open to tourists.

This remarkable transformation proves that it can be done and offers a lesson for others. It shows again the basic significance of natural resources and that damage to them can be repaired. It illustrates a site design and action process that combined input from a landscape architectural professional, a local leader, governmental agency involvement, and a great many environmental specialists. Even though there were a few stressful moments it proved that two contrasting and sometimes conflicting cultures could collaborate on the pursuit of a common goal. Best of all, it must be emphasized that the key to the success of the entire project was the vision of an individual who believed it should and could be done.

HERDADE, PORTUGAL

An example of environmentally sensitive resort design is Herdade da Comporta on the coast of Portugal (Grot 2001). Planned are several development types such as small boutique hotels and larger hotels integrated with the existing village.

Located only one hour from Lisbon, the site is 12 square kilometers in size and features a dune and beach waterfront as well as estuarine habitat. The landscape planners of the London office of EDAW considered not only the needs of a viable resort for a special market segment but also critical factors of the environment—scale, vegetation, population, and integration with the surrounding region. The developer of this project is Atlantic Development Company.

The designers set for themselves important "core values" and "environmental goals" as cited below:

Core Values

- To create a unique place known for environmental, built, and cultural assets—a place with a spiritual and deep commitment to the land.
- Proactive protection and enhancement of the natural environment.
- Preserve sanctity, vastness, and simplicity of the land.
- Enhance, develop, and respect the existing context or "flavor" of the built environment (appropriate scale/type/style).
- Diversify, develop, and optimize agricultural base.
- Ensure development is compatible with the natural capacity of the site.
- Social integration through training, education, and ownership for existing, new, and future populations.
- Re-establish strong cultural identity and awareness.
- Create balanced economic growth.
- Create a development with a high rate of return, increased asset value, attractive to a range of high quality development partners.

Environmental Goals

- Environmentally sensitive master planning to minimize land take and avoid damage to high value habitats.
- Preservation of high value wildlife habitats and appropriate management to ensure their longevity.
- Restoration of damaged habitats and the creation of new habitats.
- Solutions to conflicts between agriculture and wildlife and the promotion of agri-environmental schemes.
- Indigenous planting and landscape design.
- Indigenous planting of golf courses.
- Facilitating observation and interpretation of wildlife.
- Sensitive construction techniques and construction planning.

- Sustainable management and operation of resorts.
- Increasing environmental awareness and promotion of Herdade da Comporta as a best practice example of sustainable tourism.
- Robust control of damaging practices on the estate.

This project demonstrates the growing trend of developers and designers of tourism sites toward environmental sustainability.

TOURISM SITE CRITIQUE

Much of the motivation for better tourism site planning is to avoid negative impacts of bad planning. An example is the present concern over coastal tourism development in Thailand. An analysis of several sites reveals the need for more careful planning in the future (Sudara 1995).

Cited among the problems has been the massive high-rise service businesses that have threatened the former serenity and beauty of the once fine beaches. Tourist congestion has resulted from tourism's growth. Coastal water pollution and poor waste collection have presented health problems. In some cases, tourist development projecting into the sea has stimulated beach erosion. Flooding after heavy rains has incited excessive runoff and pollution, especially in the popular resort of Pattaya. Similar problems arose with expanded tourism at Phuket. And land speculation has allowed massive development in the wrong locations.

Fortunately in recent years both the public and private sector have initiated restorative plans and are reaching toward the goal of site sustainability. An example is the revitalization of Bangpu, protecting the mangrove forest and offering visitors an opportunity to enjoy nature, especially the local ecosystem. The negative impacts of the past may not be completely remedied everywhere but commendable is the awareness of these issues and instigation of remedies.

This critique by Surophol Sudara is cited here not to demean tourism in Thailand but rather to highlight many of the environmental problems of site developments that have been experienced worldwide. Actually, in spite of these site issues, the attractions and tourist services of modern Thailand are world-class. The new and progressive sustainable initiatives promise better planning in the future.

CONCLUSIONS

The fourteen cases of site scale planning and development described in this chapter offer a wide range of challenges and solutions. Even with their great diversity, some characteristics of planning and design are common to all. Most of these cases broke away from the traditional professional-only design process. Instead, many publics—local, environmental, governmental—were involved at several steps in the process. Nearly all demonstrated that political involvement in tourism design and development is necessary—initiation, commitment, leadership, approval.

These cases exemplify the importance of some leader having a vision. Such a person might be a designer, a government agency official or technician, a politician, a nonprofit organization, or someone else. No matter the affiliation, unless someone recognizes a need and is concerned enough to promote a project, good site design and development will not take place. Professional architects, landscape architects, and urban designers have a critical role to play in conceiving of land and building design but this also requires the spark of conviction that a tourism site project is needed and feasible.

This sampling of cases is representative of trends toward balancing environmental protection and renovation with tourism development and at a very visible scale. When good design is translated into a specific land use accomplishment, theories and principles provide tangible demonstration of their value. As more governmental agencies, business people, and representatives of nonprofit agencies increasingly put such sustainable evidence on the ground, much progress will be made for the many benefits of tourism. These examples show the need for adaptive, integrative, and creative design for tourism site development.

DISCUSSION

1. Speculate on why the overall plan and management plan for Boston Harbor Islands took so long. Why was the National Park Service a logical management agency for this project?

2. For the Blue Ridge Parkway what was the major issue in its establishment? Discuss the factor that determined its boundary and the problems inherent in this decision.

3. How would you characterize the main role of the Design Team for resolution of the planning impasse for Franconia Notch highway?

4. In the Whiteman Park project, why were major policy changes recommended for its adaptation to tourism?

5. Discuss the design implications at major tourist regions, such as the case of Michigan's Upper Peninsula.

6. What were the environmental and integrative design policies for the Framura, Italy case? Discuss how the landscape architect fulfilled these policies.

7. Discuss the advantages of interpretation, such as carried out at the Ironbridge Museum.

8. Discuss the trend toward engaging a design team for tourism site design rather than a single professional firm.

REFERENCES

Avagliano, Carmela (1995). "Tourism Development and Landscape Change." In *Proceedings, IFLA World Congress,* Bangkok: Thai Association of Landscape Architects, pp. 60–64.

Blank, Uel, Clare A. Gunn and Johnson, Johnson & Roy (1966). *Guidelines for Tourism-Recreation in Michigan's Upper Peninsula.* Cooperative Extension Service. East Lansing: Michigan State University.

Boston Harbor Islands (1994). Report of a Special Resource Study. National Park Service, North Atlantic Region.

Boston Harbor Islands (2000). Draft General Management Plan and Draft Environmental Impact Statement. Boston: Boston Support Office of the Northeast Region, NPS.

Bridging the Straits (1988). A summary report concerning the feasibility of a proposed Upper Peninsula Welcome Center/Straits of Mackinac Transportation Museum. Presented by the Mackinac Island State Park Commission, St. Ignace, MI.

DeRoo, Philip (1995). "Renovation of the St. Bavo's Square and of Other Squares in the Tower Area, The touristic heart of the Medieval Ghent." In *Proceedings, IFLA World Congress,* Bangkok: Tahi Association of Landscape Architects, pp. 112–116.

France, Robert (2001). "Leaving Las Vegas." *Landscape Architecture,* 91 (8): pp. 38–42.

Grot, Juli (2001). Masterplan for Herdade de Comporta, Portugal. London: EDAW.

Gunn, Clare A. (1997). *Vacationscape: Developing Tourist Areas,* 3rd ed. Washington, DC: Taylor & Francis.

Hall, Colin Michael (1996). *Tourism and Politics: Policy, Power, and Place.* New York: John Wiley & Sons.

Ironbridge Gorge Museum Trust (1991). *Annual Report.* Telford Shropshire, England: IGMT.

Ironbridge Gorge Museum Trust (1992). *Education at Ironbridge: Preliminary Information for Teachers.* Education Department. Shropshire, England: Ironbridge Gorge Museum Trust.

Jamal, T. and Donald Getz (1995). "Collaboration Theory and Community Tourism Planning." *Annals of Tourism Research,* 2 (1), pp. 186–204.

Jolley, Harley E. (1985). *Blue Ridge Parkway: The First 50 Years.* Mars Hill, NC: Appalachian Consortium Press.

Makopondo, Richard O. B. (2001). "Modeling Collaboration, Partnership, and Inclusiveness in Recreation and Tourism Resource Management." *Annual Conference Proceedings,* pp. 100–109, Travel and Tourism Research Association.

Orange Valley, Antigua (1991). A proposal for a Major Resort within the Antiguan and Barbudian National Park. Prepared by a development team of architects, landscape architects, golf course architect, and other specialists. Columbia, MD: LDR International, Inc.

Paolini, Kenneth W. (1982). *Boston Harbor Roundtable Discussion.* Washington, DC: Office of Coastal Zone Management, National Oceanic and Atmospheric Administration.

Phillips, Fred O. (1998). "Ahakhav Tribal Reserve." *Restoration and Management Notes,* 16 (2): pp. 140–148.

Reed (1997). "Power Relations and Community-Based Tourism Planning." *Annals of Tourism Research* 24 (3), pp. 566–591.

Rigterink, Richard (1992). (Correspondence and reports) Also: "New Hampshire Adapts Interstate," by Peter Green, *Engineering News-Record,* November 20, 1986. Also: "Franconia Notch," *Landscape Architecture,* November 1989.

Smith, Stuart B. (1989). "The Next Thirty Years." In *Dynasty of Iron Founders,* Arthur Raistrick ed., York, England: Ebor Press.

Stahlecker, Gail D. (2001). "Blue Ridge Parkway, Virginia and North Carolina." News release, Classic Award. From www.asla.org/meetings/awards/awds01/blueridge.html as of January 23, 2002.

Sudara, Suraphol (1995). "Tourism Development and Conservation of the Coastal Areas." In *Proceedings IFLA World Congress,* Bangkok, Thailand, October 21–24, pp. 226–232.

Thompson, J. William (2000). "Desert Passage." *Landscape Architecture*, 90 (3): pp. 56–65.

Whiteman Park (1989). An assessment of present and future tourism appeal. Perth, Australia: Western Australian Tourism Commission.

Williams, Michael and Lynn Blake (1990). "Wilpena Station, Flinders Ranges National Park: Planning for Cultural Tourism." In *Historic Environment,* 3 (4) pp.61–71.

Williams, Michael & Associates (1988). *Proposed Wilpena Station Resort Flinders Ranges National Park Draft Amendment to the Flinders Ranges National Park Management Plan and Draft Environmental Impact Statement.* Prepared for the SA National Parks and Wildlife Service, Adelaide, South Australia.

Wrekin Council (1992). *The Visitor's Guide 1992.* Wrekin Tourism Association. Shropshire, England; Wrekin Council.

Wood, D. and B. Gray (1991). "Toward a Comprehensive Theory of Collaboration." *Journal of Applied Behavioral Science,* 27 (2), June 13, pp. 9–162.

EPILOGUE

Those readers familiar with earlier editions of this book would have found a chapter 11, Conclusions and Principles. For this edition, that chapter has been deleted. The reasoning is based on a more thorough presentation of conclusions and the addition of discussion topics after each chapter. We believed that repeating this material at the end of the book would be redundant. We hope that readers will agree.

The book encompasses three major areas of tourism planning information—basic fundamentals of tourism as a system, planning concepts and processes at three geographic scales, and examples of planning/design at these scales. A major observation of making this revision was the dramatic increase of tourism's adaptation to areas of these three scales throughout the world in the last decade. More experienced areas are taking on new tourism tasks, although for some tourism adaptation is a new phenomenon.

We observed many improvements in planning and development in recent years. This is the first time in history that environmentalism has become a major thread of tourism concern. Tourism plans of this decade have demonstrated a great increase in public-private cooperation. Roles and responsibilities are being clarified. Never before have as many seminars, conferences, and educational meetings been held, bringing a great diversity of tourism interests together for new tourism enlightenment. Only recently has the awareness of the power of policy and its influence on tourism increased. These and many other observations prove how dynamic tourism really is, constantly in a state of flux. Better planning and design are increasingly evident.

These changes create a dilemma for the tourism planning researcher and writer. Just as soon as words are set down and published they have already entered into a first stage of obsolescence. This fact means that all tourism planners and developers have the obligation to build on the best information available with their own new concepts, principles, and applications. Although basic principles remain, planning must adapt to changes in markets and new supply development.

As long-time students of tourism, we as authors are confident that the many conclusions following chapter presentations offer a fair summary of the many planning dimensions. However, if we were asked to name the very top conclusion, the tourism planning factor that reigns supreme, it is *leadership*. Long after the concepts and principles put forth in this book have begun to fade and be replaced by newer ones, strong leadership will still be in demand for every tourism project. Such tourism planning and development leadership may emerge from any of many sources. It need not necessarily come from a titled individual, such as a tourism agency director. It may emerge from a private commercial sector or a public agency. A nonprofit volunteer organization or a professional designer may provide needed leadership.

No matter the source, the required leadership will possess several key characteristics. First, there must be dedication to the cause of better tourism planning and development. The cause must be well defined and encompassing. The individual or team must have strong convictions on the values of tourism for mankind throughout the world. The charisma of such a torchbearer must be supported by factual understandings of tourism's many dimensions. Without vision and dedication, even the most advanced technology will not produce the best tourism plans and objectives. And finally, even effective leadership may wane unless mechanisms for continuation are built into the system.

Finally, the authors would be remiss if they did not speak about the relationship of tourism to the horrendous terrorist attack on the United States on September 11, 2001. The immediate reaction was a travel depression fostered by a fear of travel, especially flying. One interpretation might reduce the value of a book on tourism planning and development such as this. One might question whether nations, communities, investors, and developers will be as interested in future enhancement of tourism. Only time will offer the full answer to this question.

But as the authors began to take an objective stance, it seemed appropriate to place this event in historic perspective. For centuries, many parts of the world have been devastated by floods, fires, volcanoes, war, disease, terrorism, famine, and other catastrophes. This does not excuse the recent event but may put it in perspective. Concurrent with all these problems, mankind's insatiable desire to travel has continued to grow, not diminish. It seems appropriate then to take the stand that wherever and whenever in the future there is a desire to develop tourism, the concepts, principles and guidelines for planning put forth here will continue to be valid.

Index